The Yearbook of Agriculture
1978

Living on a Few Acres

Finding a few acres takes careful looking. Some may remodel an old house, even use it as annex to a new house. County Extension agents help with gardening and other problems. But country living also can be rigorous, as tending sheep in winter (overleaf).

◄1

2

3

4►

6

5

6

8

Top, weed-free garden
produces better
vegetables. *Above,* com-
mercial sprayers can be
hired for small apple or-
chards. *Upper left,* family
pitches in at harvest time
to prepare garden foods
for canning or freezing.
Lower left scenes, picking
blueberries to sell at
roadside stand, family
pet watches grape
harvest.

10

Power for few acres
comes in many forms.
Above, solar-heated
greenhouse which has
turf growing on back roof
for insulation. *Right*,
home-made wood-burning
stove heats entire house
and is used for cooking.
Top right, hand-powered
post-hole digger. *Lower
right*, large gardens and
other operations call for
motorized equipment.

11

12

14

13

Below, community canning center for preserving garden produce. *Left*, store home-canned food in convenient, dry place away from light. Facing page, *lower right*, lath house offers protection for herbs from wind and hot sun. *Upper right*, herbs drying for later use, herb knot garden adds to landscape decor.

15

16

21

Above, temporary
greenhouse for flower
production. *Left,* cold
frame is convenient to
garden and protects early
plants. *Facing page,* annual flowers provide
steady supply of blooms
for sale or home use.

22

23

24

Left and facing page, roadside stands and farmers markets give growers chance to sell produce freshly picked at peak of quality. *Above,* family with five acres adapted farm wagon as roadside stand for use in front of home.

25

26

Right, beekeeper in protective clothing checks brood comb from hive. *Below,* hives placed so bees—essential for pollination—can assure commerical onion seed crop, and provide income for beekeeper. *Facing page,* smoker calms bees.

27

28

29►

31

Above, maple syrup operation. *Right,* young Christmas tree plantation. *Facing page,* mature trees in small woodlot can be harvested for pulp, lumber, or firewood.

32

33

34

35

Facing page, top, mountain ride at dude ranch can be pleasurable experience. *Below,* three fishermen admire 20-inch rainbow trout caught in fee fishing pond.

Above, campground on five-acre farm. *Right,* horses and a few acres seem to go together. *Overleaf,* early morning roundup.

36

37

Sheep, beef cattle, and
hogs are popular
livestock on a few acres.
They can be good sources
of meat for the table, and
supplement your income.

38

40

43

Goats provide milk, often become pets. *Facing page*, top, grooming cow for youth fair. *Bottom*, dairyman has 24 Guernsey cows on five acres.

44

29

Right, chickens are raised by many on a few acres for fresh eggs and meat. *Facing page,* producing show birds can be fascinating hobby. *Below,* turkeys are raised on few acres but need lots of care.

45

46

47-

Foreword

Bob Bergland
Secretary of Agriculture

Americans keep going back to the land. It is a pilgrimage that makes more sense to a lot of people than living in cities, enticing us with the promise of escape from freeways, assembly lines, and crowds.

The land offers freedom, a chance to test your mettle against nature's challenges.

The pilgrimages began with our agriculturally-minded, freedom-loving forebears of the 17th and 18th centuries. In waves since then—from the founding of community utopias in the first half of the 19th century, to the homesteads of the later 1800's, to the flight of the unemployed from a collapsing economy of the 1930's—Americans have returned to the land.

Now the tide of Americans that swept to the cities after World War II has ebbed. The flow has reversed. People are populating the countryside faster than they are cities.

I know their motives. I am a farmer.

So I know that country life can push some people beyond their endurance, can shatter illusions with a heavy and indifferent hand, and can press poverty upon the backs of the unlucky and the unprepared.

This book, *Living on a Few Acres,* is intended as a practical guide for those who make the journey back to the countryside and for some of you who are already there. It is mainly for those who intend not to gain their principal income from the land, but rather to have a job in town or live on a pension or some other source of income.

U.S. Department of Agriculture programs offer valuable assistance at the local level. In addition, this Department has a major responsibility for all Federal efforts in rural areas.

Living on a Few Acres describes both the pitfalls and the satisfactions of country life. There are plenty of both. And there is nothing quite like country living.

Preface

Jack Hayes
Yearbook Editor

Change is the only constant in farming and rural life. There have been tremendous changes in rural life since the early days of this Nation. We have seen a great exodus from the land and now we are beginning to see a substantial return.

The new trend reverses the massive rural-to-urban migration from World War II through the 1960's. Non-metropolitan counties lost 3 million people through outmigration in the 1960's. But between 1970 and 1976, rural areas and small towns grew by 4.3 million.

Many suburbanites and city people have moved to the country. Those who haven't done so and are thinking about it should study carefully all they can about what is involved before making the move. This book, *Living on a Few Acres*, should help. Read it. Visit country areas. Talk to people already there. Check all the angles.

There are others who will find this book useful. Among them are those who have been full-fledged farmers in the past but are choosing to farm smaller portions of the land, and to hold jobs in town.

We wish you good reading and good farming.

Among the many persons who contributed special skills to this book were Charles McKeown and William Rawley of the Typography and Design Division, *U.S. Government Printing Office;* Warren Bell and Paul Wertz (ret.) of the Printing Liaison Branch, USDA; and Denver Browning of the Yearbook staff, who did the Index.

Chairman of the committee that planned this Yearbook was Ned D. Bayley, *Office of the Secretary.* The Deputy Chairman was Alice Skelsey, *Science and Education Administration* (SEA).

Members of the committee were:

Jack Armstrong, *Economics, Statistics, and Cooperatives Service* (ESCS)

William A. Bailey, *SEA-Agricultural Research*

Charles Beer, *SEA-Extension*

Richard A. Biggs, *Montgomery County (Md.) Cooperative Extension Service*
Wesley Harris, *Farmers Home Administration (FmHA)*
Pieter Hoekstra, *Forest Service*
Evelyn Johnson, *SEA-Extension*
James Lewis, *ESCS*
Percy Luney, *Office of Equal Opportunity*
Charles McClurg, *University of Maryland*
McKinley Mayes, *SEA-Cooperative Research*
David Ross, *University of Maryland*
Floyd Smith, *SEA-Agricultural Research* (ret.)
Billy Teels, *Soil Conservation Service (SCS)*
Ralph Wilson, *SCS*
James D. Wiseman, *FmHA*

Contents

Part 4   How to Make the Most of It

Part 5   Disposing of Property

Pluses,
Minuses

Part One

Living in the Country— a Diversity of People

By A. Gene Nelson and Tom Gentle

Who lives on a few acres in the country? These rural residents are a diverse group—computer analysts, chiropractors, carpenters, salesmen, professors. The land on which they live ranges in size from enough for a rural residence with a large backyard to small-scale farms involving a number of acres. The locations stretch from Vermont's rocky soils to the Blue Ridge foothills of Virginia, from the prairies of Iowa to the southlands, the western deserts, and the northwest's lush forests.

The 1970's have brought renewed interest in the country lifestyle, but this rural movement is not a simple phenomenon. There are many objectives, resources and situations involved, and implications for the future are definitely complex.

Because of this diversity, there is no such thing as the average resident on a few acres in the country. Some are involved in agriculture, operating a few acres on a part-time basis. Some are willing to make the material sacrifices necessary to achieve a preferred rural lifestyle, and others have enough wealth to afford this lifestyle without income from the land.

People live in the country and own a few acres for a variety of reasons. Their objectives might be to:
* Use the acreage solely as a residence
* Pursue hobbies or recreational activities
* Reduce the family's food costs by gardening
* Provide an "alternative" lifestyle for meeting food and energy needs
* Provide an extra (part-time) source of income by selling produce from the acreage, or engaging in some other sideline.

The Rural Residents. These people sell less than $1,000 of of agricultural produce a year. Their primary interest is the

A. Gene Nelson is Extension Farm Management Specialist, Oregon State University, Corvallis.
Tom Gentle is Extension Information Representative at the university.

pastoral setting for their home. Many view the few acres on which they live as an extended backyard. They are often city folks buying up old farmhouses, plus five acres to keep a horse.

In other cases they are retired farmers, such as Ivan and Gladys living on a 120-acre farm in Illinois. At 75 and 70 years of age they are still active, but all the cropland is now rented out to a neighboring farmer and his sons who farm it as part of a larger acreage. Ivan and Gladys are primarily interested in preserving their home and lifestyle.

A California couple became rural residents by default. A real estate agent convinced them they could make a living by growing alfalfa in eastern Madera County. In reality, the land was rough hardpan and had no irrigation water. Growing alfalfa was out of the question.

The local Extension advisor convinced them to forget the idea of farming and get jobs in town because their finances permitted no alternative. Although it was difficult to accept, they took his advice and have been quite successful and happy just living in a rural setting.

It's quite a different situation for Dennis, his wife Alta, and Dave and Jean. The four, all in their thirties, recently formed a partnership to buy their farm, which has 25 acres of woodland, 10 acres of Christmas trees, and 20 acres of open pasture.

Dennis and Dave teach at a small college 8 miles away Jean is an elementary school teacher in another nearby town.

Says Dennis, "Our main reason for living here is to share a group experience and be out in the country." While they would like to earn enough from the land to pay the taxes, they are not interested in making money from farming. None of them has farming experience. They would like to rent the pasture, but have yet to find an interested party.

The partnership allows them to share the cost of the land, which none could have afforded alone. By pooling their resources they also will be able to afford maintenance on the house.

The Hobby Farmers. These farmers have income from their land as a secondary goal, but work a few acres to pursue their recreation and hobby interests. For some it might include woods and a stream for hunting and fishing. Others may have horticultural hobbies and devote their leisure time to landscaping.

Raising registered Hereford cattle is a rare avocation, but Harold, a real estate broker in a nearby city of 80,000, hopes his herd eventually will provide supplemental income when he retires.

3

Another reason he and his wife, also a real estate broker, moved to a farm was to enjoy the benefits of country living. His farm consists mostly of brushy hillsides.

He raises beef because he likes cattle and feels his land is well suited for them. His herd of 20 brood cows is as big as he wants to get. However, his small operation makes building a reputation difficult. Harold has neither the time to train his cattle for showing, nor to show them.

Limitations on his time also mean contracting out most of the heavy work to custom farmers, which adds considerably to expenses. He lost about $3,500 in 1977. Although he enjoys his hobby, he does not want to always lose money on it.

The Gardeners. These people, while they may sell some of their produce, are primarily interested in reducing their family's food bill by growing their own food. They may use their land for an extensive vegetable garden, one or two head of livestock, and some fruit trees.

Five acres have been more than adequate for Richard and Lila to supply most of the food their family of six consumes. They grow vegetables in a large garden, have a mixed orchard of fruit trees, and raise a steer and some chickens for meat.

"We save a considerable amount of money on vegetables —especially tomatoes—if you don't figure in the time I spend canning them," says Lila.

The children earn extra money for school by selling some of the harvest at a produce stand they set up next to a nearby

highway. Lila sells eggs to neighbors and uses the income to offset the cost of chicken feed.

"We have reduced our dependence on the supermarket, which gives me a good feeling. But we also enjoy spending the time at growing and preserving our own food," says Richard, who works full-time at a textile factory.

The "Alternative" Farmers. These farmers are seeking to minimize their reliance on services ranging from electricity to telephone to inside plumbing. These rural newcomers are interested in "natural" food, and the use of animal energy for farming. Many are advocates of "organic" farming, not using chemical fertilizers, and avoiding pesticides.

Keith and Carol and other members of their farm cooperative have ambitious plans for their labor-intensive organic farm. They want to supply high-quality vegetables to the nearby metropolitan area at as low a cost as possible.

Hugh Gillis, retired from his California job, is busy now with such chores as pruning and hauling poles to build a barn on a 15-acre Oregon farm. He and his wife grow fruit trees and grapes. (See text on page 6).

Scott Duff

They think their produce is fresher and tastier than vegetables shipped from other states. Moreover, the co-op's produce does not require as much energy for packaging and transportation to market.

"Right now we're selling to the smaller grocery and produce stores in the city. It's hard to compete with California growers for the big supermarkets because the Californians can supply produce all year long," says Keith.

Their commitment to organic farming methods does not make things easy. Last summer they attacked an infestation of garden centipedes with hoes and rototillers rather than use pesticides. They hire outside labor to weed their vegetable plots by hand.

The Part-Time Farmers. These farmers have definite income goals for their limited acreage. They are seeking through use of their land, labor, and financial resources to earn a significant portion of their income. They may be retirees supplementing their social security or pension income, or they may be seeking to augment their salaries or wages earned off-the-farm.

Hugh and Mary bought their 15-acre farm two years before Hugh retired from his Los Angeles job. After a thorough search, they found a ready-made farm with two acres of grapes, a cherry orchard, and numerous producing peach, apple and pear trees.

He chose small-scale farming as a way to keep active after retirement, but rates the joys of living in the country as also important. "I have a retirement income, but I want to earn something for my work here on the farm," Hugh says.

Hugh is seeking a market for his fresh cherries through selling to local stores and produce stands, or out of the back of a truck. Planning is underway to install irrigation equipment in the orchard and for a pasture to raise beef or sheep. He also is considering a vegetable crop to sell to local consumers.

Jerry, 29, grew up in Brooklyn, N.Y. Andi, his wife, is from Michigan. They came west to go to graduate school, but the classroom took a backseat to Jerry's lifelong desire to be a farmer. In 1973 they bought their 22-acre farm in an isolated area of Oregon.

"We knew what we wanted when we looked for land. We wanted to be way out in the country so a subdivision couldn't come in next door. We wanted a creek."

Jerry built a barn using wood from an old building on his property. By doing the work himself, his barn cost $1,000. He then built a shop, which cost $250 for materials.

Jerry enjoys working with his hands and repairing machinery. He bought a used tractor. "I save a lot of money by doing my own repairs," Jerry says.

Jerry and Andi grow their own vegetables and raise animals to provide meat. Their outside food expenses average about $75 a month for themselves and their two young daughters. Jerry learned how to butcher, to smoke meat and fish, to cure ham and to make sausage. They heat their house with wood.

Jerry works off the farm as a carpenter, which provides about a fourth of the family income. A trust fund provides another 50 percent and the farm pays for out-of-pocket expenses and feeds the family. The property, however, is paid for so they have no monthly mortgage payment.

Success, Problems, Failures

Many living on a few acres can be considered successful. "If peace and quiet and having a good place to raise young children are the measure of success, then most of the city folks who have come to the country have been successful," says Clayton Wills, Oregon Extension agent.

Jerry Chiappisi, who left the city life of Brooklyn to live on an Oregon farm, marks the weight on a hog's back before delivering it to market.

Scott Duff

Land can provide a measure of security and a hedge against inflation. If the family is able to furnish necessary labor and management and chooses the right enterprise, the land can supplement cash income. The family may be able to live less expensively in the country than in town, especially by producing some of its own food and if transportation costs don't become excessive.

"These successful farmers all have a love for agriculture and a determination to succeed," according to Extension agent Bert Wilcox. Determination, devotion, drive, and energy are all words used to describe those who have succeeded on the land.

Unfortunately, living on a few acres is not without its problems. Consider the following examples:

- A California couple discover walnuts and peaches won't survive southern Oregon's cold winters, but not before they had invested their life savings in 40 acres of unsuitable land
- A newcomer to country living learns too late that steep slopes and rocky soil render all but 8 of his 25 acres useless for any type of farming
- Fifteen acres of prime farmland stand idle because the owner, an attorney, finds part-time farming takes more of his time than he is willing to give

A primary difficulty stems from romantic and uninformed notions of rural life. Many people know nothing about or ignore animal and plant diseases, insect pests, and other production and marketing problems they may encounter. As a result, they often find trouble.

An almost certain way to invite disaster is to buy land before developing a realistic plan for what to do with it.

Ted Carroll, Extension agent in Virginia, points out that the plight of many new farmers is often made worse by their own mistakes. "They buy their land," he says, "and then they go out and buy a lot of new equipment that they probably don't need. And then they come in here and throw themselves on our mercy. We try to help, but it's often too late."

According to John Burt, western Oregon county Extension agent, "Failure is most often the result of mental attitude rather than being defeated by the land. The work becomes overwhelming and people lose the energy needed to succeed."

Ironically, failure to establish an economically viable unit may not necessarily spell financial failure. With land values constantly rising, rural land usually can be sold by the dissatisfied at a profit.

"Quite often, people who failed at farming have been able to sell their land for more than they paid for it. So they did make money on their investment in land even though their dreams were shattered," says another Oregon Extension agent, Marvin Young.

Ultimate success of a farm operation depends on how well the individual family members are able to cope with potential disadvantages and problems, and how strongly they feel the advantages of rural life outweigh the disadvantages.

Coping With Problems

Problems result from changes—changes in weather, goals, values, and resources. One of the changes that occurs over time is the composition of the rural family. Children are born, grow older, and eventually leave home.

The family situation has had a considerable effect on the part-time farming operation of Harold and Eldora. When they first acquired the farm, the children were young and needed attention which competed with demands of the farm.

When the four boys were old enough, they contributed to the farm operation by doing custom work for neighbors. "We bought a tractor and hay baler and put the boys to work haying for other farmers. We also planted and renovated pastures and rototilled garden plots. The boys earned $3 an hour and we were able to pay off the equipment as well," says Eldora.

Now the boys are grown and Harold doesn't have time to do the custom work. But by keeping the equipment in good repair, he still has it available for his own use.

Important prerequisites to success are planning and allowing for unexpected changes. The rural environment is capricious. Catastrophes can come in the form of low prices, unfavorable weather, disease, pests, vandalism.

Virgil was a successful small farmer in the Midwest, growing 120 acres of corn and marketing it through hogs.

"But I had some bad luck," he says. "High winds totally demolished most of my hog facilities. I had insurance, but when I figured the cost of rebuilding, I decided the cost of getting back into the hog business was just too high."

As an alternative, Virgil took an off-farm job. He still farms the cropland but on a part-time basis.

This is not an uncommon solution. The happiest seeker of the rural life may be the part-time farmer who commutes to his city job and feeds his family and his soul with a large garden and a few animals. But people living on a few acres all have their individual solutions, and, as we have seen, they are quite diverse depending on their goals and resources.

Consider the Tradeoffs Before Leaving the City

By Manning Becker, Edward Yeary, and A. Gene Nelson

Noise, congestion, traffic, pollution, a lack of privacy, high crime rates, concrete, and other things about the city are finally beginning to get to you. Peace, quiet, clean air, solitude, growing plants, pets, fresh fruits, homegrown vegetables, and open space make country living mighty appealing.

But . . . neither city life nor country life is perfect. It's a matter of tradeoffs.

A move to the country for some is a new way of life in the making. The change may be beneficial, but before making the change, families should consider several things. For example, each family must be convinced that the values and the satisfactions anticipated from living and working in the country really outweigh the inconveniences or difficulties.

Making a permanent change from city to country living for a whole family can require a big adjustment. Families have made the move and have been quite happy about it. There also have been those who wish they hadn't, or who have given up and gone back to the city. Remember that children grow up and the family's values change over time.

Spending weekends or a summer vacation in the country with good friends or relatives may well be remembered as heaven on earth, but was it the change from the daily routine that made the experience so delightful, or was it truly the advantages of country living?

It was fun to care for the livestock in the spring, but livestock have to be cared for every day of the year. Getting out of bed on a warm spring day is far different from getting out of a cozy bed to go out in the wet and cold to care for the same animals.

Manning Becker and A. Gene Nelson are Extension Farm Management Specialists, Oregon State University, Corvallis. Edward Yeary is Farm Advisor-Statewide, University of California, Parlier.

You may never be happier than when you're working in your small garden in the backyard in town after a hard day in the office, and perhaps it would be nice to have a larger space so that more produce could be raised. But will it be that much fun after the longer drive home from work and with a much larger task facing you?

Sure, country living is great, but there's no denying that some people should never move out of the city. They need the things a city offers such as convenience, entertainment, shops, nearness to people, and medical care.

Before making a decision to move or stay, you might draw up some sort of a balance sheet listing the pros and cons of what your present life offers compared with what you feel a move to the country would offer.

Air pollution may be less or different in the country, but it is not eliminated. You may be trading industrial smoke and car exhaust fumes for animal waste odors, crop and weed pollens. Dust storms can be a problem in some areas.

As a rule, the country offers more opportunities for outdoor living than the city. Your house is seldom close enough to the next one to make screens necessary for privacy. Moreover, there is usually room for a good-sized lawn that is handy for recreation. But remember that a good-sized lawn is going to take more time and care than the small plot you have in the city, thus leaving less time for recreation.

Most places in the country are closer to hunting and fishing than are the cities. In fact, one young wife said that part of the rationale her husband had used to move was that he would no longer need to spend every weekend hunting and fishing. It never occurred to her that now he would want to hunt and fish every evening.

Distance to Neighbors

In the country, the neighbor next door may be a quarter of a mile away. After the husband has gone to work and the children have gone to school, it may be more difficult to get neighbors together for a cup of coffee. Distances may be too great to walk, and the family car is in town. Then too, there are chores around the place that need to be taken care of during the day.

Are you moving to the country to save energy? A Michigan study of energy use by urban and rural families found that while energy use in the home was about the same, rural families used 42% more gasoline than did urban families. This additional gasoline was required for food shopping and driving the children to libraries, movies, and school-related events.

The food you raise yourself may be better and less expensive than that available in the city, but it may not be. Don't expect all your fruits and vegetables to look exactly like the beautiful pictures in the catalog. The directions printed on seed packets are helpful but incomplete. They don't say what to do when you see little green bugs or when the leaves start turning brown.

On the other hand, you'll never purchase sweet corn in the city market that compares with that you've picked yourself at the peak of perfection, and have on the dinner table 20 minutes later.

You will want to set some tentative goals as to how large a garden you are going to have. If it gets to be too big, you will have to incur the cost of a small garden tractor and a complement of machinery. You probably won't be able to eat all the produce when it's at prime quality. If you don't plan to sell any, you'll want to can or freeze it.

Maybe you have always dreamed of having a little more space so you could have some animals. It's a great experience for children to learn to care for and love animals, but animals need care every day.

"My children have held death in their hands and they've held life in their hands, and they've grown very self-reliant," says a Massachusetts mother. "It's an education for them that school doesn't offer."

Chickens don't require much space, neither do sheep and goats, but horses and cows do. Raising part of the feed requires more acres and time, leaving less time to enjoy the animals. Buying all the feed gives more time for the animals but less money to spend for other purposes.

If you plan to raise some animals for meat, buy all the feed, slaughter the animals, and process the meat, it is unlikely that you'll end up saving much money. Some of the food you raise yourself may cost more than if you purchased it in the store. It's hard to compete with the commercial egg producer who has 100,000 birds, automatic equipment, and the management to cut cost and maintain quality.

Remember, too, that eating pork chops from the pet hog that grew up in the backyard can make an awkward situation for some at the dinner table.

Often the hobby gets a little expensive, so people decide to expand in an effort to supplement the family income. In some cases people with small plots have done quite well financially on a part-time basis, but they have some things going for them that newcomers will not have. Most have good farming backgrounds and experience. They really enjoy their work and are willing to put in several hours of work each day or week, and usually they concentrate on some highly specialized crop adapted to top quality soil and an ideal climate.

Some small-scale farmers have been encouraged to raise specialty crops by the promise of a good market. Too often and too late, they have not found that market adequate.

Starting a new venture on your own means that no one will train you, as in the case when you are working for someone else. True, much good information is available, but when you make a mistake, there's no one else to pay for it. You become the risk taker.

While money can be made in part-time farming, there is another side to the coin. Money can also be lost.

Conveniences

Living in the country, for all the peace and quiet, has drawbacks. You may find some difficult to live with, impossible to change, and you may miss the good things an urban area offers. Let's take a look at a few things taken for granted in urban life, but usually not available in the country. Or, if they are available, the cost will be higher, because of the distances between dwellings.

In the city a reliable supply of water flows from a faucet. Water usually flows from a faucet in the country, but the source of the water is more apt to be your own well and pump. When the pump fails, it is your responsibility to get it repaired. Repair services may not be readily available.

Most rural residents are served by septic tanks which generally work well if slope and soil conditions are right, but when things go wrong, it's your responsibility to solve the problem.

Garbage is picked up in the city on a regular schedule, but in the country you often must make your own arrangements for disposal.

Schools in rural areas may be a considerable distance from your residence. Children sometimes meet the school bus before daylight and get home after dark. They may not have the same opportunities for participating in extracurricular activities or selecting subjects offered for study.

It is often the responsibility of the parent to provide transportation to and from the school for students participating in after-school activities. This may require a second car for the family.

In the country, the neighborhood store may be located several miles from your residence. Shopping habits will have to be changed or you will put many miles on the car, much gas in the tank, and more time going to the store.

Driving in the city and finding a place to park may be somewhat frustrating. Main roads in the country are generally good, but gravel and dirt roads are sometimes rough on the family car and keep it dirty most of the time. Snow or mud can close roads for several days, which requires keeping adequate supplies of essentials on hand. There's also the possibility of not being able to get to the off-farm job during these periods.

A move to the country does not guarantee that theft and some other crimes associated with cities will be completely left behind.

The nearest fire department, ambulance station, police station or sheriff's office may be miles away. Long distances

Below, Duane Karnatz and his bulk milk cooler bought with dairy receipts. Right, hand-move irrigation pipe for the Karnatz family's large garden effort. Lower right, Mrs. Karnatz starts seeds in home-made hotbed. (See text on facing page).

Jared M. Smalley

Three Years to Break Even

By Jared M. Smalley

Duane and Carol Karnatz are back on the farm in rural Minnesota, after spending 21 years in the Minneapolis-St. Paul area. Like many other persons who completed high school in the mid-1950's, they left their family farms for the growing metropolitan area to seek success and happinesss, but found mostly city congestion and frustration instead. They probably would still be trying to cope with the pressures of the big city, if Duane hadn't become disabled with a leg ailment which forced an early retirement from the postal service in 1972.

Between disability checks and Carol's job, the Karnatz family including daughters Patrice and Denise were able to make ends meet, but they yearned for a return to rural living. After much planning and looking, they found a 17-acre farm near Wadena, Minn., that met their expectations and budget needs.

"We moved on Friday the 13th of June, 1975, but we're not superstitious," Duane noted. The girls soon adjusted to their new home, and pitched in to plant an acre of strawberries and raspberries. Hopes for making the payments through raising beef and selling chicken eggs didn't work out as planned, so the family switched their emphasis to a small dairy operation and a large garden under a hand-move irrigation system.

Weather also contributed to the family's early problems, as a severe drought during 1976 was followed by an extremely cold winter that wiped out 95 percent of the berry plants. "I guess we would have become discouraged, if it wouldn't have been for our friendly neighbors," Carol recalls. She found it necessary to take a job as a ward secretary at the local hospital to supplement Duane's pension, as income from the farm fell short of expectations.

The drought passed and the garden flourished in 1977, while the dairy herd grew to seven cows and five heifers. A big planting of peas, coupled with sales of some berries and other vegetables, provided extra spending money. Between produce sales and milk checks the family expects to break even for the first time on its farm operation in 1978, Mr. and Mrs. Karnatz say.

In the meantime, they are getting along quite well by heating their home with wood, milking their own cows, raising their own chickens and eggs, canning and freezing their own vegetables, and sewing many of their own clothing needs.

Jared M. Smalley is Area Extension Agent, Wadena, Minn.

from places like these will be reflected in higher insurance rates. If roads are poor or the area is isolated, it will take a long time for emergency help to arrive.

Taxes may be lower in the country, but so is the quality of the services provided by these taxes.

Adjusting to Changes

The change from city to rural living often creates severe adjustment difficulties. Loneliness may set in for those left behind when the spouse is at work and the children are off to school.

Some people who move to the country find new financial stress. As previously mentioned, many times a second car is needed. Equipment to maintain the place becomes essential since services are so isolated. Repairs to fences where animals are present, and upkeep of buildings and equipment, suddenly take part of the monthly income. Some of the new experiences often become daily or weekly aggravations.

Chances are, if you were happy where you came from and your family has the ability to adjust, you will be happy in the country. Country living can be great, but it may not be as cheap or convenient as expected.

It's relatively simple to close up a city home and go off for a vacation, but in the country you have to consider a number of other factors. If you own animals, you must find someone to care for them. Vacations away from the home may have to be timed to not coincide with the planting, caring for, or harvesting of your crops if a small farming venture is involved. Sudden and severe storms in the winter can cause damage to properties that are left unattended.

None of these problems are insurmountable, but certainly should be considered when listing the pros and cons of city and country living.

You'll have some new and wonderful experiences even after considering the items discussed here and others you may think of. You and your family are the only ones who can determine what you are willing to put up with to gain what you feel you want from life. No one else can determine this.

Ready to Face the Realities of Small-Scale Farming?

By Edward Yeary and Manning Becker

For millions of Americans, the appealing features of life on a small farm outweigh any disadvantages. Discussions could be endless, and many questions would never be answered by words alone. Are you ready to deal with the realities of getting started?

Some questions, and answers, about your personal and family goals and resources will help you make the decisions needed to plan your entry into small-scale farming, or part-time farming on a few acres.

– Do you want to be a part-time farmer on a few acres and keep a full-time job in town?
– Can you find a farm that suits your needs in a location where you are willing to live?
– How much will it cost? If more than your savings, can you obtain financing?
– Do you have the knowledge and skills needed to operate a small farm?
– What about financing the production process?
– How and where will you sell your produce?

Part-time farming is usually the choice of people who want to "get out of town", but not too far, and to "have a little room", but not too much. Space will be available for a few fruit trees, a vegetable garden, even some poultry and animals to reduce the family food bills. A crop can be produced that will earn part of the needed family income.

Small-scale farming on a full-time basis is the goal of some urban families. Fed up with living and working in what they view as crowded, noisy, impersonal cities, they seek the more tranquil life and the satisfactory financial rewards they believe await them on a small farm.

Edward Yeary is a Farm Advisor-Statewide, specializing in farm management, University of California, Parlier.
Manning Becker is Extension Farm Management Specialist, Oregon State University, Corvallis.

Others will choose a compromise between those two extremes. Their goal is full-time farming, but they accept the reality that it may not be within reach. One or more family members will have to work elsewhere to supplement the income earned by farming a few acres of land. They can enjoy the advantages they associate with rural family life, which will be diluted by the necessity for some to commute to work, perhaps for long distances.

Some people truly want to get away from civilization, at least for awhile. They are willing to live very modestly, create with their hands, and produce with their own efforts to provide for their needs. Shunning many of the offerings of modern civilization, they choose to rely heavily on land, livestock, native materials and hard work to accomplish their objectives.

Your goal will be more easily reached if all family members, old enough to participate, have helped to make the decisions. Along with the many benefits you will enjoy, much hard work lies ahead. Problems, frustrations and adjustments will be more easily dealt with by a family united in purpose and working in harmony to reach the agreed-upon objectives.

Your Resources

If you are typical of many families who enter small-scale or part-time farming, you have three main resources from which you can expect income. These are labor, management, and capital invested in the land, equipment and operating expenses. These resources will dictate the type of small farming enterprise at which you can succeed.

You are going into competition with commercial agriculture and must face some of the realities of this competition. Most of America's agricultural production is highly mechanized.

In the growing of many crops a small unit cannot hope to compete with a typical large scale, highly mechanized farm. But this is not the case with labor-intensive crops, that is, those that require many times the number of hours of labor per acre that are needed by crops which are almost completely mechanized.

Labor provided by family members is a major source of income from small-scale farming. If you are going to sell your labor effectively on a small unit you may want to select from crops such as: fruit trees, grapes, vegetables, and others that require intensive use of labor; perhaps more than 150 hours per acre each year.

With labor-intensive crops it's easier for a small enterprise to compete.

It takes just as long to hand-prune a tree in a 5-acre orchard as it does in an 80-acre planting. By contrast, producing barley on a large-scale farm requires much machinery and only 1 to 3 hours of labor per acre each year. There is little opportunity here to earn labor income unless a large area is farmed.

So, if you are energetic, have plenty of help available, have production know-how, and all goes well, the right choice of crops will provide an outlet for all of this energy at a reasonable return. Labor and all the other production inputs must be used at the right time and in the right way or there will be no profits nor returns for your labor, management or capital. Your ability as a manager of your resources will determine whether or not you receive a return on your investment and for your labor. There is no guarantee here, it's up to you.

Getting Located

The decisions related to location and the type of farming must be made at the same time and may require some flexibility on your part. The larger the general area that suits your needs, the easier it will be to find your farm. Your choice of location and of farm enterprises may conflict with each other. Priorities may have to be given to one or the other, so weigh both carefully while you make your plans.

You are investing in both a small business and a home so this is the time to decide where you might want to live for many years. Consider all areas of interest and types of farming that appeal to you before you make a decision. It is a good idea to travel through these areas and to spend some time exploring them.

Zoning is usually a fact of life in the country as well as in most cities. Generally speaking, areas next to cities are zoned for small acreages; perhaps one or two acres or as many as five. Farther out, agricultural zoning may be in effect, and minimum sizes may be as much as 20 or 40 acres. Other choices are available, especially if you are willing to go a long distance from metropolitan areas.

If you plan to have some livestock, make sure that zoning ordinances will not spoil your plans. You may want to use chemical control measures for insect pests in your farm operation. Are there any restrictions against this in the areas that interest you?

You can learn a lot about areas and people by subscribing to community newspapers and by getting acquainted with local business and professional people. To learn about the usual high and low temperatures, annual rainfall and similar data, contact your nearest U. S. Weather Bureau office. Cooperative Extension personnel can be helpful in providing information about types of agriculture in the areas you visit.

The more time you take to just look around, to meet and talk to people and accumulate information, the more certain you will be that you picked the area best for you.

More for Less

Have you already looked at real estate advertisements? You probably have noticed that small farm units cost much more per acre than large ones. One reason is that many people like yourself are looking for small acreages. The prices asked are partly the result of a large demand for a limited supply of these units. Many are located close to urban areas and reflect the influence of these urban land values.

Does the place you are interested in have a house on it? If so, is it one you will be willing to live in during a cold, wet, windy, disagreeable winter? How about a hot, dry, windy, dust blown summer? Building or remodeling is expensive, unless you are capable of doing most of the work. Labor will be about half the total cost for most reconstruction.

No purpose would be served by attempting to quote prices since each unit will be priced by its owner and these prices change rapidly in most areas. However, small acreages of suitable farmland close to growing urban areas are generally priced far above any value represented by income that can be earned from farming.

You will need some machinery, perhaps an irrigation system, and some sheds or storage buildings. If livestock or poultry will be produced, fencing and housing must be pro-

Steadily rising land values and taxes make it increasingly expensive to produce livestock on small farms that are in the path of housing development.

Janet Yeary

vided. People are often surprised by the high cost of farming—including mortgage interest payments, equipment purchases, and seasonal crop expenses. Contrary to popular belief, land will not always produce enough to pay for itself.

Land and Water

Is the quality of the land suitable for your farming project? If it is presently being farmed and crop production is satisfactory, it is unlikely that any problems exist. But someone may have planted tree or vine crops on land not suited to their production and you could be buying trouble. If you are looking at open land, have some soil tests made and get expert opinion to satisfy yourself that your planned crop production could be successful.

You will certainly need water for domestic purposes and for livestock. In many areas of the country, water will be needed to irrigate your crops. Some areas have adequate supplies that are suitable for these purposes. In others, water is limited in both quality and quantity.

If you are planning to buy open land and develop a well, get information about the ground water supply. Check on the likelihood of having enough irrigation water if more land around you is developed. It is a good idea to get professional advice and also to check with neighboring landowners who are irrigating crops. Add their judgment to other information you gather.

Will you have the right to drill a well or to use water from a stream? Is a permit required? Ownership of water rights is a complex subject. More than just a few times the new owners of property bordering on a beautiful mountain stream have found, too late, that they had no rights to any of the water.

Laws concerning water rights differ from one state to another. Prospective buyers will do well to find out ahead of time whether or not any problems about water rights will concern them. If surface water is involved, check also to see if the babbling brook that is so lovely in the late spring is still flowing at the end of a hot dry summer.

What about labor? If you plan to develop or grow crops that demand extra labor at certain times of the year can you find that needed help? Casual labor may be available from a nearby school, community or college. You may need to make arrangements well ahead of time and perhaps furnish some transportation or temporary housing to attract the extra help.

Are you counting on family labor to get most of the work done? This can be the answer, but on the other hand after the "new" wears off some family members may find they don't like farm work and choose to sell their marketable skills where the hourly earnings are higher than on the family farm.

Even though the climate may be generally suitable for the farming program you would like to follow, there can be localized or spot problems. Check weather records for the possibility of a late spring or early fall frost if you are going to raise crops that could be damaged. Are there any peaks of high temperature that will normally occur and cause problems? If you are planning to raise crops that are not produced nearby be especially alert to climate-related problems.

Do you have a marketing program in mind? Some crops are handled by marketing cooperatives while others must be sold by the producer. Sometimes a roadside market will provide an outlet. Local retail food stores may be interested in what you plan to grow. Check on all possibilities and be sure you have available markets. You will need this information when you are arranging the necessary financing to buy your farm as well as when you have products to sell.

Knowledge, Skills

Do you have the knowledge and skills needed to operate the types of small farms that interest you? If you are counting on a reasonable amount of farm income, then you must have or acquire the ability to manage and operate this resource.

Farming skills can be acquired in many different ways. If you do not have them now, you have many sources of help available. Ways to acquire needed information include cooperative extension programs, agricultural courses in high schools and in community colleges, and working on farms.

Many of the crops most suited to small-scale farming require a great deal of production knowledge and skills. Applications of chemicals to the ground and perhaps to the crops must be done at the right time, in the right way, or total crop failure could result. A few days' error in timing can also create disastrous results by ruining the quality of the crops.

Perhaps you do not want to use any chemicals for weed or pest control; maybe not for fertilizer either.

Most crops must have some protection from insect pests if good yields of high quality products are expected. Knowledge of alternative methods of pest control must be acquired, along with the ability to use them.

Compost and animal waste may be your choices for fertilizer. Knowing how to make compost and where to locate organic fertilizer are skills and information you must have.

You are going to own a modest amount of farm machinery as well as the usual number of home appliances and these may all be located a long distance from repair shops. The cost of bringing repairmen to your small farm will be far more than you were used to paying in the city. A money-saving idea would be to learn about machinery repair and household maintenance.

Your financial resources must be considered when selecting the farming unit. You may need to borrow money to operate the farm in addition to buying it. These may influence your decision between buying an established farm or buying open land and developing your own enterprise.

Many sources of credit are available if you have a good credit rating and the lender views your investment as an acceptable risk. These include the Federal Land Bank, insurance companies, and commercial banks.

It is usually easier to finance an established farm than to finance open land, housing, and crop-developing costs. If your decision is to buy open land and develop permanent crops such as trees or vines there will be about 3 to 6 years of spending before you have very much farm income. Your financial resources may be adequate to carry you through the period. If not, then it may be necessary to find employment for one or more family members.

Financing will be an easy or difficult task, depending on the amount of cash and other resources you can use for this purpose. Borrowing large amounts of money requires plans and budgets to get the needed credit.

Plan and Budgets

In your preliminary planning you have considered several alternatives and now is the time for detailed planning on the specific farm you are considering for purchase.

What has to be done? When? What will it cost? How much labor is needed? You need detailed information and answers to these and to other questions before you actually make an offer.

A well developed plan of operation will help both experienced managers and novices to get a farming venture started. The necessary operational and business plans include a cash cost of production schedule for each crop or livestock enterprise, a total farm labor requirement schedule by months, and a total farm cash flow budget.

Start with the cash cost of production schedules. Charge for everything you will buy and use. Project the labor required during each month. The total for each month can then be measured against the amount you and your family can accomplish. The labor schedule will help you plan for the times during which you must hire help or arrange to trade work with neighbors.

The cash flow budget will show accumulated total farm cash costs and income received during the crop and livestock production cycles. These budgets should include an amount for family living and an operation loan, if one is required.

Money for operating expenses can be borrowed throughout the production cycle as needed and for the shortest possible time. As part-time farmers you may want to include your income from off-farm work in the cash flow budget to show your total family income and expense schedule. This will show the amount that is left over to pay any long-term debts that may exist.

There are many reasons why you and your family may choose life on a small farm unit. These include the complexities of urban living, a need for greater freedom and independence, and a desire to be more in control of your own lives. There is no one formula for success in meeting these expectations. Still, an energetic family working together on suitable land with a good plan to follow, a willingness to work hard, and the skills to manage all of its resources, has the greatest opportunity to achieve its goals.

Changing to a New Lifestyle: the Little Things Do Add Up

By James Lewis, Ed Glade, and Greg Gustafson

Many people in urban areas have contemplated relocating to a relatively more rural environment. In fact, since 1970 the nonmetropolitan population has increased by about 6.6 percent compared to 4.1 percent for metropolitan areas.

The transition from an urban life to a rural environment can be a most rewarding and personally satisfying experience. It offers opportunities and relationships not readily available in the urban setting. However, along with all the benefits go certain responsibilities and adjustments which should be understood and carefully considered.

For many persons thinking of relocating "back to nature", these considerations are only of minor concern and may actually be part of their preconceived desires. To others, however, the necessary adjustments and occasional disappointments may be more than they bargained for. Nevertheless, an awareness and careful thought about all aspects of rural life will greatly enhance the chances of a successful transition.

This chapter is intended to point out some of the transitions in lifestyles which should be anticipated. Often we discount the personal importance of living adjustments due to the excitement and anticipation of changes.

For most people moving to rural areas the adjustments are welcome changes, but some have found the transition too much of a sacrifice. A number of people have become disgruntled, viewed the initial decision to move as a big mistake, sold their error at a loss, and moved back to an environment similar to the one they left for the rural area. This is partly because there are certain amenities in everyone's life style that don't appear to be very important—until they are given up.

James Lewis and Ed Glade are Agricultural Economists with the Economics, Statistics, and Cooperatives Service, Washington, D.C. Greg Gustafson is an Agricultural Economist with ESCS at Oregon State University, Corvallis.

Perhaps this is nitpicking about a lot of little things but the little things do count and when everything about advantages and disadvantages to living on a few acres are added up, those little things can make the difference in whether or not you are satisfied and happy with rural living. The problem is that we seldom take the time to sit down and consciously evaluate the things that make us happy and the value of things we must either give up or put up with in a different environment before a change is made.

Many of the activities which are part of the urban community are not as accessible or abundant in rural areas. There are fewer choices available in rural areas for theaters, restaurants, bowling alleys, shopping centers, grocery stores, repair shops, physicians, medical facilities, etc. Some people get a great deal of satisfaction out of shopping and comparing from

Spinning wool supplied by your own sheep can be one of the satisfactions of a country lifestyle.

Scott Duff

one store to another. In many rural areas these choices are limited. The time and distance traveled to get places where these services are available is greater in rural areas. Fast food and all night drug stores with selected grocery items are conveniences in urban areas but are unlikely to be available in rural communities.

Radio and television signal reception may be relatively poorer in rural areas due to greater distances from the stations. An annoyance though a minor one, could be fewer stations with programs to your liking. Private telephone lines may not be available.

Public services such as police, fire protection, water, sewer, road maintenance, and garbage collection are not as abundant or accessible in rural areas. Many of these services are expected and readily available in urban areas but in rural living some become your responsibility. Garbage and trash disposal are good examples of something you must take the time to resolve.

Some Make Mistakes

There are examples of where a vacation or weekend place is acquired and then at some later date the decision to make it a permanent residence is made. When you first buy the place it is a delight to visit. Indeed there may be an abundance of things to do in the little time you have to spend on those long weekends and vacations. But you also find time to sit back, relax, and enjoy the peaceful serenity of your few acres and the sweet country air.

Upon making the place a permanent residence you may find yourself catching up on all the improvements and renovations that were planned. The lifestyle in general is at a slower pace than in urban areas. What once appeared to be peace and tranquility can become boredom and dissatisfaction for some people. This is not to imply that there is nothing to do in the rural area, but that the things there are to do may not be particularly satisfying for some on a routine basis.

Caring for livestock, for example, is a reasonably minor chore in good weather. When the cold hard winter comes and pipes freeze and burst and you have to go chop ice so the livestock can get water twice a day regardless of wind, rain, sleet or snow, all those fun things become a job. It's a good job if you like it. But if you're not sure, find out and seriously think about it before you get involved.

Other important considerations that can make a difference in how happy you are on your few acres include the availability of off-farm employment opportunities, neighbors, friends, play-mates for children, and the family's consent on making the change. There will be fewer neighbors, friends will be farther away as will be playmates for children. One unhappy member of the family can make all the rest uncertain about the new lifestyle. Getting back and forth to social, civic, and athletic activities will require planning, coordination, time, and sacrifice by some of the family members.

A family's desire concerning its mobility and leisure time must be balanced against the requirements in both work and money involved in alternative farmstead activities. Seasonal and daily patterns of life must be adjusted to accommodate these variations. The decision to raise a small garden or a few chickens is one thing, but the keeping of milk cows or dairy goats which must be milked twice a day about 300 days a year is something entirely different. All the pleasures and enjoyment of country life can be quickly lost when faced with a never ending cycle of chores. Poor choices of enterprises or combinations of enterprises can lead to trouble and frustration.

Therefore, to help avoid the danger of "getting in too deep", or the feeling that you're tied down to the place, be aware of the full extent of involvement required in each new undertaking. It's better to start a little slow in developing your "few acres" than to rush into something unprepared or un-suspecting.

By carefully planning, coordinating, and controlling the various farmstead enterprises and activities, rural living can provide a healthy, wholesome mixture of productive work and recreational opportunities. Local county agents and Extension personnel, along with information contained in this Yearbook of Agriculture, can offer significant help along the way.

Realistic Goals

No family should get its hopes too high in any new ven-ture, and this is particularly true of adjusting to the rural scene on a full-time basis. While beginning stages of the transition can inspire one to try to be as self-sufficient as possible, the facts remain that the cost is just too high. Although making your own butter, bread, and preserving most of your foods may sound gallant, healthful, and natural, you must remember that only the experience of many years makes your dream come true.

Limiting the amount of necessary work, and adding more variety to your country place gradually, can ease the transition and prevent overwhelming frustration.

When you can handle a small garden with a few chickens, for instance, you may then consider the addition of a family milk goat, or even a cow. However, for the sport minded, happiness may be a horse or pony instead, or for the specialized hobbyist, a few sheep for wool to process, or a rabbit project for furs to work and sell. It's up to you, but proceed with caution.

Getting away from the routine of rural life may be necessary from time to time, and vacations should be part of one's established priorities. While it may be more difficult to get away than for urban dwellers, it is not impossible. With a little planning and preparation, various arrangements can be made.

For example, most people who relocate to a rural area still have urban friends who enjoy regular visits to observe the rural scene. They might dream of a future venture of their own. These families are usually more than happy to take over your feeding chores when given specific, written instructions, and emergency phone numbers, such as your veterinarian, an experienced farmer, and where you can be reached if necessary. While you may see this as an imposition, most people consider it a welcome learning experience for their children and themselves.

Other choices of arrangement might include paying a teenager from a nearby farmstead, or swapping chores during vacation with another family with a similar setup. An extended trip may mean using a combination of several people to help at different times.

Whether you attempt to become totally self-sufficient and live entirely off your few acres, or decide to just raise the family's fruits and vegetables in a backyard garden, be prepared for certain facts of rural life: water freezes in winter—crops fail—fences fall—and livestock get sick. Realizing that sickness, injury and disappointing harvests can occur, and being able to cope with these events, are all part of the rural experience.

So even though the best laid plans don't always pan out, more often than not it is an educational experience which provides an opportunity to understand your environment and to constantly learn new ways to use resources of nature to your advantage.

Will you be happy?

Not everyone would be happy on a few acres in the country. But, would you? This is not an easily answered question, and, in any event, only you can answer it.

If you are honest with yourself and know yourself well enough, you can find out a lot by taking a personal inventory. An appreciation and tolerance of nature, of course, is essential. The quiet of the countryside, the smell of freshly mowed hay, the sight of livestock grazing in a lush green pasture, and the taste of vine-ripened tomatoes are there to be enjoyed. However, nature can also be harsh and unpredictable.

Being resourceful is important. If you have to call in outside help whenever a fan belt in the car breaks or the lawnmower needs a new sparkplug, a large amount of time and money will be spent at the repair shop. Minor problems, such as when Japanese beetles infest the vineyard or the calves get scours, are less bothersome when you can solve them yourself.

Patience is a must. Life in the country is slower paced. City dwellers seldom build or grow anything at the same scale as you will in the rural area. It takes time and planning to plant a garden or build even a modest-sized greenhouse. In the garden it takes from four to six months of careful nurturing before you can taste the fruits of your labor. In the orchard it may take four to six years before the first apple harvest from those newly planted saplings.. An appreciation for and acceptance of nature's time schedule is a prerequisite to enjoyment of life in the country.

Country living offers some dramatic differences in lifestyle relative to the city. In general, rural life is more family oriented. There is less anonymity in the country—most rural residents know quite a lot about their neighbors. Some may find this openness to be a refreshing change while others might deplore any loss of privacy. Rural social activities also present a contrast to typical urban activities—church-related events, farm organization meetings (Farm Bureau, Grange, etc.), the local auction, county fair, 4-H events, garden club meetings, and local high school athletic events are traditional mainstays on the rural social calendar.

The bottom line, however, in evaluating your own suitability to country life will depend on how well you like the farm routine. On your few acres you will spend most of your time doing chores and performing seasonal tasks. Much of the time you will be working alone or with another family member. Hence, genuine enjoyment of this life requires that you like the work and that you enjoy your own company.

A final consideration, but one that's usually beyond the control of most individuals, is that of local ordinances and zoning laws. In many areas, the keeping of livestock and poultry is restricted to certain size acreages, and incorporates strict boundary line buffer areas. These requirements, however, may vary widely from area to area.

In some counties at least 2 acres are required before livestock can be kept, and animals must be at least 100 feet from any property line. Restrictions are also placed on the number and kinds of livestock allowed. On the other hand, adjoining counties may have no restrictions at all and if they do, restrictions may be loosely enforced. Therefore, in order to avoid trouble which can affect plans and hamper efforts, check all aspects concerning local restrictions before engaging in any new enterprise.

People moving from urban to rural environments are looking for changes. However, the transitional adjustments may be too much of a change for some while not enough for others. Regardless of any promises or shortcomings of the rural climate in tangible terms, do not overlook close reflection on how much habits, preferences, or desires can weigh in whether or not a decision to live in a rural area is wise. Many people have moved to rural areas, become dissatisfied with their decision, lost time and money, and moved back to an environment similar to the one they left for the rural scene. Hopefully, you will not become one of those statistics.

You are the only one who knows what makes you happy. So before you decide to move and invest your resources, sit back and carefully reflect on the transition to a rural lifestyle.

One Family's Satisfactions With Home on Few Acres

By Ned D. Bayley

For more than 20 years, Wedge and Candy O'Donnell have made their home on a few acres. They raised a family there. They hope to spend most of their retirement there.

They are satisfied with the home life provided on a few acres for a combination of reasons. Both Wedge and Candy have rural backgrounds. They like to work outdoors. They believe the country is the best place to raise children. They have had a stable income outside the home.

Their first three efforts to acquire a few acres aborted. They bought an inexpensive acre in Wisconsin that contained a tiny cement block house and bare land. They hoped to enlarge their holdings. After a few years, the State decided to abut their property with a major highway. They sold out at a profit.

They bought the second place at the peak of a price cycle. It was a few miles from the first and had 40 acres of rich farmland. The old farmhouse was sound but run down. The barn leaned precariously away from the prevailing winds. They tackled the house first, modernizing the furnace, repairing and redecorating the high interior walls and ceiling, and making rich red drapes for the tall windows.

After six months, Wedge accepted a job in another State and they moved. The price cycle had gone past its peak. They lost all the profit from the first sale.

The third place contained 40 rolling, gravelly, partially wooded acres in Minnesota. They cleared brush and trees to make a driveway. They dragged a brooder house onto the property as a weekend shelter until they could build a home. For their building site, they selected a high knoll looking towards

Ned D. Bayley is Staff Assistant, Office of the Assistant Secretary for Conservation, Research and Education.

the east and a small lake. They worked with a local builder on house plans.

The State of Minnesota requisitioned their building site for a gravel pit. At a high profit but with shattered plans, they had to let the place go.

Fortunately, the third abortion occurred as Wedge was being offered a promising professional opportunity in the East. They moved to Maryland and immediately scoured the hills and dales for another place in the country. Their ability to buy only 5-1/2 acres in the East as compared to 40 in the Midwest was symbolic of the difference between the two areas.

The place they bought lay between two ridges and had about an acre and a half of woods and a small twisting creek cutting through the front yard. A 400-foot driveway descended from the road, skirted the front yard, and crossed a bridge into the garage. The woods surrounded the cape cod house. Open pasture fields radiated out from a four-stall stable between the house and the road. A white-board fence encircled the fields.

One of the first things to be changed was the white-board fence. One season of whitewashing and constant repair was enough to induce the O'Donnells to replace it with 4 strands of barbed wire on steel posts.

Damming the Creek

The creek was the next major target for renovation, and continues to demand attention. Two small dams were added to the existing one in order to control erosion. The family (then containing five members) spent nights and weekends moving rock piles from the fields of a neighboring farm. They fused the rock into layers of hand-mixed motar as armor for the sharp bends in the creek banks.

Several years later hurricane-caused floods ripped the armored banks apart. The family started once again with heavier rock. Still later they dug a new channel to straighten out a bend that threatened to undermine their wildflower garden.

They continued their landscaping with a new terrace in front of the house, protective screens of white pine between the front yard and the pastures, a screen of closely planted hemlock between their place and the neighbors, large numbers of flowering shrubs and evergreen bushes, and curving border plantings of annuals and perennials.

They liked everything about the house except the kitchen. It was only 8 by 10 feet. When money became available, the O'Donnells had it enlarged. Candy designed the entire addition and built all the shelves, doors, and drawers. She used equipment at a night-school woodworking class to supplement what they had at home.

Later the O'Donnells improved the basement, and enclosed the porch with insulated glass. Recently they added two bay windows to Candy's sewing center to take advantage of the winter sun and summer shade from the room's southeast exposure.

Yes, the youngsters had horses. Before getting them, however, each child raised a steer, learned how to feed and care for an animal, and used the money from the sale of the steers to help buy a horse or a pony. They had four at one time, including a foal born on the place.

Mainstay of the stable was a 25-year old, scruffy, white welsh pony with one good eye. All the children learned to ride on him. Another equine often in use was a big palomino gelding with more draft than saddle blood in him. He harrowed the gardens, plowed snow, and pulled fence posts, besides being ridden. The horses and pony were sold when the daughter left for college.

Arming the creek bank with rock.

Robert C. Bjork

Besides the horses, the O'Donnells had cats and boxer dogs. At one time there were 32 does and several bucks in an enclosed rabbitry. The rabbits served as 4-H projects and as subjects for anatomy study in school. The children learned the meaning of mating, birth, growth, and death in a home laboratory.

Three Gardens

In some years there were as many as three vegetable gardens: two for the household and one as a 4-H project. The garden produce regularly filled a huge deep freeze and the shelves of the root cellar. In recent years, a cold frame has made lettuce possible in the early months of spring.

One of the gardens eventually was used by the younger boy to grow nursery plants—the only 4-H project that actually made money. The vegetable and flower gardens produced materials for Candy's floral and dried arrangements.

Birds dropped red-cedar berries from their perching on the fence rows. The rapidly growing seedlings provided the O'Donnells with Christmas trees for several years.

The boys' enthusiasm for horses waned after a few years and their interests turned to machinery. One of them renovated two horse stalls into a machine shop for small engines—still another 4-H project. Later, both boys used the former hay shed as a garage for overhauling jeeps.

Right, lettuce crowds cold frame as Candy tills the garden. Below, weeding is a never-ending job.

Robert C. Bjork

The 5-1/2 acres have raised the O'Donnell family and entertained their friends and relatives. The pastures not only maintained livestock but also provided a go-cart track, a hill for sleds, and a place to play football and baseball. The yards were often used for badminton, volleyball, croquet, and horseshoes. The woods furnished the fuel for outdoor barbecues and for cozy nights by the living room fireplace.

The O'Donnell children are gone now as well as the horses, dogs, cats, and rabbits. An acre of the pasture has been sold for development. With the help of machinery, Wedge and Candy keep the lawns growing and mowed, the flowers blooming, and the vegetable garden productive. The community has grown up around them but the perimeters of the acres are protected by woods and screening plants.

A green heron, a bittern, a hawk come occasionally to the creek hunting for minnows; smaller birds bathe in it and drink from it; crows soak their food in it. When the snow comes, cardinals, titmice, woodpeckers, bluejays, sparrows, chickadees, and other small birds crowd the outdoor feeder. A few squirrels and an occasional raccoon frequent the grounds.

The O'Donnells plan to retire on their few acres—at least until their energy runs out, or high taxes drive them out. Whatever happens, their home on a few acres has been full of living and, to them, worth all the effort.

Robert C. Bjork

Left, Wedge fertilizes
grass near dam.
Above, thick spears of
asparagus shoot up.

Acquiring
That Spot

Part Two

William E. Carnahan

Selecting a Region, Community, Site

By *William H. Pietsch*

Upon deciding that you want to move from your present situation to a "few acres in the country," choosing the location emerges as a crucial question. By now you should have determined the objectives an acreage will help you attain more satisfactorily than your present location.

Based on the range of options available as you make the decision, you may need first to determine the general geographic region in which you will relocate. Your second step will be to select a community that satisfies the needs of your family. This community might range from a suburb of a major city that allows "a few acre" lots, to a rural "community" where your next door neighbor is several miles away. Your third step will be to select a specific site that you can afford from among those available for sale.

The first and third of these steps will be influenced by the objective you wish to achieve by moving in the first place. The second step will be influenced primarily by your family situation and desired lifestyle as well as what you wish to accomplish on a few acres.

For each of the rural living categories examined, this chapter will take up the unique aspects of geographic region (Step 1) and specific site selection (Step 3). That presentation will be followed by a discussion of selecting a community (Step 2) which is less influenced by the type of operation than by individual lifestyle preference.

For purposes of discussion, the following categories of rural living objectives will be considered: a) a rural residence; b) a rural residence with expanded gardening and orcharding opportunities; c) a "mini-farm" aimed at food self sufficiency;

William H. Pietsch is Extension Economist and Associate Agricultural Economist, Cooperative Extension Service, Washington State University, Pullman.

or d) a retirement or part-time farm. As each of these primary objectives is evaluated, factors to consider include the geographic region, climate, and local topography that underlie achieving the objectives.

The Rural Residence

If your objective is to have a rural residence, choice of a geographic region will be based primarily on climate (your personal comfort), social and cultural preferences, and employment opportunities.

Since other people considering relocation are likely to use similar criteria in their decision process, areas generally desirable on all these factors attract a lot of new residents. In many areas this means strong competition for the available "few acre" locations. Much of the westward migration in the United States during this century can be attributed to climatic conditions viewed as favorable by large numbers of people.

As you consider geographic areas, your age and thus your planning horizon is important. If you are nearing retirement, factors such as cost of living and climate will likely outweigh employment opportunities and long-range water and energy availabilities. If you have a longer planning horizon, you will give more weight to expected employment potential, environmental quality changes, and direction of social and cultural changes.

Selecting your specific site for a rural residence will be influenced by such factors as the view, surrounding vegetative growth, and the general appearance of immediately surrounding areas.

Factors easily overlooked are availability of a water supply, soil characteristics that influence septic tank construction, surface drainage and flood hazard, and wind exposure. These are difficult or impossible to change once you've committed yourself to a site. Unless a common water supply is available, an agreement to purchase land for a homesite should be contingent on availability of an adequate water supply.

Expanded Gardening. This living situation is similar to the rural residence, but with the added objectives of growing vegetables and fruits primarily for home consumption. Most gardeners learn to limit their selection of crops to those climatically suited to their specific situation, or else find innovative ways to overcome the challenges posed by nature. While choice of a geographic region will likely hinge on several other factors, a serious gardener should give some consideration to climatic restrictions on crop alternatives.

Many factors contribute to successful fruit and vegetable production. The overriding elements is the combination of environmental characteristics we think of as climate. One of Webster's definitions of climate is "the prevailing or average weather conditions of a place as determined by the temperature and meteorological changes over a period of years." Individual characteristics such as day-length, sunshine and temperature intensity, frost-free period, and relative humidity combine to determine the quality of the environment for vegetative growth.

Choice of a specific site for the garden can be extremely important. Besides the characteristics desirable in a rural residential site, you need to consider soil depth and texture, frost hazard, and direction and degree of slope.

Although certain modifications in soil characteristics can be achieved, you will get a headstart on a successful garden by starting off with a deep, well-drained soil that contains substantial organic matter.

Texture can be a critical soil erosion factor when combined with certain slope characteristics. Few experiences are more disappointing than to find your topsoil (and perhaps the plants that had been growing in it) at the bottom side of your garden after the summer rain you had awaited with so much anticipation.

Within any general climatic area, frost hazard can vary greatly. Elevation of your garden relative to its surroundings can spell the difference between an ideal site and one that is "frosty".

The principle underlying "frost pockets" is that cold air is heavier than warm air. This weight differential causes cold air to displace warm air in low-lying areas. Frost pockets occur where cold air cannot drain to still lower areas. Site selection that embodies an awareness of this principle is critical if you plan to grow crops that require the entire frost-free period to mature.

Direction and degree of slopes on the site you are considering can be important indicators of your gardening success. As indicated above, the degree of slope has an important influence on soil erosion potential. Excessive slope also makes preparing and tending the garden more difficult—especially if you use power gardening equipment. Desirable slope direction is influenced somewhat by the crops you intend to grow.

It is commonly recommended that the vegetable garden have a gentle south or southwest slope. This allows maximum use of solar energy to warm the soil in spring as well as to speed crop growth throughout the season. The recommendation holds generally for those crops whose early season growth is important to their maturity.

For certain tree fruits, however, warming the soil leads to early spring growth that may bring the trees into blossom while freezing temperatures are still likely. Thus, if orchard production is an objective, and if frost is a hazard, slope direction and air drainage can be important in your site selection. As with soil conditions, certain modifications of other natural situations can be achieved, but normally only at a cost—either in terms of time, effort, or money.

The "Mini-Farm"

Objective of this operation is to provide a wider array of homegrown foods for the family. Food items to give away, barter, or to sell in small quantities are incidental to the primary purpose of this operation. Primary additions to the gardening situation described earlier include producing meats, fish, eggs, milk and other dairy products, and honey.

Climatic conditions are generally less critical for livestock production than for fruit and vegetable crops. Thus, the evaluation of climatic factors for the mini-farm is similar to that of the residence-gardening option.

Perhaps the factor that most distinguishes the mini-farm from the expanded garden rests in the differing public perception relative to crop and livestock production. Zoning restrictions developed to protect against potential annoyances (sounds and odors) or damage (digging, waste disposal, or chewing) keep animals out of many neighborhoods. Such restrictions are not normally regional, but local options. Thus, selection of a community takes on an additional dimension in the case of a mini-farm.

As you consider the community, take careful notice of emerging changes in settlement patterns. Regardless of the present situation, if rural residences and/or expanded gardening operations are increasing, certain kinds of livestock on your mini-farm may become increasingly unwelcome in the neighborhood. If your serious gardening neighbors observe your goat pruning their azaleas or your pig sampling their strawberries, you can be assured they will question whether more restrictive zoning requirements are in order.

As regards specific site selection, requirements described for the rural residence and expanded gardening situations are appropriate.

Due to the animal wastes produced on your mini-farm,

drainage (surface and underground) will be important. The site must allow for locating your livestock activity so that waste materials cannot be carried to your water supply or your neighbor's. The site should also assure that odors from your livestock will not be brought to your home or neighboring ones by the prevailing wind.

Part-Time Farming

Unlike the three situations discussed above, the part-time farmer plans to use the production from a few acres as supplemental income. The primary source of income may be from off-farm employment, retirement benefits, or other sources.

While not necessarily the case, many part-time farms (especially on limited acreages) are merely expanded versions of the mini-farm. The primary difference may be that production of crops and/or livestock in excess of family needs is intentional.

The extent to which you intend to supplement your primary income with farm earnings will dictate the size and intensity of your farming operation. That will, in turn, influence both the geographic area and specific site that contributes the most to reaching your objective.

Since the part-time farmer is likely to spend more time operating the farm than the "mini" farmer, location relative to primary employment may be an important factor. Time spent commuting is not available for the primary work activity, or operation of the farm.

Certain types of production (because of their seasonal or relative labor and management intensiveness) are more suited to off-farm work than others. Determining types of farm production to be undertaken will help you decide the characteristics the individual site must possess to be used in meeting your part-time farming objectives.

Since the part-time farm is operated to produce goods for sale, the location factors include access to markets, availability of services (financial and informational), and periodic labor if needed.

Unlike the "mini" farmer or gardener, if the part-time farmer intends to have farming contribute to family income, a site will be needed with natural characteristics suitable to production.

Moderately expensive soil or climatic modifications are likely to be more than the part-time operator can afford without violating the criterion that the operation add to the income available for family living.

Picking a Community

Your choice of community will be based more on personal living desires than on the way you intend to utilize your few acres. Your background and lifestyle will have influenced your expectations so far as public services and facilities are concerned.

The importance you attach to the various services and the ease with which you can use them will largely determine your choice of community.

When selecting a community, check out services provided. Since your tax dollars will be a part of the basis for support, you may want to assure that other taxpayers have priorities for services similar to your own. Generally the variety but not necessarily the quality of available services will be related to the number and diversity of the people that support and demand them. Some public services to investigate are:

For families with school or preschool children, the availability and quality of public schools in the area may be an overriding locational factor. If transportation is provided, the duration and nature of the ride can have definite impacts on the child's schooling. In many rural areas, the school is an important social institution and convenient access is especially important.

Reasonable access to adequate health care, while of some importance to everyone, is likely to be of high priority to families with small children as well as older people. The ability to gain access to health care in a timely manner may be more than simply a matter of distance. Availability of ambulance service may be of critical importance for certain types of emergencies.

Safety of your family and property is of primary concern regardless of where you live. Most rural areas have high quality —but thinly spread—law enforcement capabilities. If you're accustomed to a feeling of safety resulting from highly visible police activity, the infrequent sight of a county sheriff's car may leave you feeling insecure.

Perhaps of more common concern is the availability of fire protection. Before you choose a site, find out if it is in a fire control district. This will not only determine the availability of firefighters in an emergency, but also help establish the rates you will be charged for hazard insurance.

Access to your home can be a major obstacle in some parts of the country. While roads made impassable by mud or snow can be an inconvenience if you were planning a social engagement, they can spell disaster should you have a medical emergency or a fire.

Many rural roads are well maintained. But those most likely to receive care are mail routes, school bus routes and fire access roads. Access to an all-weather road can be an important factor in overcoming the feeling of isolation sometimes experienced by residents of remote areas.

Availability of public transportation may be an important factor for rural residents who have jobs a considerable distance from where they live. Apparent continuing increases in energy costs will force some commuters to find less costly transportation alternatives or settle closer to their place of employment.

A narrower range of opportunities for activities provided by the community is likely in rural settings. Many people who migrate to those rural settings, however, view their new living situation as providing a private recreational activity.

Beyond the availability of services, the cultural, political, and religious environments of an area can heavily influence your enjoyment of living there. In the same manner as new residents evaluating an urban neighborhood, the potential rural resident should examine whether the community offers a setting in which family members can enjoy functioning.

Time spent visiting residents of the neighborhood can be of great value in determining the attitudes and values of the people you are considering joining in the community. If your relocation is to an area remote from many of the public services you feel important, an increasing feeling of dependence on those people around you is likely.

Information Sources

Preparation for acquiring the land for your new living experience requires a great deal of information. Many public and private sources at both the state and local level may be useful to you.

At the state level, the Land Grant College can be a valuable source of information. The combined efforts of research and extension within these agricultural colleges provide information on practices required to successfully produce the "commodities" common to that climatic region. Most states also have an Agriculture Department that can provide additional information relating to local agriculture.

Once you have selected a specific location in a state, many sources of information are available to help you locate the specific site most suited to your objective. Several agencies of the U.S. Department of Agriculture (USDA) have offices in nearly every county.

Detailed information on soil characteristics and methods of

resolving soil problems is available from the Soil Conservation Service.

Nearly every county has a County Extension Office. This is a cooperative effort involving county government, the state Cooperative Extension Service, and USDA. Educational programs for youth, family living, community development, and all aspects of agriculture are conducted.

Weather bureaus are important sources of data concerning average growing conditions.

County officials can provide needed information on tax levels, zoning restrictions, and building requirements that may influence your decision on location. This information is available either free or at nominal charges.

Many private information sources are also available as you look at a specific location. Real estate brokers can be a valuable source of information on property values and proposed land use changes or restrictions. They will normally provide this information without obligation. Garden clubs and other social and professional organizations can be additional sources of valuable local information.

Unless you are familiar with the area you are considering, time spent personally evaluating the area is essential.

Several cautions are in order for anyone seeking a new experience on a few acres.

First, define your objective clearly. Make certain you are moving toward a new well defined experience, not merely running away from your present situation.

Second, be careful of over-enthusiasm. The thought of new experiences and challenges is exciting to nearly everyone. Don't let that excitement cloud your vision.

Third, be cautious of hasty decisions. Advance planning of your land acquisition will be well rewarded when both the location and site characteristics are carefully matched to the objective underlying your relocation decision.

Can You Get What You Want? Homesteaders and Others

By Daniel G. Piper

There appear to be four distinct groups of people interested in living full-time on a few acres in a rural area. First, those interested in subsistence homesteading. Second, retired persons desiring to live out their lives on a few acres. Third, those who wish to live in the country but want to commute to city jobs.

Fourth, there are those who wish to escape urban pressures but whose resources, either financial or emotional, or both, do not permit an independent residence. These are the people who live in communes, most of which are owned by a single owner, with permission to build and to live usually granted free to commune members.

The homesteaders range from those who want to obtain as much food and energy independence as possible to those who want the best of both worlds and can afford it.

Energy independence is expressed in some by rejection of available utility power, by erecting windmills, using oxen or horses instead of tractors, adopting a raw food diet, heating with solar energy and wood-burning stoves.

Organic farming methods are frequently adopted for perceived health benefits as well as the fact they require less purchasing of off-farm supplies such as fertilizer.

Homesteaders vary regarding their attitude about cooperating with local health and building regulatory authorities.

At one extreme are those who learn to play by the local laws, do the required paperwork, and construct their improvements under existing building codes. They say there already is a resident hostility to homesteaders and they don't want to alienate anyone. This group is usually better off economically and can afford the considerably higher costs of going by the book.

Daniel G. Piper is an agricultural economist with the Natural Resource Economics Division of the Economics, Statistics, and Cooperatives Service, stationed at Berkeley, Calif.

'Doing Their Thing'

At the other pole are homesteaders who don't recognize land deeds or inspectors. They argue that the current legal system has made it illegal to be a pioneer, and that the manufacturers and trade unions have written the building and health codes to make it more difficult for the owner-builder.

A much larger number, including many who are buying their own land, believe they should be allowed maximum freedom in building design as long as they hurt no one.

Rural building inspectors have had much difficulty in enforcing local building codes because of the strong desire of many homesteaders for freedom to "do their own thing." In some areas the people housed in unrecorded, unlicensed structures outnumber those in regular housing.

Pressure from this movement in California has led to legalization of some kinds of rural buildings called Class K which permit non-flush toilets among other things.

The commune population frequently locates in more remote areas, sometimes using a bad road as a form of protection. These people develop many different organizational formats dealing with decision-making, income-sharing, gardening, animal-raising, and child care.

Retired persons are more apt to cooperate with local regulations. Because of their age and often limited energy they are less likely to take shortcuts on utilities and sanitation.

The retired are probably used to living in an urban setting. Hence they are willing to continue to purchase professional services such as plumbing even if they must pay a premium. However, inflation is forcing a number to sell out.

Some settlers who maintain close urban ties through their jobs are not attempting to undertake a subsistence homestead. Depending on their individual situation, they will seek to be within commuting distance and locate near a hard surface highway. They will not be able to personally devote as much time to improving their site as homesteaders or retirees.

Living Arrangements

There are a number of possibilities regarding the organization of those occupying a site. An individual can do it by himself or herself. A couple can settle on the land. A group can form a legal partnership on an acreage.

The last arrangement may be advantageous if a larger parcel can be obtained at a lower cost per acre. Funds can be pooled and the land subdivided later with each living unit taking separate title to an individual parcel.

Check local zoning regulations to be sure this is allowed. Some areas have minimum acreage laws for parcel splits, and limits to the number of residences per parcel. For example, the recently enacted California Coastal Zone Act stipulates a minimum tract size of 20 acres for certain coastal areas. Large lot zoning is becoming increasingly more common.

All people involved in settling on a tract of land should understand their goals. It is often useful to draw up an owner's agreement or an agreement between occupants to specify the group's goals. They can then select the form of legal arrangement best suited to their situation.

Many groups are competing for land. They include commercial farmers, loggers, miners, speculators, land developers, those desiring a recreational or second home, and "dropouts" seeking to break away from the mainstream of American urban life. Farmers, speculators, and land developers are primarily interested in accessible, large acreages of high quality level land for crops or urban development. They desire the land to be accessible for ease of equipment handling and transportation.

Recreational users are interested in small accessible tracts but less concerned about agricultural features. "Dropouts'" want at least a certain amount of seclusion while a few seek almost inaccessible land for its isolation value.

The best strategy for persons seeking to live on a few acres is to locate in areas where there is little competition from commercial interests, land developers, speculators, and recreationists.

The price of land will be lowest where there is less active competition for the available tracts, all else being equal. But the very act of settling brings gradual growth of demand and increased land values and taxes.

Site Evaluating

There are a number of strategies for obtaining your site. The simplest way is to go directly to the local real estate dealer. The disadvantage is that available sites may not satisfy your desires.

A better strategy is to first develop criteria for evaluating both the area of the country in which to settle and the desirable features associated with a specific site. Once these are established, you can either go on your own or seek professional help in the evaluation process.

Site features of concern to one group may not interest others. For example, access will be very important to the

48

retired and those with jobs in the city but a minor consideration to homesteaders and communes. By contrast, soil fertility
will be important to homesteaders but of minor interest to city
workers, with retirees going either way.

Some general factors to consider when evaluating a par-
ticular piece of real estate include:
- Nature of physical improvements, if any.
- Accessibility of the site to the owner, to others.
- Fertility and drainage state of the soil.
- Adequate exposure to wind and sun for a homesite and gar-
  den.
- Water supply.
- Building code restrictions and degree of enforcement and
  their translation into building costs and structural limitations.
- Distance from medical, educational, and other facilities.
- Congeniality of neighbors.
- Length of growing season.
- Opportunities for local employment.

Determining the value of rural land is a complex process
which essentially boils down to a function of supply and demand. If the piece of rural land has potential for conversion
to higher valued uses such as recreational or commercial uses,
its value will be enhanced.

If large numbers of people are interested in homesteading
in a desirable area, the price may be driven up. For example,
in one area of California, identified as desirable by many homesteaders, the price for parcels of land under ten acres has been
bid up from about $1,500 per acre in the late Sixties to about
$4,500 per acre at present.

Even though the county's sloping land is normally used for
sheep grazing and its level land for marginal dryland barley
production, the high demand has pushed prices far beyond the
range of many potential buyers.

Building Costs

Although purchase of raw land is an important step in
becoming established on a few acres, the high costs of building
on that land usually far exceed the cost of land for people
living on a few acres.

There are both advantages and disadvantages to buying
land with improvements. Even though there may be improvements on the land, extensive costly remodeling may be necessary.

Homesteaders generally are owner-builders, while commuters to urban jobs may not have the time to build themselves.

Retirees may or may not have the energy or interest in becoming owner-builders.

Construction costs may be reduced by people who do at least some of their own building, especially if they use materials from the site or from nearby.

Paying for land and its improvements generally cannot be done in a single cash transaction because of the low cash position of most buyers. Most persons have to plan on arranging to pay for the purchase over time.

There are three possible financing approaches to follow: have the seller extend credit to the buyer, borrow from friends or relatives, or obtain a loan from third party commercial lenders.

In California, as in some other areas, state or local restrictions on subdividing land may make impossible the long used process of paying for one's land by selling off a piece of it.

Two basic alternatives involve direct loan agreements between buyer and seller. These are land contract financing, and exchange of a mortgage or deed of trust in return for cash or credit.

Key difference between the two approaches is the point at which the seller conveys title to the land to the buyer. In the land contract, title is generally conveyed only after all payments are made. When the seller gives a mortgage, on the other hand, the buyer receives title to the land immediately.

The land contract is both the most prevalent and also the most risky way to buy rural land. The method is risky because during the payment period the seller gives the buyer only possession of the land.

Sellers often like land contracts because they offer an advantage if the buyer can't make his payments. Since the seller already has title to the land, he does not have to go through a regular court foreclosure. If the buyer leaves, the seller then has the land and also all the money paid by the buyer.

Use of a mortgage for financing offers much greater protection for the buyer than a land contract. The buyer immediately receives title and promises in writing to completely repay the seller.

Reluctance to Loan

Many commercial lenders are reluctant to loan money on rural non-commercial farmland for several reasons. First, the market is thin compared to serving urban and suburban dwellers, and commercial farmers. Second, homesteading appears to lenders to have a less certain income flow than for urban residents or commercial farmers.

In some states, homesteading protects the property from being used to pay off debts. Thus, if the land is officially homesteaded, the commercial lending institutions become even more hesitant to become involved in financing.

For these reasons, loans granted for non-commercial rural property will often have high interest rates, shorter maturity dates, and require that a higher percentage of appraised value be paid in a down payment.

There are three basic ways for a group to purchase and occupy a tract of land. These include tenancy in common, joint tenancy, and incorporation. Each form of joint-ownership has advantages and disadvantages.

In tenancy in common, each purchaser shares an individual interest in the total parcel of land and it may be in an unequal interest relative to the others. Each party's land is specified in a contract among all the owners. If any co-tenant dies, his interest in the property goes directly to his heirs.

In joint tenancy, each person has an equal interest in the land. If an owner dies, his interest in the property goes to the surviving owners, not the heirs. This form is not permitted in 14 states.

A corporation has all the legal aspects of a single person. The advantage is that shareholders do not have personal liability for acts of the corporation. The main disadvantage is the high cost of incorporation.

Making Your Final Choice and Following Through

By D. David Moyer

This chapter gives guidelines on how to make the final selection as to which parcel of land to purchase. It suggests specific information needed for the selection process and where to obtain these facts. It also discusses what you can expect regarding the transfer process itself, outlining the steps and costs typically involved.

I assume that the broad kinds of analysis suggested in previous chapters has been completed. Therefore, knowing what you want ideally, what you can obtain realistically (or can afford), and having evaluated the advantages and disadvantages (in terms of dollars, esthetics, etc.) of several general areas, you are ready to select and purchase rural property.

First of all, let's focus on three sets of factors that you should include in the process of selecting your few acres in the country. Naturally, specific factors about the parcels you are seriously considering are important. Factors concerning your family and factors concerning the community in which the land is located should also be considered.

It is best to evaluate two or more parcels using these factors. By evaluating several parcels, you are also less likely to fool yourself by overestimating the advantages and underestimating the disadvantages of a particular site.

Family factors include evaluating each potential site in terms of physical needs as well as esthetic desires of the family. To evaluate the family factors, you have to rely on your family for answers. Therefore, it is important that you be honest with yourself and take your time in making the decision of where to locate.

First consider the purpose that the property is to serve.

D. David Moyer is a Statistician, specializing in studies in land ownership and land record systems, with the Economics, Statistics, and Cooperatives Service, stationed at the University of Wisconsin, Madison.

Whether it is to be primarily a full-time residence, recreational retreat, retirement home, or an investment property has an important bearing on what weight is assigned the various factors and, ultimately, which property to acquire. For example, if the primary purpose of the property is a second home for recreational purposes, you should consider the following kinds of questions.

- How often can you realistically expect to visit the property?
- Will acquisition of this property result in a major change in recreational patterns of your family?
- Based on past experience, will the changed recreational pattern be for better or worse?

If the property to be acquired is to serve as the primary family residence, questions such as these should be considered.

- Will job-holding members of the family still continue to work in the city?
- Will there be an increase in transportation expense for commuting to work?
- What impact would another doubling of energy costs (as has occurred in the last 5 years) have on the family budget?
- What will be the effect of the move on the time jobholders spend with the family? (Many exurbanites who move to the country, hoping to spend more time together and strengthen family ties, have experienced the opposite result.)
- Do the children have special education or medical needs? If so, are these available in the community? What will be the additional commuting time required on special buses or chauffeuring required by other family members?
- Does the family have automobiles and time to chauffeur children to and from after-school activities when bus service is not available?
- How will future changes in the family cycle affect the impact of school commuting problems?
- Do you plan to supplement your income with part-time farming or from a home occupation or business?

Dr. Walter McKain warns against undue reliance on the chicken-farm myth. Vegetable gardens, fruit trees, chickens, a few cattle are often assumed to be good ways to reduce food costs. However, potential part-time farmers should carefully evaluate their capabilities before starting any agricultural enterprise. Past farm background and a good mechanical aptitude are important attributes to help insure a successful part-time farm.

- What is the effective rate of interest you will be paying on any mortgage on the property? Remember that our income tax structure works to reduce a 9 percent interest rate to about 6 percent for someone in the 32 percent tax bracket. Therefore, consider your income and wealth positions when evaluating the impact of your land purchase on family finances.

You can compile a similar list of questions if you are looking at retirement property or other special situation purchases. Comparing your present situation with several alternatives will help ensure that all family factors are considered, as well as the relative importance of each factor.

Parcel Factors

A number of parcel factors should be considered, including regulations affecting land description, zoning, subdividing of land, other land use controls, septic disposal and water systems, plus miscellaneous factors related to specific parcels.

One of the first items to check is the parcel description. Later, at the time of closing, you will probably want to obtain the services of a title attorney or other professional title examiner. Initially however, examine the recorded plat, certified survey map, or last deed recorded on the parcel.

Compare the legal description with the property itself. Familiarity with the legal description of the parcel will be helpful as you compile information from various government offices about specific parcels you are interested in.

The county recorder's office usually can provide you with the legal description of land parcels. Commercially prepared plat books also contain property descriptions and can be found in the local library, assessor's office, or the county Extension office.

Having obtained the legal description, your next stop should be the local zoning office. This will usually be a county or township office.

Some counties may not be zoned and this stop can be skipped. However, evaluate very carefully what the lack of zoning could mean regarding the parcel you are considering, especially as to future development on adjacent lands. A major purpose of zoning is to guide development so conflicting uses are not permitted on adjacent lands.

Assuming the area is zoned, discuss your plans for the parcel with the local zoning administrator. Attempt to obtain answers to the following questions.
– Is the property zoned for the use you intend?
– Is zoning on adjacent properties likely to result in conflicts with your interests?

- If the land is in an agricultural zoning district, is the zoning used of the cumulative or exclusive type? Since almost any land use is permitted in cumulative agricultural zoning districts, this type of zoning is relatively ineffective in dealing wth many land use problems.
- What are the applicable zoning regulations affecting building location on the parcel? Especially note required setbacks from the street or road, side yard and rear yard lines, as well as from septic field locations.

This is particularly important if you have a specific house plan in mind. Early efforts to consider fitting a proposed house to the parcel, rather than fitting the parcel to the house, will result in a house and lot that complement each other, as well as being compatible with garden, patio, driveway, and recreation areas.

Are there special zoning regulations that may affect use of the parcel? (Such as restrictions on mobile homes, livestock, residential development adjacent to navigable streams and lakes, and areas that are subject to flooding.)

Many areas of the country also are covered by separate ordinances specifying requirements governing the creation, transfer, and permissibility to build on a parcel. Information on these matters can be obtained from the zoning administrator, local or regional planning commission, and the recorder's office.

A recent incident in Dane County, Wisconsin, is a good example of the kinds of things to watch for and the problems that arise. Two brothers, residents of Michigan, each bought

Zoning restrictions on livestock may affect your plans.

20 acres (totaling 40 acres) of land-locked property, upon which they each intended to build residences. They were assured by the real estate agent that a half-mile access easement was adequate for their purpose.

After the purchase was completed, however, they discovered the county subdivision regulations require that all parcels on which buildings are to be placed must front on a public road or street.

It is technically possible for them to buy a 66-foot-wide right of way and build a half-mile road to the property. However, the expense makes it uneconomic and in addition the town board refuses to accept the road, even if the owners could afford to build it.

The net result is a dream turned into a nightmare, with the brothers facing substantial economic loss, besides not being able to use their few acres in the country as they intended.

Septic Systems

While talking to the zoning administrator, also inquire about waste disposal (septic) systems. Additional information sources on septic systems and wells and water systems are the county sanitarian, county health department, local office of the U.S. Soil Conservation Service, State Department of Natural Resources, county Soil and Water Conservation District, and county Extension office.

If there is an existing residence on the parcel, ask about the septic system's age, likely condition, and size. Adequate size is important if useful life of a septic system is to be maximized.

Septic systems are "sized" based on number of bedrooms and type of plumbing fixtures in the home. For instance, the Wisconsin Plumbing Code requires a 500-gallon tank for a two-bedroom home, and an 800-gallon tank for a four-bedroom home. If a garbage disposal, automatic washer, or dishwasher is added, a four-bedroom home must have a 1,200-gallon tank.

Information on soil type and soil permeability, slope, depth to ground water, and depth to bedrock is useful in determining how well an existing or new septic system will operate. The offices mentioned above can tell you whether the parcels you are considering are good, marginal, or severely limited as to septic systems.

Before you make an offer to purchase, determine that the parcel has passed any required percolation (perc) tests. If this has not been done, any offer to purchase should be contingent on the parcel passing such tests.

Actual installation often requires additional permits and inspections by local officials. Becoming familiar with the operation of your septic system and planning for the periodic pumping of sludge from the septic tank will extend the life of the system and save money in the long run.

Most rural areas obtain their water supply from a well. With increased well costs and greater concentration of people in some rural areas, group wells, serving 3 to 10 homes, are being installed.

Regardless of whether you are on your own well or part of a group, obtain information on typical costs of installing new wells. Costs of $1,000 are not uncommon. Be prepared for periodic maintenance costs on pumps and related equipment. Also, periodic testing of group wells is often required by health officials and this testing cost is often charged to the user.

Several additional parcel factors—while not as important as zoning, subdivision, and sanitary regulations—should be considered for each parcel, particularly as they might affect your use or the future value of the property. See table.

Miscellaneous Parcel Factors

	Possible Problem Areas
Easements	• Across your parcel?
	• Access to your land over another parcel?
Restrictive Covenants	• In deed or on recorded plat?
Fences	• Ownership patterns, how will they affect fence maintenance?
Topography	• Compatible with planned house, garden, crop areas?
	• Access road ever impassable due to severe slope, snow, floods?
Drainage Patterns	• Effect on you and your neighbors as to house site, garden, crops?
	• Standing water problem?
Weather-Climate	• Frequency of tornadoes, fog, drought.
Recreational Water Quality	• Is water adjacent or nearby, acceptable for planned use?
Property Value	• Consistency of present and past assessed values, past sale price, and asking price.

Community Factors

Having completed a review of the family and parcel factors, the final step is to consider a number of factors about the community that are important in selecting a specific parcel for acquisition. In general, this final review step is to determine what kind of community the parcel is located in, what changes have occurred in the community during the last few years, and what changes will likely occur in the next several years.

This information can be acquired during conversations with many of the information sources contacted about parcel factors. "Windshield surveys" while driving around the community and visiting with potential neighbors and other local residents are also good ways to evaluate how well the community fits the ideal location for your family.

Ask zoning officials what kinds of rezoning changes are being made, where the changes are typically occurring and the frequency of such changes. Similar information regarding residential plats and certified surveys can be obtained from the local subdivision review official or the local planning commission.

Review with local planning officials what changes have occurred in the past and what land use is called for by plans for the area. Also ask about techniques available for implementing local land use plans, and how effective these techniques have been in obtaining the desired results.

Remember that if you have little difficulty in acquiring your few acres, others will probably have similar experiences. Therefore you may soon have several (or many) new neighbors.

Evaluate carefully the impact of the political subdivision location of the property. Often there are significant differences in level of services provided as well as tax rates between neighboring townships or municipalities. Keep in mind that many services such as garbage collection and snow plowing which are included as tax-paid services in urban areas become the responsibility of each landowner in rural areas.

Useful Site Selection Information Sources

County Zoning Administrator
County Recorder or Register of Deeds
County Extension Agent
Property Tax Assessor
Subdivision Review Official
County Soil and Water Conservation District
Certified Soil Tester
County Sanitarian or County Health Department

Township, County or State Highway Department
Township Clerk
U.S. Soil Conservation Service
U.S. Geological Survey
State Department of Natural Resources
Local or Regional Planning Commission
Local Library

Getting the Parcel

The second half of this chapter discusses what you can expect, in terms of procedures and costs, once you decide which property you wish to purchase. Included are discussions of the offer and acceptance, financing and title examination, and the procedures and costs you can expect at the meeting where the closing takes place.

Your major concern should be to avoid any surprises, during or as a result of the purchase and closing process. With proper planning and preparation, the two most important kinds of surprises—those relating to quality of the title and costs of transfer—can be avoided.

In many cases it is best to obtain the services of an attorney to assist in preparing the offer. Once an offer is signed by both buyer and seller, it becomes a binding contract on which the entire transfer process rests.

In relatively simple transfers or if the buyer is very familiar with the transfer process, he can prepare the offer himself. Usually, if a broker is handling the sale, he will be willing to assist or to prepare the offer. If you intend to hire an attorney for the closing, the additional charge to prepare the offer is usually well worth the cost.

The offer must be in writing to conform with the Statute of Frauds. Amount of detail in the offer will depend on the transaction's complexity. Purpose of the offer is to specify what is being bought (and sold), what the price and other terms of sale are, and who is responsible for the various costs that result from the transfer process itself.

Which party typically pays each of the specific transfer costs is discussed later. However, you can specify in the offer what costs you believe the seller should pay. Whether he will agree will depend on a variety of factors including how long the property has been on the market, the price, and the quality of the property.

Following is a list of items included in a typical real estate offer:

- Legal description of property. This usually is not as detailed as the description in deed, but should be complete and unambiguous.
- Names of buyer and seller. If married, include both husband and wife's name, regardless of how title is held. This protects against problems with the statutory rights of spouse that are not necessarily a matter of record.
- Price. Includes such items as amount of interest money, any additional payment to be made on acceptance, if existing mortgage is to be assumed, and total price offered.
- Personal property—fixtures. Note any items you specifically want included or about which you believe there could be disagreement later. Items legally defined as fixtures (personal

property that because of its use or close association with the parcel has become real property) often are involved in litigation, so it is always best to avoid later problems by listing them in the offer.

- Form of deed. This specification can take one of several forms. For instance, the seller may be required to "provide title, free and clear of all encumbrances except those to which the buyer agrees." Alternatively, the buyer can request a warranty deed as opposed to a quit claim deed.

  In a warranty deed, the seller guarantees the title to be free and clear, whereas a quit claim deed only transfers the rights the seller actually has, whatever they may be.

- Form of title assurance. The seller usually is required to provide evidence of a good and marketable title some reasonable time before the closing. For example, in Wisconsin this evidence is almost always in the form of an abstract or title insurance policy. Also, the Wisconsin seller has the option of which to provide, unless the buyer specifies a particular option in the offer.

  Regardless of the form of title assurance, the buyer's attorney will want to review it before the closing.

- Date and place for closing.

- Date buyer to receive possession. (Usually on or near the date of closing.)

- Conditional clauses. Most common conditional offer is contingent on the obtaining of a mortgage of specified amount, specified length, and for a specified rate of interest. Obtaining a change in zoning is another condition sometimes used.

- Miscellaneous clauses. The buyer can specify who is to pay particular closing costs and how items such as taxes, insurance, and water bills are to be prorated. He can also request certificates assuring property is free of pests (such as termites) and that the utility systems and appliances in the house are in good working order.

- Deadline for acceptance of offer. This will prevent an undue delay before you can make another offer on this property or an offer on another property.

Remember that the time to negotiate is with the offer. Once the offer is accepted by the seller, it is a binding contract and most of the conditions of closing are bound by it.

After the offer is accepted, there are two major items to be taken care of before the closing. First, financing will need to be obtained (unless it's a cash deal) and, second, the buyer's attorney will need to complete an examination of title.

For the title examination, the buyer's attorney will typically evaluate the evidence provided by the seller. For instance, he will review the abstract to assure that the "chain of title" is complete. If a title insurance policy is provided, the attorney will determine if it is adequate, noting particularly any exception to insurance coverage noted in the policy.

Financing arrangements may be set by the accepted offer. This would be the case if the seller is providing financing by a land contract or mortgage. Otherwise, the buyer will want to contact several lenders to obtain the most advantageous financing.

While interest rate is an important factor, also give consideration to other factors, particularly the charges the lender may assess at the closing. Interest rate differences of one-half percent and more can be offset by charges for loan application fees, credit reports, property surveys, and lender lawyer fees that the borrower may be required to pay.

A recent U.S. Department of Housing and Urban Development publication entitled *Settlement Costs* provides excellent tips on how to compare lenders, as well as a detailed discussion of closing procedures and settlement costs.

The Closing

Function of the closing is to bring all interested parties together and permit them to execute and deliver the necessary documents, make payment of the purchase price, and settle the costs of the transaction itself. If all parties have made proper preparation, the closing is a relatively simple procedure.

In a typical property purchase, the following people will be present at the closing: buyer and wife, seller and wife, an agent of the lender, a real estate agent, and one or more attorneys. Often the buyer and lender each have an attorney and, while less likely, the seller may also be represented by an attorney.

At the closing, the seller will deliver to the buyer a signed and notarized deed. Ideally, the buyer's attorney will have helped the buyer decide whether the title should be held in joint-tenancy, tenants-in-common, or as a sole owner. The type of title is important because of its effect on future transfer of the property, inheritance of the property, and State and Federal inheritance taxes that will accrue when the buyer or his wife die.

The buyer's attorney will also require proof, in the form of written documents, that any defects he discovered in the title examination have been corrected. He also will want assurance that the seller's mortgage has been paid, that there are no outstanding mechanic's liens, and that all taxes are paid up-to-date. These assurances may include documents such as affidavits, mortgage satisfactions, and tax receipts.

The seller will also provide a copy of the abstract or a title insurance policy, the insurance policy for any casualty insurance being transferred to the buyer, and copies of other relevant documents (for example, a property survey).

The buyer will be required to provide any further cash down payment (usually in the form of a certified check) and a note and mortgage that is signed and given to the lender. The note specifies the amount of debt and terms of repayment and the mortgage pledges the property as security for the debt. In some parts of the country a trust deed is used, which is similar to a mortgage.

Final step in the closing is settlement of the various pro-rata charges and costs of the transaction itself. It is usually handled by a series of checks among the various parties (broker, buyer, seller) and is based on the Settlement Statement. This is the step where buyers most often find they must pay charges they were not aware of or not prepared to pay. Recent Federal legislation is designed to reduce the number of surprises at this point in the closing.

The Real Estate Settlement Procedures Act (RESPA), a Federal statute, provides that certain information about settlement costs be provided to the seller before closing. For instance RESPA requires that if you apply for a loan, the lender must supply you an estimate of the settlement costs.

Also, one business day before the closing, you have the right to inspect the Uniform Settlement Statement. This standard form is filled out by the person who will conduct the closing meeting, and contains a complete listing of all closing costs, with a designation for each item as to whether the buyer or seller is paying the cost.

While RESPA helps reduce surprises concerning costs the buyer must pay at closing, most buyers need to have a good idea of the amount of these costs long before closing. This is necessary so realistic estimates can be made about the amount of money needed to complete the transfer. As John Payne has noted, the gross purchase price of a house often is not as important to buyers as the amount they will have to pay at the time title passes and the financing charges that must be paid each month.

The amount of closing costs that must be paid by the buyer varies depending on the area. However, it is usually a good idea to plan on buyer closing costs ranging from 2 to 5 percent of the sale price of the property.

A recent Wisconsin study on typical buyer closing costs in the Midwest identified 23 costs paid as part of the real estate transfer process.

These 23 costs are grouped into six major categories in the table.

Buyer Closing Costs—Percent Reporting and Average Amount Paid, in Midwest

Cost	Proportion of Buyers Reporting	Average Cost —All Buyers
Sales (commission, advertising)	2%	$ 14
Financing (credit reports, appraisal fees, loan origination fee, points, mortgage insurance)	79	255
Establish Title (title examination, title insurance, title abstract, survey, document preparation, attorney fees)	92	120
Statutory Charges (recording fees, transfer taxes)	87	8
Prepaid Charges (taxes, special assessment, insurance)	7	8
Miscellaneous (escrow fee, closing fee, etc.)	22	46
TOTAL	100	451

As is to be expected, buyers seldom pay any of the sales cost. On the other hand, about four-fifths of the buyers incurred one or more of the seven costs associated with financing the purchase.

Loan origination fees were a major cost and usually paid by the buyer. Origination fees typically amounted to one percent of the loan.

The costs of loan discount payments (or points) were paid about equally by buyer and seller. However, if the seller pays, the buyer should be aware that the purchase price often has been adjusted upward to offset this cost.

Financing-related costs such as application fees, credit reports, inspection reports, and mortgage insurance premiums are sometimes absorbed by the lender. Therefore, the buyer should consider whether these costs are extra when comparing rates among lenders.

The six costs to establish title were shared about equally by the buyer and seller. For instance, the seller was more likely to pay the title insurance premium, abstract fee, and cost of having documents drafted. The buyer was more likely to pay title examination fees and for a land survey, while buyers and sellers were equally likely to incur lawyer fees.

For statutory charges, the seller usually pays the transfer tax and the buyer is more likely to pay recording costs.

Real estate taxes and special assessments are typically paid after the time period incurred. This was true in the Wisconsin study where sellers often reported they had to pay part of the real estate taxes at the closing. (These costs are usually credited against the amount the buyer owes, thereby reducing the amount due at closing.)

The opposite is true for casualty insurance, with most Wisconsin transfers reporting this item paid by the buyer at closing (that is, a credit to seller for insurance premium he paid, with coverage now transferred to the buyer.)

Miscellaneous costs such as escrow fees and closing fees are typically paid by the buyer. For instance, many lenders require that a sufficient deposit be made in an escrow account to cover real estate taxes for the first year. Similar charges may be incurred for casualty insurance premiums.

Buyers in the Wisconsin survey paid fees of about two percent of the purchase price at the closing. These results will vary throughout the country with details available to the buyer on the settlement statement required by RESPA regulations.

If you follow these guidelines, your purchase and use of a few acres will more likely live up to your expectations.

Further Reading:

U. S. Department of Agriculture *The Mechanics of Land Transfer,* 1958 Yearbook of Agriculture, 206-217, on sale by Superintendent of Documents, U. S. Government Printing Office, Washington, D. C. 20402. $7.60.

U. S. Department of Agriculture, *The Exurbanite: Why He Moved,* 1963 Yearbook of Agriculture, 26-29, on sale by Superintendent of Documents, U. S. Government Printing Office, Washington, D. C. 20402. $7.00.

U. S. Department of Agriculture, *What and Where of Community Services,* 1971 Yearbook of Agriculture, 254-258, on sale by Superintendent of Documents U. S. Government Printing Office, Washington, D. C. 20402. $5.80.

U. S. Department of Housing and Urban Development, Office of Consumer Affairs and Regulatory Functions, *Settlement Costs,* HUD-368-F(2), August 1977.

Improvements
for your
Place

Part Three

Remodeling a House— Will It Be Worthwhile?

By Gerald E. Sherwood

An existing house on a small acreage may offer comfortable living conditions at a moderate cost. Such a house may be generally adequate, but to meet your needs and desires it will probably require some remodeling. Perhaps it is old and run down, too small, or otherwise not quite what your family wants. Remodeling may be the answer.

Retaining an existing house has real advantages. It may save money, because rebuilding is expensive. Remodeling is also a conservation measure, for it doesn't use all new building materials. If the house has a unique and desirable character, saving this example of our architectural heritage is another plus.

But every existing house isn't worth remodeling. How do you evaluate the house, and how do you assess your needs and plan to satisfy them?

Determining whether a home is worth remodeling requires a thorough inspection and analysis. Besides your own observations, professional help is usually necessary in some respects. Regardless of who performs the actual evaluation, these points should be useful to the homeowner:

The Foundation. The foundation is vital because it supports the entire structure. Any foundation failure may distort the house frame, resulting in problems with doors and windows, loosening of siding and interior finish, and cracks that allow air to blow through the house.

Most foundation walls of poured concrete have hairline cracks that have little effect on the structure. However, large open cracks indicate a failure that may get progressively worse, so some professional guidance may be in order. Crumbling

Gerald E. Sherwood is an engineer in the
Engineered Wood Structures Project, Forest
Products Laboratory, U.S. Forest Service,
Madison, Wis.

mortar in brick or stone foundations can be repaired, but if most of the mortar has deteriorated, a major repair is required. Complete replacement of the foundation may also be necessary.

Localized failure or minor settling may be corrected by releveling beams or floor joists.

If pillars are used under the house, or under porches, check to see that they are sound. Sometimes replacement is more difficult than it appears.

Standard Wood Frame. The building frame should be examined for distortion from failure of the foundation or from inadequate framing. General condition may not be too apparent, but key points should be checked for any damage from decay and insects. Instructions for recognizing such damage are given later.

Floor supports are easily observed in a basement, but may be difficult to check in a crawl space. In a basement, wood posts should be examined for decay at the juncture with the floor.

Girders resting on these posts should be checked for sag by sighting along the girder. Some sag is quite common under heavy loads such as bathtubs, heavy appliances, or partitions. This sag usually affects appearance more than strength, but it can be critical if the floor above slopes noticeably.

Sill plates or joists and headers rest on top of the foundation and thus are exposed to moisture from the concrete. Wherever possible, examine these contact points for decay and

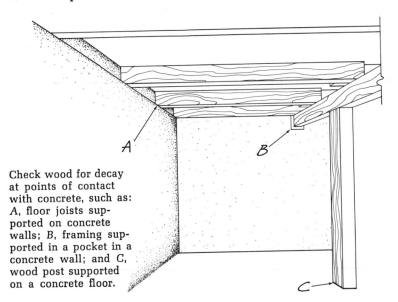

Check wood for decay at points of contact with concrete, such as: A, floor joists supported on concrete walls; B, framing supported in a pocket in a concrete wall; and C, wood post supported on a concrete floor.

insect damage. If the basement or crawl space seems very damp, the entire floor framing system should be examined.

Sag in floor joists is not critical unless it is readily apparent. A more common problem is to note springiness as you walk across the floor. The floor can be firmed up by adding extra joists or girders to increase stiffness.

A point of particular concern in the floor system is framing around stair openings. Check floors around the opening for levelness. Where floors are sagging, the framing should be leveled and reinforced.

Wall framing usually has more than adequate strength, but may become distorted if the foundation settles too much or floor framing is inadequate. Check doors and windows for squareness to make sure they do not bind. Also check for sag in headers over wide window openings or wide openings between rooms. Headers that sag noticeably will have to be replaced.

Look at the roof for sag at the ridge, in the rafters, or sheathing between rafters. If the ridge line is not straight or the roof appears wavy, some repair may be necessary.

Uneven foundation settling in this old farmhouse resulted in a building badly out of square. Evidences include A, crack in foundation; B, eaveline distortion; C, sagging roof ridge; and D, loosefitting frames or even binding windows.

Siding, Windows, Roof. Exterior wood on a house will last for many years with reasonable care. Paint failure is often caused by excessive moisture, but it may also result from poor surface preparation, poor paint, improper application or incompatible successive coatings. Regardless of the cause, you may have to completely remove the existing paint before repainting.

In examining the siding, look for space between horizontal siding boards by sighting along the wall. Where warped boards leave big gaps, new siding may be required; however, if boards are not badly warped, renailing may solve the problem. Check the ends of siding boards for decay where two boards butt together, at corners, and around window and door frames.

Good shingle siding appears as a perfect mosaic; worn shingles appear ragged, and individual shingles often are broken, warped, and upturned. New siding will be required if these shingles are badly weathered or worn.

Cracks in brick or stone veneer can be grouted and joints repointed, but large or numerous cracks may be unsightly even after repairs. To prevent water from entering masonry walls, examine the flashing at all projecting trim, copings, sills, and intersections with the roof. Plan to repair any of these places where flashing is not provided or where it needs repair.

Check all windows for tightness of fit and examine the sash and sill for decay. Weather-stripping can better seal the window, but if the sash or sill is decayed, that part must be replaced. If you plan a replacement, however, measure the window and determine if it is a standard size. If not, the opening will have to be reframed or a custom sash will have to be made —both of which are expensive.

In cold climates all windows should be double glazed or have storm windows. Again, if the windows are not a standard size, making up storm windows may be expensive.

Severe roof leaks should be obvious from damage inside the house. But general condition of a roof that is apparently not leaking will be more difficult to determine. If possible, look carefully in the attic for any signs of problems.

On the roof, scrutinize the condition of the shingles. Asphalt shingles show deterioration by becoming brittle and losing surface granules. More important is wear in the narrow grooves between the two shingles, which may extend completely through to the roof boards. Where such signs of wear exist, new roofing is required. Wood shingles need to be replaced when numerous individual shingles are broken, warped, or upturned.

Built-up roofing on flat or low-sloped roofs shows wear by

Watch for sag at *A*, ridge; *B*, rafters; or *C*, sheathing. Rafters are frequently tied, as at ceiling joist, *D*, to prevent them from spreading outward. Flashing, *E*, is used at intersections of two roofs or between roof and vertical planes.

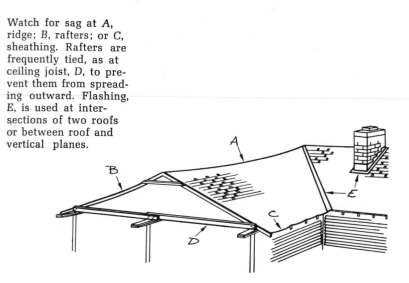

bare spots in the surfacing and separations and breaks in the felt. Bubbles, blisters, or soft spots also indicate the need for major repairs. Examine the condition of the flashing; corroded flashing should be replaced.

The Interior. Interior surfaces may have deteriorated due to wear, distortion of the structure, or the presence of moisture.

Wood floors should be checked for buckling or cupping of boards. If cupping or buckling is not too severe and boards have not separated excessively, the floor may be returned to good condition by refinishing. But first, be sure the floor is thick enough to withstand sanding.

Floors with resilient tile should be examined for loose tile, broken corners, cracks between tile, and chipped edges. If any tile must be replaced, the entire floor will probably have to be redone because new tile will seldom match.

Plaster on walls and ceilings almost always has some minor cracks, which can be patched. If large cracks or holes are numerous, a new wall or ceiling covering may be required. Bulging or loose plaster also indicates the need for new walls or ceiling.

If walls have more than two or three layers of wallpaper, the wallpaper should be removed before applying new paper or paint. Paint also may have built up to excessive thickness on walls and ceilings, or be badly chipped. In either case, complete removal of old paint is required, so application of a new panel material may be a better solution.

Trim can be refinished. However, if the old trim is badly chipped or checked, the finish will have to be completely removed. Ornately carved designs are the most difficult to refinish even though they may provide the most rewarding results. If replacement of some trim is required, new sections can be custom-made. However, this may be costly.

Decay and Insect Damage. Look for decay in any part of the house where wood has remained wet for a time, such as close to the ground. Decayed wood can be identified by its loss of sheen, abnormal color, and sometimes fungal growth on the surface. A test for extent of decay is to prod the wood with a sharp tool to see if it mars easily. Pry out a splinter. If toughness has been reduced, the wood may break across the grain with little splintering and lift out with little resistance. Sound wood will lift out as one or two relatively long silvers, and breaks are splintery.

Termite damage is sometimes harder to spot. Look on foundation walls for earthen tubes. The tubes are evidence of subterranean termites, which use them as runways from soil to the wood above. These termites follow the grain of the wood as they eat their way through, leaving galleries surrounded by an outer shell of sound wood.

By contrast, nonsubterranean termites, which are found only in warm coastal areas, eat freely across the wood grain. The nonsubterranean type requires no connection to the ground.

If any termites are in evidence, get the opinion of a professional exterminator.

Insulation. Older houses are often drafty and cold, but insulation, storm windows and doors, and weather-stripping can help make them warm and comfortable.

First, check the ceiling for insulation. If there is little or none, your first priority would be to add ceiling insulation. Ask your local utility company for guidance on how much to add, as well as for general guidelines on energy conservation. But as insulation is added, good attic ventilation becomes critical. Floor insulation is also a good investment.

Storm windows help by reducing heat loss through glassed areas by up to one half, and also reduce air leakage around the windows. Weather-stripping doors and windows, as well as caulking all cracks, costs little for materials but does require considerable labor.

After these simple corrections are considered, the next step would be the walls. Where insulation is added to the walls, a greater expense and effort is involved because this must be professionally applied. The results may be worth it, however, in comfort. In areas of cold winters, some precautions are necessary to prevent moisture problems in the walls.

Moisture Control. When moisture-laden air can leak through cracks to the outside of an old house, moisture is seldom a problem. But as the house is tightened by weatherstripping and storm windows and doors, the air is no longer carrying off that moisture. A vapor barrier, placed on the warm side of the wall, prevents that moisture from simply moving into the walls and condensing there as it meets the cold outside wall. Lack of a vapor barrier commonly results in exterior paint peeling problems.

Several coats of oil-base paint on plaster provide some resistance to water vapor. If relative humidity in the house is kept low, this paint may be an adequate vapor barrier. Additional vapor resistance can be added by applying a highly moisture-resistant paint or vinyl-coated wallpaper. Where new panel material is planned as interior finish, a good vapor barrier such as polyethelene film can be placed over the plaster first.

Crawl space moisture often migrates up through walls and living space. Moisture levels under the house can be reduced by good ventilation, but a vapor barrier ground cover is even more effective because it prevents ground moisture from coming up into the crawl space.

Utilities; Heating. Utility systems have generally been improved in recent years, so the older home may require updating. Of course, some rural homes, particularly those that have been vacant for years, may have no central heating, plumbing, or electrical system. This should not be a deterrent from remodeling the house that is structurally sound.

In this time of concern for saving fuel, the efficiency of the heating system compares in importance to insulation for the house. Utility companies can advise you on this. If the heating plant is quite old, perhaps a new one will save enough fuel to pay for itself. Observations can be made of the general condition of the furnace or boiler, but professional help should be sought for a thorough inspection.

Where firewood is available on your acreage, it can be considered for heating or at least as a secondary source. Wood-burning units are available that can be attached to the furnace, and many wood-burning stoves are on the market. Fireplaces are good for removing the chill during mild weather. However, in cold weather they generally draw warm air from other rooms and put an added load on a central heating system.

Plumbing; Electricity. Check several faucets to see if the flow is adequate and there is good pressure. For a private water source, the gage on the pressure tank should read a minimum of 20 pounds, preferably 40 to 50 pounds. Also have any private well tested for purity.

Look for leaks in the water system. Rust or white or greenish crusting on pipe or joints may indicate a leak. Check for clogged drain lines by flushing the toilet and observing any sluggishness. Also run water for a few minutes to determine if the drain lines are clogged. Where a new drainage field is needed, some codes require percolation tests of the soil.

Electrical service should be at least 100 amperes for a moderate three-bedroom house, and probably 200 amperes for a larger house or if air-conditioning is included as an electrical load. New service with a larger capacity will be a major expense.

Some wiring is usually exposed in the attic or basement where it can be checked. If cable insulation is deteriorated, damaged, brittle, or crumbly, or if armored cable or conduit is badly rusted, wiring should be replaced. Also notice if there is at least one electrical outlet on each wall of a room and if ceiling lights have wall switches. If these are lacking, consider the expense of bringing the wiring up to current standards.

General Considerations

Consider the arrangement of the house and the changes that may be required for convenience. Look at the rooms in terms of furniture placement and adequate circulation space. Also consider storage requirements. Many older homes have few, if any, closets. Do not be bound to traditional uses of rooms, but look at spaces in terms of your living requirements. Sometimes moving a door or eliminating a partition may do wonders for livability. However, keep in mind that partitions may be supporting the floor or roof above, and that changes will add to cost.

Appearance of a house is largely a personal matter, but some general guidelines can be given. The well-designed older home often has a desirable character and changes should be kept to a minimum under this condition.

Unity of design is generally a governing factor. Windows and trim should be of one type and in keeping with the house's style. Porches and garages should blend with the house rather than appear as attachments. Too many types of siding, or ornamentation that appears stuck on as an afterthought, may present a confused appearance. Also consider what can be done with paint and landscaping. No house looks good in a rundown condition or with unkept grounds.

Summing Up. The first two items mentioned, foundation and framing, are the most critical. If the foundation is good and framing is generally square, the house is probably worth remodeling. However, if numerous other repairs are required, or the house generally does not meet your needs, remodeling may be questionable. In the final analysis, make a detailed cost study or get a contractor's bid to help you decide if the remodeled house is worth the cost.

Besides a scenic view, this "two-into-one" house provides plenty of living space for the Harmons, and opportunities for creative and useful work such as building bookshelves and a fold-out kitchen pantry. (See text on facing page).

Harmon-y in a Double Home

By Jimmy Bonner

Leon and Mildred Harmon began dreaming of a country home back in 1938 after they were married. And after 39 years of city living they finally made one–by putting two houses together.

Their unusual country home is near Water Valley in Yalobusha County, Mississippi. The couple joined two houses built in the early 1930's that had been on the farm of Mrs. Harmon's mother.

The Harmons, who are both retired Monroe, Louisiana, schoolteachers but originally from rural north Mississippi, had dreamed for years of a home back in the country with plenty of room. So last year they left the city and returned to Mississippi.

"Leon and I both wanted the privacy and peacefulness that country living offers," said Mrs. Harmon. "And we wanted our grandchildren, who live in large cities, to have a chance to visit a place that hasn't been changed by man. Out here, we can see nature at its best–trees, birds, squirrels, fish and all kinds of wildlife. There's just nothing like it."

The Harmons put together their home during a period of about five years, working during weekend trips from Louisiana. By the summer of 1977, they were able to move in and add the finishing touches.

A self-trained carpenter, Mr. Harmon connected the two houses himself and did all the carpentry work and most of the other work needed. Mrs. Harmon helped with the interior decorating and landscaping, with assistance from Mrs. Mamie Shields, Extension home economist in Yalobusha County. The two houses combined gave them nine rooms and about 3,200 square feet of floor space.

Mr. Harmon estimates the total cost of building the house at about half the price of a comparably sized and equipped new home, or about $15 per square foot. As an added advantage, the house reflects their own taste in design and decoration.

The house is completed now, but the work hasn't stopped. "We plan to have azaleas and a nature trail in the woods surrounding the house," said Mr. Harmon, who enjoys fishing and hunting on their land. "We also will have a garden and orchard."

Adapting to a new, quiet country life hasn't come without problems, said Mrs. Harmon. "We miss the shopping conveniences and the culture of the city," she said. "It also took quite some time to complete the house. But we would do it all over again because we enjoy the work and living in the country."

Jimmy Bonner is Assistant News Editor, Mississippi Cooperative Extension Service.

Assess Family Needs

Your present family needs go beyond the number of bedrooms in the house. Consider your living habits, work, relaxation, and entertainment; and think of the needs of each in terms of space, privacy, storage, access to other areas of the house, and comfort level required for each activity.

Entertaining large groups means a need for large rooms. If you frequently have overnight guests, a spare bedroom may be important. If you plan to raise your own food, the kitchen and utility area should provide space for the processing involved, and storage of canned or frozen food.

Each family is unique. These are only suggestions that may stimulate your thoughts in assessing present family needs.

All families change and consequently family needs will not be the same in the future. A young, growing family undoubtedly will need more space in the future. Look at expansion possibilities. Consider using presently unfinished space such as attic, basement, or attached garage.

Generally there should be room for an addition to a rural house. The main thing is to plan such an expansion or addition at the time remodeling is being done. The heating system and electrical service should be sized for add-ons if expansion is planned. Rough plumbing may also be planned for expansion.

If your family is presently large, it will decrease in size as children establish their own homes. Perhaps see if an area of the house can be closed off when unused, but be available for guests. Or could a part of the house be converted to an apartment for rental? By making the division between units with removable panels, the house could be changed either way as needed by your family. Of course, the feasibility of this arrangement depends on being in a location where there is a demand for such rentals.

Getting Ideas and Help

Your local library should have books and periodicals on ideas for remodeling as well as how-to-do-it booklets. Many of these can also be purchased at newsstands.

The lumberyard or building supply dealer has many free brochures on application of various building products, and in larger stores you may find slide-tape instructions on how to accomplish specific aspects of remodeling. County Extension offices have samples of available government publications.

State Agricultural Extension offices are usually located at the state university and have a professional staff to advise the homeowner who is considering remodeling.

For the house that has historical or significant architectural interest, specific assistance is available in most states from the state preservation officer.

Services of an architect may be a good investment where extensive remodeling is planned or where the house has particular architectural significance. A good architect can assure satisfactory results as well as control costs through good planning. Education and experience requirements for licensing architects are quite stringent, so technical competence is usually assured.

The best way to locate a good architect is through references from acquaintances. Another possible source is through observing work that you like and finding out who the architect was. Where an architect must simply be picked at random, ask to see examples of his work and follow up on references of previous clients. Regardless of how the architect is selected, have a written contract with a clear understanding of responsibilities of both parties.

Remodeling contractors will generally work with you in inspecting the house to determine what needs to be done. Many will give free estimates that can be used in determining the feasibility of remodeling. Some contractors provide a planning service. As with the architect, the best source for finding a contractor is through recommendations of acquaintances. References of satisfied customers should also be checked out.

A possible source of contractors is the National Home Improvement Council, 11 East 44th Street, New York, N.Y. 10017, which can refer you to local members. However the selection is made, a written agreement can save a lot of misunderstanding.

Finance, Construction

Conventional as well as government-insured loans are available through banks and savings and loan companies. Long-term mortgages similar to new house financing are available for major remodeling jobs. It may be possible to extend or refinance a present mortgage.

Home improvement loans can be obtained for less extensive remodeling. They generally must be paid off in 10 years or less.

It may be best to discuss possibilities with several loan companies for a broad understanding of options.

Federal low-interest loans are available to elderly and low-income people. Check with the local Federal Housing Authority for requirements. The Farm and Rural Development

Administration also makes loans for home improvements in rural areas.

To accomplish the actual construction you have the option of hiring a contractor, doing the work yourself, or a combination of these two. Much depends on your abilities, available time, and extent of work required.

Hiring a contractor does offer some advantages. The work can be completed more quickly by an experienced crew working full time. Financing may be more easily arranged under these conditions, and you can be enjoying your completed house at a much earlier date.

Do-It-Yourself. It may be quite feasible to do much of the construction yourself. You will have to evaluate your own abilities and make the final determination.

There are some precautions to take if you plan to do all of the work. Projects go very slowly when worked by one individual in his spare time. If the house is to be occupied immediately or at the earliest possible moment, do the necessary items at once and plan to work in "projects" with a breather space between them. Nobody wants to live in a mess continually and nobody can work continuously without risking having the project "go sour."

Be as realistic as possible. It will increase enjoyment of doing the work and satisfaction with the finished home. And, finally, realize that your ideas may change.

Old house, being remodeled into guest quarters, will be part of a new house to be erected at right.

William E. Carnahan

A Personal Experience

By Richard A. Biggs

Having grown up on a family farm, I knew that when I chose a place of my own it would have to be in the country. When you're used to the sight of a fox darting across a snowy, moonlit landscape, or the sweet aroma of new hay being packed away in the barn, or the sound of frogs croaking at the arrival of spring, it is simply impossible to consider any other way of life.

Soon after we were married, Nancy and I had the opportunity to purchase an old house and a few acres. The house had not been cared for properly for about 15 years, and needless to say, was in rather poor condition. There was no heating or septic system, the water pipes had all burst during previous winters, only three electrical outlets worked, windows had been broken, and the front porch floor had rotted into oblivion. The lawn had become a tall pasture and the shrub border consisted of lilacs supported primarily by honeysuckle, brambles, and poison ivy.

In thinking back, I believe the main reason we fell in love with this place was because it had been neglected so long. We realized that every single project we undertook would be a fantastic improvement. By squinting our eyes very hard, we could picture what the house and lawn must have looked like at one time, and we became determined to bring it back to life again.

Three years went by while we both worked to save money for the much needed remodeling. On weekends we would drive up to our "new" home and work out in the yard. At one point, I recall thinking that maybe we ought to slow down a bit, as there wouldn't be much left to do when we actually moved in. What a joke! (We've lived in the house for six years now, and I've come to the conclusion that we'll never really finish restoring the place.)

At the end of our three-year wait and save period, we decided it was foolish to delay remodeling any longer. The cost of building materials was rising much faster than our savings balance!

We contacted a local builder who gave us much needed advice on soundness of the structure, major jobs that must be done, and approximate cost. Nancy drew up a rough plan of the

Richard A. Biggs is Extension Agent-Horticulture,
Montgomery County Cooperative Extension Service,
Gaithersburg, Md.

changes and additions we desired, and my father, being an architect, drew up the final plans.

Saturday Meetings

To help keep remodeling costs down, our builder agreed to meet us each Saturday morning and advise on how and what to start removing. This consisted of such jobs as tearing out walls, small closets, and the four brick chimney flues. These various tasks took several months, and finally we were ready for the carpenters, electrician, plumber, roof man, and heating system installers.

Before long we had a five-zone heating system, new electrical wiring, a septic system where the garden used to be, insulation and storm windows installed, a new front porch, and new wallboard tacked over the existing horsehair plaster mix.

Unfortunately, things did not go exactly as we had anticipated, and our remodeling costs began to soar way over the initial estimate.

When the roof man came to repair broken and missing portions of our slate roof, he found that each slate had weathered so much that it was impossible to replace one without splintering each surrounding one. Naturally, the only solution was to install an entire new roof.

On another occasion, we accidentally discovered a large colony of termites when I fell through the kitchen floor. This led to a rather extensive wood replacement bill.

The major remodeling took four months and we were finally ready to move in. There were still several inconveniences to contend with such as the eight-foot hole in the wall where the fireplace would eventually go. Fortunately it was springtime and these conditions were corrected by mid-summer.

We have been here for six years now and have found that this particular abode is not always the "dream house in the country" we first envisioned.

Our white frame house sits on top of the windiest hill in the area and heating costs are phenomenal. We thought we had taken enough heat conservation measures, but it turns out we were sadly mistaken. Additional insulation, plastic over the storm windows, and 100 pine seedling as a future wind break all help, but we still have a long way to go.

Termites have been a constant problem, and after unsuccessful trench and treating of the house myself, we finally had a professional exterminator in. We still have termites in the outbuildings that will have to be controlled this spring.

4-Year Paint Job

The idea of living in a large frame house, with various shingle designs and intricate scroll work, appealed to me until I started painting the outside. It took four summers to scrape, sand, and put three coats of paint on the entire house. When I finally finished the last side, the first side had already begun to peel.

Our driveway is about 1/4 mile long and consists of dirt, rocks, and gullies. It has gotten to the point that when it rains, friends call first to check on our driveway condition before they come to visit.

When you live in an old house surrounded by fields of wheat and corn you have to be prepared for occasional unwanted visitors. During our first winter, a rather large rat decided to spend the cold months with us. After several unsuccessful nights of trap setting, and Nancy's firm decision that it was either the rat or her, I set out a poison bait.

According to the directions, the rat was supposed to go outside to expire. Unfortunately he didn't read the directions

The "gingerbread" trim on an old house may look beautiful, but can be very hard to scrape, sand and paint.

Susan Griffin

and it took several weeks and many cans of air freshener before our living room was enjoyable again.

While there have been several disheartening experiences in our "new" home, the happy times far outnumber the unpleasant ones. Each year we add a few more plants to the landscape, finish off another room in the house, add a bit of gravel to the driveway, and harvest a little more fruit from our young orchard.

Occasionally, we even have time to stand back and admire our home and realize that we have something very special indeed.

Refinished interiors.

Susan Griffin

Further Reading:

Abt Associates, Inc., *In the Bank or Up the Chimney? A Dollars and Cents Guide to Energy-Saving Home Improvements*, U.S. Dept. of Housing and Urban Development, GPO-023-000-00297-3, on sale by Superintendent of Documents, U.S. Government Printing Office, Washington, D.C. 20402. 1975. $1.70.

Anderson, L. O., and Sherwood G. E., *Condensation Problems in Your House: Prevention and Solution*, USDA AIB 373, on sale by Superintendent of Documents, U.S. Government Printing Office, Washington, D.C. 20402. 1974. 85¢.

DeGroot, Rodney C., *Your Wood Can Last for Centuries*, USDA, FS, GPO-001-001-00419-7, on sale by Superintendent of Documents, U.S. Government Printing Office, Washington D.C. 20402. 1976. $1.00.

Forest Products Laboratory, *Wood Siding—Installing, Finishing, Maintaining*, USDA HG 203, on sale by Superintendent of Documents, U.S. Government Printing Office, Washington, D.C. 20402. 1973. 35¢.

Haverty, Michael I., *You Can Protect Your Home From Termites*, USDA FS, GPO-001-001-00420-1, on sale by Superintendent of Documents, U.S. Government Printing Office, Washington, D.C. 20402. 1976. 60¢.

Sherwood, Gerald E., *New Life for Old Dwellings*, USDA AH 481, on sale by Superintendent of Documents, U.S. Government Printing Office. Washington, D.C. 20402. 1975. $1.70.

Stephen, George, *Remodeling Old Houses Without Destroying Their Character*, Alfred A. Knopf, New York, N.Y. 11022. 1972, $3.95.

U. S. Department of Agriculture, *Renovate an Old House?*, Home & Garden Bul. No. 212, on sale by Superintendent of Documents, U. S. Government Printing Office, Washington, D.C. 20402. 35¢

Building That Dream House: Don't Be Caught Napping

By James D. Wiseman

Building a new home can be one of the most rewarding experiences of living on a few acres. What could be more enjoyable than seeing the dwelling which you have carefully planned and designed take shape and become the home that you and your family will enjoy and be proud of for years to come.

However, without careful planning and work on your part, your experience can deteriorate into a trauma, resulting in a house that is less than your "dream"; poorly planned, too expensive to build and maintain, or one for which you have signed binding legal instruments not in your best interests. Your entire investment could be jeopardized by neglecting to adequately protect yourself through careful selection of site, builder, and lender.

The first step in planning a home is to determine the size and design needed for your family. Unless you have unlimited funds or a rich uncle, space not used is an unnecessary initial expense and will require time and money to clean, maintain, and heat over the years. The size of your family is probably the primary factor used to determine how large a home you need.

In addition to size, you should also be concerned with the age of the family. A young growing family will need room to expand, while older children will soon be leaving the home. Special family interests or hobbies should also be considered when planning the size and design of your home.

Special needs of elderly or handicapped members of the household should be given careful consideration. Just a little effort in this area, such as wider interior doors, single-level designs, special features in kitchens and bathrooms, and ramps, when necessary, can certainly enrich the lives of the elderly or

James D. Wiseman is a Rural Housing Loan Specialist with the Farmers Home Administration.

handicapped without adding appreciably to the overall cost of construction.

Proper room arrangement will enhance the livability of the home by allowing for adequate traffic flow and separation of activity areas from sleeping space.

Some sections of the nation tend to favor specific home styles such as spanish, ranch, cape cod, or colonial. Design, therefore, can definitely influence the resale value of your home; a fact to keep in mind when choosing your house plan.

You may prefer to consult a professional architect when designing your new home. He can provide valuable technical expertise in design as well as structural features. He can also help with selecting a contractor and preparing the contract. The cost of architectural services will vary according to the dwelling's size and cost and the extent of the services you request.

Standard Plans. Standard building plans may be used as an alternative. Many firms produce and sell such plans at a moderate cost. Booklets containing floor plans and sketches of many models may be purchased or obtained at building supply dealers or in the magazine section of your local store. Standardized building plans are normally available in practically any size or design you might prefer.

Factory-built houses are another alternative you may wish to consider. A factory-built unit is simply a dwelling which is built in sections in a factory, transported to the construction site, and erected on the foundation. The components, which are built under close supervision in the factory, range in size from 4-foot wall sections up to one-half of a modest-sized ranch house.

Companies that manufacture these units usually have several models a purchaser may choose from. These models normally reflect different room arrangements, size, exterior and interior finish, as well as different styles of architecture.

Another important factor to consider in designing a home is the amount of energy required to heat and cool it. With today's skyrocketing fuel costs, the added initial expense of constructing an energy-efficient dwelling will be returned several times over in utility savings. For suggestions on how to make your dwelling energy-efficient, contact your local utility suppliers or your county or state Extension Service.

Special-purpose features should be carefully considered in planning the home. A carefully designed, well-laid out kitchen can literally save miles of walking over a period of a few years. In addition, a bright and cheerful kitchen will make some of the routine kitchen work much more pleasant. The addition of kitchen built-ins, such as dishwashers, garbage disposals and compactors, and ranges are a special treat if they do not stretch the budget too far.

Young, growing families on limited budgets may be interested in unfinished basements or second stories as a means of expanding to meet their future needs. Design of the dwelling may also allow for adding bedrooms, baths, or a family room without upsetting the basic room arrangement or causing unnecessary expense.

The Right Site

Selecting the right site on which to build your new home is just as important as selecting the style or size of the home. If you already own "a few acres" then you must choose the best site available, provided your land includes one or more acceptable sites. If you do not own land, or if your acreage does not offer an acceptable home site, there are some important factors to consider before buying land.

Local zoning and building codes may have an effect upon your choice of a building site. In some areas, land zoned for agriculture cannot be used for residential purposes unless the tract exceeds a minimum size of 10 to 20 acres or more.

Availability of water and sewer taps may be a limiting factor. In some areas, wells and septic tank systems may not be economically feasible. Other restrictions such as health ordinances, sanitary codes, flood plain regulations, and deed restrictions; must be carefully considered when choosing a site.

You might obtain assistance and advice on these matters from the local zoning commission, mortgage lender, attorney, or home builders association.

Soil conditions of a home site are extremely important. Heavy clay soils are subject to swelling when wet, and shrinking when dry. This swelling and shrinking can break foundation footings, cave in basement walls, and buckle walks and driveways. Unstable soils may slip, if on a slope, causing a building to literally pull apart. Soil scientists from the Soil Conservation Service can examine the soils on your site to advise you of any limitations the soils may impose.

Many people have made the mistake of not matching the site and design of the dwelling. Ranch style homes are sometimes built on hilly terrain, resulting in excessive land leveling costs. Split level homes built on flat sites require artificial hills, which are expensive and look out of place.

Excessively high water tables can cause a multitude of

problems in constructing a home. Occasionally, high water tables are not noticeable except during a rainy season; so check with the adjoining property owners if you are making your site selection during a dry period. High water tables may eliminate any plans you have been considering for basements. An inoperable septic system, probably the most frequent problem encountered when building in rural areas, is often the result of a high water table or heavy clay soil.

Privacy and traffic noise are two factors worthy of your consideration in choosing a site; however, building and living in a dwelling located far off a public road may have several disadvantages. Driveways, even if not paved, are expensive to build and maintain. Snow removal may be a problem and an added expense you haven't counted on. Many utilities such as water, sewer, gas, and electricity are more costly if the dwelling is located too far from existing services.

On the other hand, properties located on paved roads with existing water and sewer lines may become heavily trafficked and lack some esthetic value of rural living.

Final Plans

Once you have selected your site, you can turn your attention to developing final construction plans based on your family needs and design features mentioned earlier, selecting materials, and developing construction specifications. Since the process of developing final plans may vary, depending on whether architectual services, standard plans, or factory-built homes are used, I will elaborate on the steps to be taken as well as the pros and cons of each type service.

If you choose to purchase a factory-built unit, the job of developing final plans will consist primarily of selecting which of the models offered best fits your needs and desires. After you select a model, the company representative can help you decide on the options offered. While you may not have as many choices when selecting plans associated with a factory-built house, you do have the advantage that construction can probably be completed sooner than with a custom-built dwelling.

If you are planning to use standard plans you should do some further checking before spending your money on a set of plans which may be impractical for several reasons.

Unless you are intimately familiar with construction details and building costs, you may wish to take the floor plan and sketch to a builder or building supply firm for their comments and approximate cost estimates.

Although a builder cannot make a definite bid without complete plans and specifications, he can give you a rough estimate as to how much it will cost to build the house you have chosen. Based on his estimate, you may wish to look for a plan which more closely fits your budget. If you expect to borrow funds to construct the dwelling, you may also wish to show the floor plan to your prospective lender to see if it meets his or her approval.

When you have made your choice, order at least three complete sets of building plans. You will need one each for the contractor and lender and should retain one set for yourself. If any revisions have been made, be sure they are made on all copies.

Plans drawn by an architect allow you the most freedom in choosing the style and type home you desire. He can give you valuable advice concerning cost and durability of your home; however, he must understand what you really want before he can do his best job. Make sure you have carefully thought things through and give him as much specific information as you can, unless you prefer and are willing to accept his creative concepts.

Materials. Selecting the building materials and completing the specifications can be an exciting experience; however, it does take a lot of time and exploration. The basic building materials, such as concrete footings and framing members, are fairly standard. Finishing materials, however, represent your taste and should conform with the general style architecture you have chosen. Begin your search for materials at your local building supply dealer. He has samples of the various materials he handles and can probably tell you of completed homes where these materials have been used. If he does not handle a particular item, check with other dealers.

Be cautious about specifying items which require special ordering as they may be hard to get and may delay construction of your home.

The building supply dealer can furnish you with cost information which you should consider before completing the specifications.

"Specs" Sheet

Do not attempt to complete the specification sheet without technical assistance unless you are familar with and understand construction terminology and technique. The specifications will become part of the construction contract which is a legally binding instrument between you and the contractor.

Since the contractor bases his bid on these specifications, any omissions cannot be corrected without an increase in cost.

Each item in the specification form should be completed. Give careful attention to special items such as kitchen appliances and fixtures, bathroom fixtures, floor and wall coverings, etc. Check to see what sources of heating fuel are available, such as natural gas, electricity, L.P. gas, or fuel oil. Depending on local prices and availability, some fuels may be significantly cheaper to use than others. Solicit the advice of local builders, heating contractors, and even the utility companies before making a final determination.

Be sure you understand what is on the building plans and in the specifications. Misunderstandings between the owner and contractor due to incomplete plans and specifications have turned many dream homes into nightmares. Although the average person is not expected to be familiar with all the details and terminology associated with residential construction, you should make every effort to be sure you and the builder fully understand what you want and have included everything in the specifications and plans.

"Lock and Key." There are several ways your new home can actually be built. A "lock and key" contract is by far the most popular means of construction and for most of us, the best way.

A "lock and key" is simply a total contract in which the contractor completes all work from site preparation and footings to completion, generally including basic landscaping or at least the finished grading. Upon completion and final acceptance and payment, the contractor will present you with the keys to your new home which is ready for immediate occupancy.

Selecting a Contractor

The most important job a prospective homeowner has when building by contract is selection of the contractor. There are several things to consider besides cost. If you plan to solicit bids from several contractors in your area, ask them for references and inspect homes they have built in the past. You may want to talk with previous customers and consider their recommendations. Find out if the contractor guarantees his work and materials, and if he does, see if he honors this warranty.

The contract itself should be specific. It should refer to the plans and specifications already prepared and specify the exact complete cost of the project. Cost-plus contracts, where the contractor gets paid for all material and labor plus an additional percentage (such as 10 percent), may become much more expensive than you planned.

The contract should specify the date construction is to begin and a completion date. A daily penalty, for noncompletion after that date, should be included which will cover your costs such as interest, rent, storage, etc.

Payment terms also should be included in the contract. Payment terms could consist of a lump sum payment at completion or partial advances of 60 percent of the work completed at any stage.

The work should be inspected during construction. Inspections are the owner's responsibility and not necessarily the responsibility of the lender or local building commission. If you need assistance, contact an architectural firm, local builder's association, or real estate broker.

Inspect the work at least three times during construction; just prior to the pouring of concrete footings or slabs, prior to installation of wall coverings when all rough plumbing, electrical ductwork and framing members are exposed, and at completion.

The dwelling should be occupied and final payment made only after all items of construction have been completed, including the removal of construction materials and debris.

An alternative to a "lock and key" contract is to build your home yourself with the help of carpentry labor and subcontracts for some work, such as electrical, plumbing, heating, and cooling, etc. Before you seriously consider building your own home without the benefit of a general contract, you should be aware of several absolutely essential factors. First, you must have sufficient expertise in the construction business to at least supervise the construction.

You should have enough time available to spend several hours each day at the construction site when the work is in progress as well as time to select and purchase material. In addition, you should have experience and/or assistance available in purchasing materials and contracting with subcontractors. Keep in mind that contractors are regular customers of material dealers and subcontractors and can usually buy cheaper than you or I can.

What Affects Cost

Building a new home is probably the largest investment or purchase that most of us make in our lifetimes. The cost of such a purchase is therefore of paramount importance. Since the cost of construction varies significantly from one section of the country to another, I will not attempt to provide cost guides, but instead will discuss the items which strongly influence cost of construction.

Primary factors affecting cost are size, design, and amenities (or extras). The fact that size affects cost is obvious; however, many individuals make the mistake of building a dwelling larger than needed, which not only costs more initially, but results in higher than expected maintenance and utility bills. Basic adequate housing for a family of four can be built with a heated or living area of 1,000 square feet or less. Additional space can be considered depending on individual needs and financial resources.

Design has a tremendous effect upon the cost of construction. Normally, simply designed homes with straight walls and straight roofs are less expensive than houses with offsets, wings, and various roof lines. In areas where multi-story homes are common, the cost of building a second story is usually cheaper than spreading the same amount of living area out in one story.

The room arrangement may also affect cost. If the kitchen and bathrooms are in opposite corners of the dwelling, the plumbing cost will be much higher than in a home where these facilities are grouped together.

Cost of an average-sized, basic, adequate dwelling can be doubled by adding extra items such as fireplaces, extra baths, kitchen appliances, expensive floor and wall coverings, etc. Remember that any item you specify in your house that requires specialized labor, such as custom cabinets, brick work, and ceramic tile, adds greatly to the cost. If you are on a limited budget, get an estimate of these costs before making your final decisions.

Financing a Home. Borrowing money to finance a home is not only a necessity for most of us, but may be just good business. Inflation—which tends to reduce the purchasing power of savings—and the rapid rise in the cost of construction make it impractical to save enough to build a new home.

Many sources of financing are available for prospective homeowners. These sources include, but are not limited to, local banks, savings and loan associations (known as building and loan associations in some areas), mutual savings banks, mortgage lenders, Federal Land Bank associations, and Government agencies. All of these lenders offer long term real estate loans at comparable interest rates.

91

Down payments normally range from 5 percent to 20 percent of the value of the security property. In some cases your land may substitute for the down payment provided you do not already have a mortgage on it.

Most lenders, other than Government agencies, are simply businessmen like the grocer or merchant. They are selling the use of their money. Shop around in an effort to buy your money from the one who offers the best terms including not only the interest rate, but length of repayment, etc.

Yes, building a new home is a big job, requiring much time, many decisions, and a lot of hard work. The challenge, however, can be fun and rewarding if you approach the job in a logical, sensible manner following the suggestions outlined. The pleasures of living in a well constructed and attractively designed home are well worth the time and effort you will expend. I hope you will enjoy your new home on a "few acres" for many years to come.

Further Reading:

U. S. Department of Agriculture, *Construction with Surface Bonding,* Information Bul. No. 374, on sale by Superintendent of Documents, U. S. Government Printing Office, Washington, D.C. 20402. 45¢.

U. S. Department of Agriculture, *Fire-Resistant Construction of the Home . . . of Farm Buildings,* Farmers Bul. No. 2227, on sale by Superintendent of Documents, U. S. Government Printing Office, Washington, D.C. 20402. 35¢.

U. S. Department of Agriculture, *House Construction: How to Reduce Costs,* Home & Garden Bul. No. 168, on sale by Superintendent of Documents, U. S. Government Printing Office, Washington, D.C. 20402. 35¢.

U. S. Department of Agriculture, *Use of Concrete on the Farm,* Farmers Bul. No. 2203, on sale by Superintendent of Documents, U. S. Government Printing Office, Washington, D.C. 20402. 45¢.

Family Work and Storage Areas Outside the House

By Theodore Brevik and Marion Longbotham

Families occupying a home on small acreage in the country will use many work areas outside the house. Storage spaces are needed for equipment and products the family uses in outdoor work and recreational activities.

This chapter has general guidelines for such work and storage areas. Families can make adjustments to fit their special needs.

Work and recreational areas may include a garage, shed, yard and garden, outdoor cooking area, play and recreation areas, and others. Some areas may be dual-purpose.

Storage spaces are needed at or near the work and recreational areas for many items. Storage may be needed for items listed below as well as for other items:

* Autos and/or a pick-up truck (8 by 18 feet)
* Yard and garden equipment—garden tractor & equipment, power mower & gasoline, leaf or grass push rake, rakes, brooms, shovels, hoes, axes, lawn trimmers & edgers, wheelbarrows, extension ladders, stepladders, hoses, sprinklers, weed sprayers, insecticides, herbicides, fertilizers
* Home maintenance equipment—paint, brushes, ladders, caulking materials, window-washing equipment, carpenter tools, storm windows & screens
* Electrical, plumbing and other home maintenance tools
* Auto servicing and repair materials—oil, waxes, cleaners, grease, spare parts, anti-freeze, filters, car ramps, jacks, spare tires, oil cans
* Recreational equipment—bicycles, wagon, sleds, swings, slides & other toys, skis, canoes, snowmobiles, boats, skates
* Workbench

Theodore Brevik is Professor and Extension Agricultural Engineer, University of Wisconsin-Extension, Madison. Marion Longbotham is Professor and Extension Specialist/Housing and Household Management, at the University.

- Miscellaneous equipment—powersaws & other wood-working equipment, lumber, garbage cans, empty bottles
- Food products—root vegetables, canned and frozen food

A plan that provides adequate, handy work and storage space requires careful thought. To start, develop a list of work area and storage needs including automobiles and/or other vehicles.

Keep in mind that needs change over the years and equipment items often are switched to newer and sometimes larger machines. At other times, new interests develop and are added to or substituted for initial interests and activities. Open flexible areas can allow for these changes.

One useful approach to help decide on size and arrangement is to study available plans and to visit other families who have already built and are busily engaged in living on small acreages.

Storage space for keeping products when they are not being used should be convenient, safe, and adequate in size and shape. These qualifications apply to new or remodeled storage areas.

Judging Convenience

Convenience of storage can be judged in terms of the stored item being easy to see, reach, grasp, remove, and replace. Don't stack articles unless they are of similar nature, such as firewood or bags of mulch.

A half hour of garden work can be done at a convenient time of the week if the needed machines or tools can be easily removed and replaced. If the machine or tools are hard to reach the job is apt to be put off.

Bicycles, wagons and other toys should generally be easy to get at because of frequent use.

Storing articles near where they are first used also adds to convenience.

Machines removed from storage areas often require routine maintenance before being taken outdoors for use. This implies storing the maintenance products so they can be used on the machine between the storage area and outdoor work area.

Some machines and tools require space for maintenance or repair. In these cases there should be a space that can be cleared so they may be worked on, preferably in the workshop area.

If the workshop is in the garage, the car may have to be put outside to leave floor space for repairing a yard machine.

94

Tools associated with the workbench should probably be stored in a locked cabinet above the bench. Seasonal equipment, for warm or cold weather use, can be stored in a less convenient area in the "off" season and in a more accessible area during the season of use.

Items can sometimes be interchanged seasonally in the same storage spaces. An example: lawnmowing and snowblowing machines. Skis and croquet sets are seasonally used and could justify more security in their storage.

Utility connections are needed to operate or service some machines, tools, or products. Locate these conveniently in relation to the area where they are used. Adequate lighting at work and service areas is recommended. Utilitarian lighting fixtures need not be expensive.

Storage should not be too deep. Most items requiring storage do not require more than 12 inches in depth.

Safety Aspects

Planning for safety in work and storage areas is important. Several aspects of safety should be considered both for the regular workers and other persons who use a work or storage area occasionally.

The storage area should provide for safety and security of the articles being stored. Some articles require protection from the weather. Others need to be protected from extreme heat or cold. Some machines are heavy and require an adequate base for support. Outside and garage storage areas may need special security measures to deter theft and vandalism.

The operator should be able to remove and replace machines and articles without personal injury or damage to the item being handled or serviced.

Electrical lighting and connections should enhance safety. Beyond general lighting needed for safety while handling equipment, special lighting is needed at work areas such as a workshop.

Adequate ventilation is needed in work areas where carbon monoxide, dust or noxious fumes may be generated. An exhaust fan can be mounted in the wall or ceiling.

Children are unaware of hazards which exist in such work and storage areas. Responsible adults can keep hazardous items in locked or inaccessible storage spaces and use other means to protect children. Hazards which cannot be kept secure may require parental training and guidance of the children for their own protection.

Types of Storage

Types of storage areas include:

* Enclosed storage areas—cabinets, closets, tool chest
* Horizontal surfaces—counter and bench tops, table tops, shelves
* Vertical surfaces—wall-mounted hangers or pegboards on which articles such as rakes, hoes, and shovels are hung for storage with easy access
* Floor surfaces—for freestanding, heavy, or bulky items; for machines which need to be permanently mounted to a strong base
* Ceiling surface—ceiling-mounted hangers from which items like a canoe or bicycle may be suspended
* Special storage areas—a place where items often used together are stored together, such as workshop or garden pest control materials

Each family must make decisions about storage. Options are to fit the items into the available space, improve the efficiency of the available space, or add space for storage. When storage areas are crowded, consider disposing of some seldom-used items. This is often practical for a family of decreasing size.

Families should provide for flexible use of space and for efficient use of storage to avoid the cost of overbuilding storage structures.

The Garage. The garage will probably be important not only for the car, but for the variety of tools and power equipment associated with living on a few acres. The garage needs to be handy, large enough, and designed and built for safety and security. A totally enclosed structure that is not only large

enough for one or two cars, but built to accommodate other chores associated with small acreages, would probably work best for many families.

Most homes in recent years have been built with attached garages large enough for two cars. A common size has been about 20 by 20 feet, which leaves little room for a work bench or for storing power equipment popular for yard work, gardens, and snowblowing, and for bikes and the like. Many garages of this size usually become filled with these things as well as wagons, snowmobiles, and other recreational equipment on one side, leaving no space for one or sometimes two cars.

For most homes on small acreages, a well-planned attached garage would be convenient and economical. Two to four feet added to the width of the garage and eight to ten feet added to the length or depth can provide for storage and work areas and still leave room for cars.

A garage 24 feet wide and 28 feet deep would include useful work and storage areas, as well as ample room for driving in and out. This added space is less expensive than building the same space in a separate structure since the same door, lighting, drains and the like are used as in putting up a car garage alone.

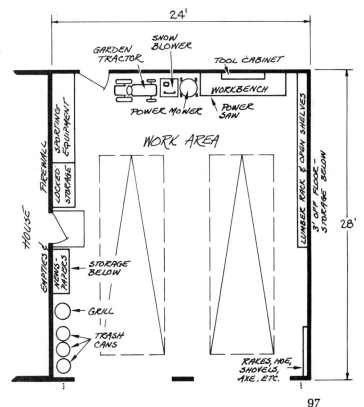

PLAN VIEW

97

Ideally, though, a plan should be developed that would provide for convenient storage of all items to be kept in the garage, a handy work area, and convenient access to and adequate storage for the automobile.

An attached garage large enough to accommodate the above has many advantages. The convenience to the house has advantages such as easy access, maximum security, and economy if well planned when you are building a new house.

Frequently it is necessary to provide a separate garage workshop where no garage now exists or the house has only a single car attached garage. If the small acreage operation requires a limited amount of field machinery it may be practical to design and build a combination garage-shop and machine shed. Considerations for the location of such a structure include:

- Storing the family car
- Convenience from the driveway, the highway, and the fields
- Distance from the house because of security, convenience, frequent access to stored items, and exposure to the weather
- Heating and insulating the workshop area
- Storage of fuel for tractors or other power equipment
- Nearness of electric power
- Drainage and protection from wind
- Storing boats, camping trailers, and snowmobiles

If the family car is stored in a separate building along with bikes, wagons and other toys, the structure should be within 50 feet of the house. If it is primarily for a truck, field machinery and garden tractors a more convenient location may be 75 to 100 feet from the house.

Drainage

Pick a well-drained site for year-round, trouble-free access in all kinds of weather. If good drainage does not exist naturally, it will generally pay to fill the area to provide drainage. The driveway and walkway leading to the building should be well-gravelled or preferably hardsurfaced, with concrete or hot-mix blacktop.

If a heated workshop is to be included, select a structural system easy to insulate. Also provide an interior liner that is fire-resistant and easy to maintain. Plywood is often a good choice because it is easy to attach fasteners for storage shelves and cabinets or for other items that are convenient to store on walls. Cover the insulation to protect it against mechanical damage and against flammability if one of the expanded plastics in board form is being used.

The floor of the garage-workshop area should be 6 to 8 inches above the surrounding grade and sloped to drain either to a floor drain or toward the garage door. A slope of 1/8-inch per foot is usually adequate to provide good drainage.

A garage door should be at least 9 feet wide and 7 feet high. Field machinery may require wider and higher doors depending on the type of equipment used.

Heating with wood may be practical on many small acreages—especially if you have your own woodlot or there is an adequate supply of low cost wood nearby. Allow plenty of room for the wood heater and install it with adequate clearance. Make sure the connection to an all-fuel chimney is done in accordance with recommended practices. Do not store any combustible materials near the heater.

Store the wood you plan to burn nearly a year so it is thoroughly dry. This means storage for wood should be large enough for a two-year supply. Obtain circulars which explain woodcutting, storing, and use for heating.

Store flammable liquids and gases in approved containers where there is no likelihood of flames or sparks. Some liquids produce explosive vapors which if allowed to accumulate could create a hazard.

Obtain circulars from your local Extension service which describe recommended safe storages and use of the proper type of fire extinguishers.

Storage Shed

If the garage with an existing house is too small, it will likely be more economical to build or buy a low-cost storage shed. Small storage buildings are economical and easy to erect. They offer low-cost storage but may not provide as much security as an attached garage.

These buildings are generally of metal and vary in size from about 6 by 8 feet to approximately half the size of a single-car garage. They can be built with no floor or placed on a concrete slab. Their most practical use is for storing yard tools and equipment.

Check the strength and durability of doors when buying these low cost metal sheds. Items requiring shelf storage or wall hangers are less conveniently stored in such sheds. Generally electrical service is not provided, so after dark use is not convenient.

The Basement. Many one-story homes have generous basement areas, a portion of which could be used for a workshop—

especially a hobby workshop. Storage of outdoor items in a basement is generally not convenient unless there is a walkout entrance. The size of the door may be important, depending on the need to move stored items in and out.

One disadvantage to a basement workshop is the dust created—especially if any great amount of woodworking with power saws is done. It may be more practical to locate power-saws in the garage area to minimize dust problems in the house.

A basement workshop should have ready access from the outside. The stairway should be convenient to the back door and planned so there is a straight approach to the stairway as you enter the house. The landing between the basement door and the outside door should be at least 4 feet long and planned so door swings do not conflict with each other or restrict movement to the basement. There also should be adequate space at the foot of the stairs.

Stairways to the basement should be at least 3-1/2 feet wide. Provide a handrail for safety and light switches and lights to illuminate the stairs. Be sure the stairs are well-built, with a "rise" and "run" that provides for easy movement up and down. A riser of 7-1/2 inches and a tread of 10 inches makes for a comfortable stairway.

Food Storage

Space for canned goods. A 5-1/2 to 6 inch vertical clearance between shelves will hold No. 2 and 2-1/2 metal cans but a 6 inch clearance is needed for glass pint jars. Allow a 3-1/2 inch depth for a single glass pint jar or No. 2 or 2-1/2 can, or 6-1/2 to 7 inches if placed two deep. A 7-1/2 to 8 inch vertical shelf clearance will hold quart glass jars. Allow a 4-1/2 to 5 inch depth for a single quart jar or 9 to 9-1/2 inch depth for two jars.

These dimensions allow for top and side clearance of the cans or jars. Storage space needed for canned goods varies, so plan space allowances according to your family needs.

Frozen foods require a freezer for storage longer than three weeks. Either chest or upright type freezers are heavy when filled, so be sure the floor will support them. Do not place the freezer where the room temperature goes down to freezing or below, and avoid locations where the sun or a heat source will cause the motor to run often. Avoid locations of high humidity. Try to have a counter or table nearby to use while loading or unloading food packages.

Frozen foods may also be put in a rented, commercial frozen food locker and packages brought to the house for brief storage prior to use.

Fruits such as apples and pears, and vegetables such as potatoes and onions, can be stored in a small basement room. This room should be on the north side of the house, walled-off and well-insulated from the rest of the basement. Properly built, such a room can also serve as a disaster shelter—unless you have a separate shelter room located on the side of the house from which storms approach.

A window is not needed but a vent to the outside with a damper is useful in maintaining the desired temperature, as well as to get rid of vegetable odors.

The humidity should generally be moderately moist—moist enough to avoid drying the produce and dry enough to avert spoilage. Natural air movement or use of a fan in the storage room also helps prevent mold growth.

Various fruits and vegetables require different temperatures, humidity, ventilation and other conditions for optimum storage. Your county cooperative Extension office will have information about methods and conditions of storing produce in your geographic area.

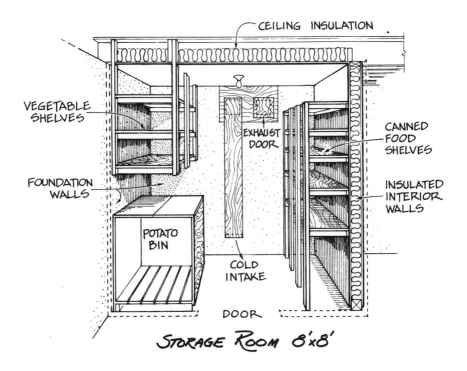

STORAGE ROOM 8'x8'

Determine your needs for produce storage and provide bins, shelves, and hangers for mesh bags. Check the produce regularly and dispose of items starting to spoil.

Canned goods can be stored in a basement or pantry. Temperatures of about 50° to 60° F are desirable but should not go below freezing or higher than about 110° F.

Properly constructed outdoor cellars are excellent for storing many vegetables, especially root crops. They also can serve as a storm or other disaster-type shelter. Built below ground they offer a uniform temperature but must be strong enough to support the weight of several feet of soil above and soil pressure on the sidewalls.

Build where there is likely to be no water problem. Or else construct a tile drainage system that will keep water out.

It is likely a root cellar would be more economical to incorporate into a new house if you are building with a basement.

Lighting

Planning the electrical requirements, including lighting, is important in any workshop area. Lighting is of two types: general lighting and task lighting. A lighting intensity of 10 foot candles is adequate for general overall lighting. Task lighting varying from 20 to 50 foot candles is recommended at the workbench, depending on the detail of the tasks at hand. Some work may require more light.

Ceiling lights are commonly used for general lighting. A reflector above the light bulb will help direct the light downward. Light colored walls and ceiling also will help reflect light and aid in general lighting. Place lights about 10 feet apart and about 6 feet from the side at a ceiling height of 9 to 10 feet. One hundred watt bulbs generally will provide the necessary overall lighting

A higher intensity light is needed over the workbench. If incandescent lights are used, two are recommended to avoid shadows on your work. Two 150 watt lamps placed 4 feet above the workbench and 4 to 5 feet apart will provide about 50 foot candles on a 3-foot-high workbench. Two 40 watt fluorescent tubes will provide about the same light level. A fluorescent fixture requires a higher initial outlay but the cost of operation is less.

It may be desirable to plan for portable lights in case the workbench is to be moved in the workshop area. When a closeup light is needed at a saw or drill, a portable lamp holder with a 60 watt bulb will generally do the job.

Plan convenience outlets for flexibility of use. Outlets should be no more than 10 feet apart. This means that you won't have to reach more than 5 feet along the wall to any convenience outlet. Circuits should be properly fused. Provide 15 ampere service for lighting outlets, and 20 ampere service for electrical tools. Be sure outlets are properly grounded and the equipment you use is fitted with the three prong plug.

In a large garage-workshop it may be desirable to provide ceiling drop cord outlets to avoid long extension cords from wall outlets. If larger than 1/3 horsepower motors are used it may be best to plan for 230 volt outlets. The electrical service entrance must provide for this, of course.

Lighting above the entrance doors is recommended, with a 100 watt bulb 10 feet high above each garage door. Plan for switches in the house as well as in the garage. For added convenience and security, a radio-controlled garage door opener may be installed.

Costs, Financing

Improving existing storage space will cost less than adding storage space. If you can do the work yourself, installing pegboard and hangers on walls may range from $10 to $50. Adding shelves will probably cost more, and adding closets and cabinets will likely be the most expensive.

Costs of improvements or additions can range from a few dollars to several hundred. Small costs for occasional hardware and materials purchases can often be handled as part of the weekly or monthly family spending.

Before you proceed with a moderate or large-size job, determine probable costs for the improvements or addition of workspace and related storage space. To do this, first get your plan on paper. This gives a basis for estimating the cost of construction and materials if you do the job entirely yourself or subcontract some parts of it. If you have the work done the plan will be a basis for obtaining bids.

Families may assume the financing from their savings. Or they may seek financing from outside sources. Conventional lenders include savings and loan associations and banks.

Some families are eligible for government-insured loans. Sources include VA loans for veterans, FHA and the Farm and Rural Development Administration for moderate and low-income families.

Loans may be available from private sources such as parents, relatives or local citizens. Also consider financing through renegotiating an existing mortgage, borrowing on insurance, or using an asset as collateral for a loan. Check also on items that will affect financing costs such as interest rates, length of loan period, and other conditions of the loan.

Further Reading:

Complete Do-It-Yourself Manual, The Reader's Digest Association, Pleasantville, N.Y. 10570. 1973. $13.95.

Fix-It-Yourself Manual, The Reader's Digest Association Pleasantville, N.Y. 10570. 1977. $12.99.

How Things Work in Your Home (And What To Do When They Don't), Time-Life Books, New York, N.Y. 10020. 1975. $14.95.

U. S. Department of Agriculture, Storing Vegetables and Fruits in Basements, Cellars, Outbuildings, and Pits, Home & Garden Bul. No. 119, on sale by Superintendent of Documents, U. S. Government Printing Office, Washington, D.C. 20402. 40¢.

Landscaping Around Home—
Get Help, Plan Carefully

By Fred Buscher and Jot Carpenter

Landscape development for a home on a few acres can be approached differently than for a small lot. The house on a small city lot has little space for plantings in the front or back yard. A suburban house site has more room for plantings, even hobby gardening such as roses, flowers, vegetables, and fruit, or even sports such as tennis.

But the opportunities to develop a landscape for a large lot or small acreage are almost unlimited. With planning, it's possible to find the space for most outdoor activities, such as gardening hobbies and small farming ventures.

The house and other buildings on the property can be visibly enhanced with plants. The landscape around the home should provide pleasure and convenience to the family but still be easy to maintain. With good planning, the landscape should be useful, add value to the property, and provide beauty for the family and community to enjoy.

Although it's possible for an owner to develop a landscape plan, you would be best advised to consult with a landscape architect or landscape designer to prepare a master plan. You could hire a consultant for a preliminary study and, combined with reading and self-study, complete the plan yourself.

A great deal of garden literature deals with landscape design of small properties. Many people have accumulated a library of information and through experience and observation know what they want in the landscape and garden.

Each residential planting plan can be unique. No two houses, sites, or families are exactly alike. It follows that no single landscape plan will fit all properties or answer the requirements of all families.

Fred Buscher is Area Extension Agent, Horticulture, and Professor in the Cooperative Extension Service, The Ohio State University, Wooster. Jot Carpenter is Professor and Chairman, Department of Landscape Architecture, at the University.

A landscape design needs to be developed for the architectural style and lines of the house. The building materials, colors, and entryways can offer clues and suggest landscape ideas for plants and structures. But most important, the planting should be designed for the people who will live there and use the land.

The landscape design process is a procedure used to develop a landscape plan that will be both useful and beautiful. First, information must be gathered and recorded on the conditions of the site and the needs of the people who will use the land. This information serves as the basis for development of a landscape program.

The extent to which a landscape program is developed will depend on the owner's or the family members' attitudes towards gardening and the outdoor environment. A desire for comfortable and beautiful surroundings can influence the amount of time and expense devoted to grounds maintenance.

A limited budget should not prevent you from developing a landscape plan. Most of the preparation, planting, and construction can be completed by the homeowner and the family.

Larger projects or extensive plantings can be phased and budgeted over a period of years. This is why a landscape plan is an essential first step. It permits all parts of the total landscape to be fitted together at later times like parts to a puzzle. When carefully planned, the finished landscape will be a complete and pleasing picture, rather than a jumble of plants and accessories unrelated to each other.

Site Analysis

First step in landscape planning is the orderly and logical recording of conditions and facts on the buildings and land area. The most useful landscape design will depend on how well the landowner or consultant can overcome or modify site restrictions, or enhance and protect the property's good points.

An analysis of the site should include a list of the existing conditions, natural or manmade, that have immediate or potential effects on the property. These can include anything that is heard, seen, smelled, or felt.

Size and shape of the land, direction of the sun, winds, and views all present restrictions and/or opportunities to landscape a small acreage. The land and buildings each express some characteristics, beauty, advantages, and limitations. The owner or designer of the landscape needs to get the feel for the land to understand what the site has to offer, suggest, or express.

Success or failure in producing a functional landscape plan often depends on how well the designer understands the site's characteristics. Each site, no matter how small, offers some unique opportunities.

Organization of the land and outdoor spaces is critical so that all the use requirements of the owner can be met. The landscape should look good 12 months of the year, not only in spring and summer. Buildings, plants, and structures can be planned to strengthen each other. No amount of planting can overcome the lack of good organization.

Natural forces of sunlight, rainfall, winds, frosts, and temperature cannot be eliminated. However, they can be modified by the landscape design. Broad categories that must be considered are: climate, topography, land, soil, vegetation, house, utilities, and community. In the site analysis, these conditions should be located, described, and evaluated. A value judgment is needed on each condition. Is the condition useful? Is it good or bad?

Climate and weather affect outdoor activities more than any other factors. Plants or structures can be used to create shade, trap heat, redirect or slow wind movement. Minimum temperatures determine the kind of plants that can be grown. Temperatures also determine the range of outdoor work, gardening, and recreation activities. Landscape design is influenced by the effects of rainfall, frost-free periods, and wind direction.

The changing direction of the sun from winter to summer creates a whole different set of sun and shade patterns on the land and buildings. Knowing where there is sun or shade at different hours and seasons helps solve problems for plants, gardens, and outdoor activity areas.

Road noise, traffic, glare, and street lights affect the landscape. The pattern of street and auto headlights on the windows and outdoor areas impose a set of restrictions or benefits that may be modified with plants or structures.

Topography must be considered, too. Is the site sloping, or rolling? Will the planned activities work on the existing grade? Perhaps the activity should be changed. An alternative would be to modify the site by grading if the hobby or activity has a high priority.

Esthetically, sloping or rolling land is more dynamic and has more advantages for the design of the house and landscape. A level site has neutral and minor landscape interests. So more interests can be planned with fewer restrictions. On a flat site, bold colors and exotic materials are possible. Level areas have less protection from the wind so more climate control elements such as trees (windbreaks) or structures may be needed.

Drainage and grading are closely related problems. If a plant-growing activity is planned, drainage or grading may be needed. Even with only a few acres, a landowner should be concerned not to allow "brown water" to run off the property. Help on drainage or grading problems can be obtained from an engineer, landscape achitect, the U.S. Soil Conservation Service, the Cooperative Extension Service, or a landscape contractor.

Soils information is needed for the site analysis to determine what plants (shrubs, flowers, vegetables, fruit) will grow best on the land or if changes must be made. A soil test can be obtained through the county Cooperative Extension Service. Results of the test will indicate the soils' lime requirement, fertility status, and if corrective or maintenance fertilizers are needed.

Many counties have a detailed soil survey made by the U.S. Soil Conservation Service. This gives more detailed data on the soil texture, structure, plant nutrient, and drainage characteristics. Knowing this helps you forecast the potential for growth, development, and success of a landscape planting, horticultural venture, or hobby.

Native plants on the property enable you to "read" the landscape and they provide clues to the local environment and soil conditions. Trees, such as red maple and sour gum, and shrubs, such as arrowwood viburnum or red-stemmed dogwood, can indicate wet or poorly-drained soils.

Identify and evaluate why the existing plants are growing on the site. Are they worthwhile? Do they add to the landscape? If they are removed, will this adversely affect or improve the area? Some native plants may be an asset, some a liability.

The House Plan

The house exerts a strong influence on landscape design. The house plan will affect the relation of the house to the outdoor areas and gardens. Rarely are the house and garden designed together.

A door from the kitchen or dining room is the logical place for outdoor cooking and eating. A door from the living room to an outdoor patio is a logical place to sit or entertain. A door from a bedroom could lead to a private garden.

Assess the views from inside the house looking out and

from outside looking in. Should they be screened, hidden, or used? To be able to look into a neighbor's attractive landscape is like owning the land without the taxation.

Consider the impact of movement of people and vehicles about the place. Dimensions for outdoor use areas—such as walks, steps, patios, and furniture—are larger than inside. Walks should be wide enough for two people side by side, at least four feet wide.

The house tends to dominate on a residential property and be less dominant on an acreage. Most houses are geometric and the land around them can be developed with the same geometric or formal pattern.

The formal plan is still the easiest and safest for a landscape—especially for smaller areas. On larger properties, the informal free-form design works best. The formal design can transition out from the house to the large informal space.

The value of trained landscape designers is their ability to integrate the site and program in a functionally good and esthetically pleasing way. Owners often are not able to express why they like what they see—but will admit they find it attractive and enjoyable.

Utilities (water, sewers, electricity, telephone) need to be identified on the landscape plan. The location of meters, height of wires, sewer clean-outs, and tile lines all influence the placement of plants and gardens. So do walks, driveways and property easements.

In site analysis, circulation deals with people and vehicles on the property. Landscaping can enhance what people will feel, see, or experience as they walk or drive on the site. Are there enough lights for entering at night? Are the entrances visible?

One criterion used to evaluate a community is the appearance of its homes and grounds. Zoning and building regulations will control the type of building and homes to be constructed and neighborhoods that will be created. They can protect or restrict the building of a swimming pool, privacy fences, or boundary line plantings.

Zoning regulations, however, should not be considered a hard and fast contract between local government and property owners. Regulations and conditions have to change with the times and can be tested by the appeal process.

Nothing to Hide

One of the most inappropriate landscape approaches in use (or misuse) today is the foundation planting. This technique has been based on a virtually abandoned house form. Unfortunately, the use of plans in this manner in residential landscape design has not changed with the alterations in architectural design.

Today's houses are built with little visible foundation. Standard construction brings the facing material of wood, brick, or stone close to the ground. There is generally nothing to hide—yet the hiding process continues.

Many rules and recommendations have been written on foundation plantings. But the authors believe it would be better to forget the foundation planting concept and to consider the entire property—house and site—as one living environment.

Plan for People. Any success for a useful landscape plan is more assured when the project is well thought out and not forced onto the site. A key to planning the landscape for the people who will use it is to thoroughly understand their goals—what they need, what they want, and what they can afford.

Family characteristics (number, ages, sex, interests) will reflect the type of landscape desired. Children's play areas and outdoor cooking, eating, and entertainment areas could be important. The hobbies and garden ventures for fun or profit need to be provided in the best locations.

Special interests, attitudes on beautiful surroundings, outdoor activities, and maintenance will suggest the size and extent

Typical planting in front of a 1900 period house, planned to overcome the architectural styles of that era and hide the foundation.

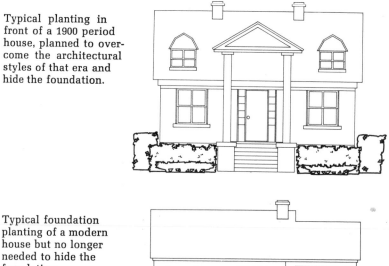

Typical foundation planting of a modern house but no longer needed to hide the foundation.

110

This planting design emphasizes a more functional and important entry approach.

of gardens and plantings. What will be the needs for storage, parking, roadways, service areas, animals, or pets?

Few people can afford to complete a landscape all at once. Financing can finish the job sooner. The extra cost to borrow money can often be justified by the immediate use and enjoyment plus the property's added value.

On the other hand, once you begin developing the landscape plan you may be forced to change your dreams by the realities of a budget and the available space.

In the landscape design process, your next step is putting together a program based on the information gathered on the site and your goals.

This information can be both written and in graphic form recorded on a plot plan of the property. Three parts of a program that designers find helpful are:

· A plot plan to record the facts of the site
· A site or environmental analysis plan to record the physical and environmental facts on the site
· A functional diagram to show how the land will be used

The plot plan is drawn to scale (usually 4, 8, 10, or 20 feet to the inch). The plot plan should show all existing features—such as the house, drives, walks, trees, boundary lines—as they appear on the lot as if seen looking down from an airplane. Suggested further reading at the end of this chapter can provide more detailed information on developing a plan.

The next step is to place tracing paper over the plot plan. Information gathered in the site analysis can be recorded on this paper overlay. All the physical elements, good and bad, that affect the site can be noted.

Examples of elements to be listed are the shadow patterns of the winter and summer sun, good and bad views, direction of slopes, drainage areas, winds, and existing vegetation. These items all add to potentials of the site or identify something that needs to be modified. Possible solutions can be noted on the plan. All this information is useful and helps identify the areas or elements to change, eliminate, or retain.

Last step in developing the program is to figure and note the interrelationship of the outdoor use areas in terms of rough size, abstract form, and sizes by drawing bubble or

Plot Plan

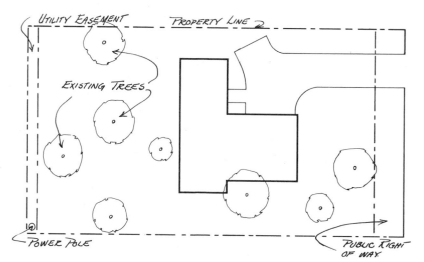

Environmental Analysis

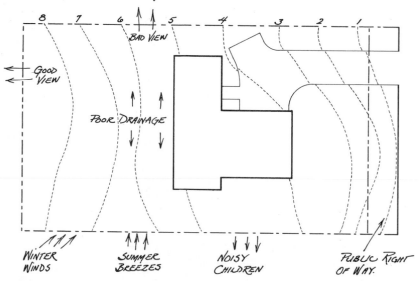

functional diagrams. Functional diagrams are drawn on another sheet of tracing paper placed over the previous drawing. These diagrams help visualize the connection of the outdoor use areas to the interior rooms of the house.

Functional diagrams show areas of separation between activity and use areas. They help you evaluate an outdoor use area and compare it to the house, circulation links, or identify new problems. Functional diagrams help you double check whether or not the use or activity area fits your needs or desires. They help in evaluating the impact of the use area on the soil, vegetation, neighbors, or community.

Elements and activities of the functional diagram should be organized to give the optimum relationship to each other. For example, the outdoor eating and cooking area (patio) should be in an area for easy transfer of food and dishes from the kitchen. Storage facilities for patio furniture, garden tools, or play equipment need to be related to the appropriate outdoor use area.

Sometimes a compromise must be made on whether to modify the site to the landscape program or to adapt the program to fit the site conditions. By developing the best possible functional diagram, you can decide if it is worth giving up these relationships on the site or to modify the site to achieve the optimum relationship. For example, a slope might be graded to provide a terrace or a patio relocated to preserve a tree providing shade for a kitchen window.

Outdoor Rooms

With completion of the most acceptable functional diagram, the outdoor "rooms" are located on the site. These areas may need to be divided, separated, screened, or connected with plants or structures. Grades could be changed. Existing plants may not be in the right location for an activity or area. The functional diagram is the germination of an idea or concept for the start of a landscape plan.

The analysis phase provides two major clues for developing the final landscape plan. First, it indicates the best places for specific needs. Second, it suggests the best form or shape for the site. For example, in the analysis the best place for a garden is determined and the form it should take to fit more perfectly with the house, roadway, or equipment storage.

The next step to the design process is development of a preliminary plan to locate the plants and structures used in the design. These are the elements which create in an abstract way the outdoor rooms.

Indicate all plants and structures in terms of their width, height, length, and functional purpose. For example, shade could be provided by either a tree, an awning, or an arbor. A privacy screen could be either a hedge, wall, or fence. In this phase, the approximate sizes are shown in an abstract fashion for the structures, paved surfaces, locations for plants, areas to be shaded, and changes of grade.

This form of preliminary plan gives a picture of the proposed design, not detailed enough to work from, but detailed enough to test the program. Examples and pictures from books and magazines can supplement the design. In this stage, the

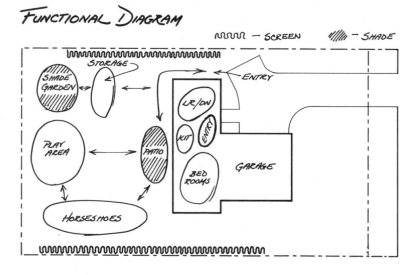

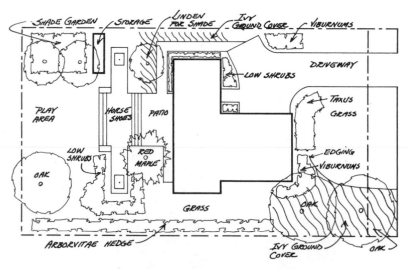

plan can be reviewed, evaluated, and tested to determine if it will work.

The Blueprint. The final design is the master plan. This plan is the "blueprint" showing to scale the exact location for all structures, pavements, and plants, plus their names and/or building material. At this point the arrangement and form of the plants and structures are determined.

Understanding the principles of landscape design and plant composition will help you complete the landscape plan. More in-depth information on this phase can be found in the further reading at the end of the chapter.

Effective landscape design is not as simple as it may sound. Certain environmental and space design problems must be solved by considering the functional spectrum of plants and applying the design principles of simplicity, balance, scale, sequence and focalization.

The time to begin a maintenance program for the yard is when the landscape is first designed. With careful thought, maintenance does not have to be drudgery. Many ambitious landscape designs can make unnecessary work for the owner if planned for maximum effect rather than the minimum mainte- nance desired by most people today.

As the landscaped area begins to develop, change, and mature, so does the family. Their requirements must be antic- ipated in a landscape design that should be flexible enough to adapt to your family's changing needs.

Further Reading:

Buscher, F. K. and J. D. Carpenter, *The Landscape Design Process for Residential Properties*, Area Center, O.A.R.D.C., Wooster, Ohio 44691. 1973. Free.

Buscher, F. K. and J. D. Carpenter, *Planning for Maintenance in the Landscape Design Process*, Area Center, O.A.R.D.C. Wooster, Ohio 44691. 1973. Free.

Buscher, F. K. and J. D. Carpenter, *A Brief History of the Foundation Planting*, Area Center, O.A.R.D.C., Wooster, Ohio 44691. 1976. Free.

Nelson, W. R., Jr., *Landscaping Your Home*, Circular 1111, Coopera- tive Extension Services, College of Agriculture, University of Illinois, Urbana, Ill. 61801. $4.00.

Robinette, G. O., *Plants, People and Environmental Quality*, on sale by Superintendent of Documents, U. S. Government Printing Office, Washington, D.C. 20402. No. 2405-0479. $4.35.

U. S. Department of Agriculture, *Landscape for Living*, 1972 Year- book of Agriculture, on sale by Superintendent of Documents, U. S. Government Printing Office, Washington, D. C. 20402. $6.85.

Land Improvements—
What You Need to Know

By E. F. Sedgley

Land improvements take many forms. In general, they are features added to the land to improve its productivity or meet special needs of the landowner.

Examples of common land improvements include water supply systems, drainage systems, pasture improvements, windbreaks, ponds, fences, roadways, and conservation measures.

Land improvements serve a variety of purposes. They may be necessary to protect property, to make possible various income-producing enterprises, or to provide recreation opportunities.

Whether you've already purchased a place in the country or plan to, an assessment of existing and potential land improvements is important.

Chances are you will not find the ideal country place with all the features you desire. Land improvements are usually expensive. Sometimes it is impossible to develop the improvements you would like because of adverse site conditions or legal constraints.

If you are seeking a place in the country, it's important that you be able to recognize the value of existing improvements and assess the feasibility and cost of improving submarginal land to meet your needs.

If you are looking for a country place with the desired improvements already established, learn to evaluate the quality of the improvements. Are the fences in good repair? Is the dam that creates the farm pond sound? Is the horse pasture well grassed? Are drainage or irrigation systems functional? Are there signs of serious erosion?

If you intend to start from scratch on an undeveloped piece of land, you need to be especially cautious in assessing

E. F. Sedgley is the State Resource Conservationist,
U.S. Soil Conservation Service, Denver, Colo.

the potentials for improvements. This is particularly true if you plan to move to a region with which you are unfamiliar.

Never buy a homestead unit "site unseen." Avoid the plight of a Virginia family that recently purchased a 120-acre undeveloped tract in Colorado's San Luis Valley through a mail transaction. The family envisioned a modest investment in land improvements since they knew the tract was undeveloped. Their mistake was equating unfamiliar conditions in Colorado with familiar ones in Virginia.

What the family did not know was that the land had no irrigation water rights, the area received less than 7 inches of annual precipitation, and the soils were too poor to grow almost anything. Local ranchers say this kind of land is good only to hold the world together. There is absolutely no way to improve this land to make it livable or productive.

The following sections are intended to provide basic information to help you assess the quality of existing improvements and to evaluate the need, feasibility, and cost of the improvements you may want.

Water Supply

In the country there is no city water system to supply you with treated water that is safe for domestic use and livestock.

Fortunate indeed is the country dweller who has a good well that supplies quality water to his home and a live stream or spring for watering his livestock.

If you intend to establish a country home on an undeveloped tract your first concern should be a water supply.

Availability of water varies considerably in different parts of the country. In areas of high precipitation, water-bearing aquifers frequently are close to the surface and wells shallow, dependable, and relatively inexpensive.

In more arid regions this may not be the case. Water has become a scarce commodity in much of the American West. Many western states are concerned about receding water tables; permits are required to construct all wells, and in some areas permits for domestic wells are being denied.

Even if you have the required permit, the cost of drilling a well may be prohibitive. Depth to water and the kind of geologic materials between the surface and the water generally govern the cost. Under normal soil conditions a well may be drilled and cased for around $15 a foot. If drilling through hard rock is required, the cost may reach $30 to $40 a foot. In many areas, wells reach reliable water sources in less than 100 feet. In others, the closest water may be at 500 to 600 feet.

In evaluating a recent application for wells by a land developer, the Colorado state engineer estimated the wells would require drilling to a depth of 2,000 feet and cost $30,000 to $50,000 per well.

Other costs associated with wells include pumps, power supplies, and storage facilities. Under favorable conditions you will need to figure a $2,000 to $3,000 investment for a well. In unfavorable situations the cost may be prohibitive.

Another concern in developing wells is water quality. In some areas ground water is so saline that it is unfit for human or livestock use. In mountainous areas ground water is frequently polluted by poorly designed waste disposal systems installed upslope.

A good source of information about the cost and feasibility of wells is the local well driller. He usually maintains drilling logs and is familiar with local conditions. The state agency that issues well permits may also be able to supply useful information. It may even be worth your while to hire a consulting geologist if you have serious doubt about the availability of water.

Undeveloped springs are sometimes a good source for both domestic and livestock water. Look for seepy areas that

Seepy area on a ranch developed into an excellent source of livestock water. Duane Scott

might indicate a near-surface release from a water-bearing soil layer. Even if ponded water is not evident, you may be able to develop a good water source by installing a collecting wall and gravel-packed perforated pipe. With proper storage facilities a flow of a gallon or two per minute should meet your needs.

Be sure to check the quality of water from a spring. Your county Extension agent can usually arrange water-quality testing for a small fee. Developed springs should be fenced to prevent pollution from livestock.

Drainage Systems

Excessive soil water can be a serious limitation to most proposed land uses. Severe drainage problems may go unnoticed by the inexperienced. Land that is dry during the summer months may turn into a marsh during spring rainfall and snowmelt.

Learn to associate certain kinds of plants with soil moisture conditions. The appearance of sedges, rushes, or other water-loving plants is reason to suspect a high water table.

In areas of limy or alkaline soils, presence of salt crystals on the surface also indicates a high water table. These are brought to the surface in solution by excess water and precipitate out as the water evaporates.

Land with a high water table can sometimes be converted to highly productive uses by installing a drainage system. You will need expert advice to determine the feasibility of drainage.

An investigation should be made to determine the depth and direction of flow of the ground water. Drainage ditches running in the direction of flow are usually ineffective. A good system starts with an interceptor ditch or tile line across the direction of flow above the affected area. To function properly, all drainage systems must have an outlet at lower elevation than the area being drained.

Underground tile systems are more expensive than open ditch systems but have advantages that may make them worth the extra cost. They do not take up land and can usually be farmed over. Open ditch systems tend to clog with vegetation and can be breeding grounds for mosquitoes.

Before you drain your land, check local land use ordinances. It may be illegal to drain certain wetlands. You may also be denied government assistance if you plan to drain certain wetlands.

In some areas you may have to join an established drainage district to obtain access to an outlet system.

The cost of drainage is highly variable and must be determined through onsite study. You will need to compare the cost with the benefits you expect to receive.

An excellent source for assistance on drainage problems is your local soil conservation district. Through an agreement with the U.S. Soil Conservation Service (SCS), each district has a staff of SCS technicians to help them carry out a conservation program.

A request to your district is often all it takes to get technical assistance in designing and laying out a drainage system. You will have to make your own arrangements to have the system constructed.

Site was dry when house at bottom was built, and owner was unaware of flood plain hazard. Other building was lifted off its foundation and moved 300 feet by flood plain ice flow.

H. J. Lyford

R. B. Dean

You may be able to get adequate advice and assistance from a drainage contractor. Some contractors rely on SCS for technical services; others have the expertise needed for the entire job themselves, including design and layout.

In areas of high rainfall, a good surface drainage system may be necessary to carry off water from high intensity storms. If you are considering purchasing a small unit subdivided from a larger farm or ranch, make sure your lot is not located near the outlet of a terrace system or in an unprotected flood plain. Structures to protect you and your land may be expensive, impractical, or not permitted because of local restrictions.

Pasture Improvement

Overestimating the carrying capacity of land for livestock is perhaps the most common mistake made by new country dwellers. A saddle horse will eat 30 to 40 pounds of forage a day. A typical non-irrigated pasture in many parts of the country can produce only 500 to 1,000 pounds of usable forage per acre per year. It would take 30 to 50 acres of this land to pasture one horse for a year without supplemental feed and without degradation of the forage and soil resources.

Pasture-carrying capacities can be significantly increased by irrigation, rotation grazing systems, fertilization, or establishment of better species. But even with an improved pasture system, one acre of land will support a horse for only about four months.

Overgrazed pastures soon become barren or weed-infested areas that are erosion and health hazards. Local ordinances may require you to install erosion-control measures and curtail grazing on such areas.

Feeding hay or grain to your livestock, or leasing additional pasture to sustain them through the year, can be a significant unexpected expense.

Windbreaks. A mature windbreak of trees and shrubs around the homestead or along field boundaries is a valuable land improvement. It may take 20 years to grow an effective windbreak, but once established it will reduce fuel costs, protect property from high winds, reduce soil blowing, keep snow from drifting against buildings, attract wildlife, provide shelter for livestock, and add beauty to the landscape.

To be successful windbreaks must be properly designed and consist of adapted species. The most effective windbreaks contain three to five rows with the lower growing species on the windward side and the tallest species to leeward.

Individual plants must be properly spaced in the rows and the distance between windbreak and homestead carefully computed so that wind-blown snowflakes do not bury the homestead.

Windbreaks are relatively inexpensive to establish. Young plants are available from commercial nurseries, forestry agencies or other conservation agencies for around $15 to $30 per hundred. Planting may be done by hand or with planting machines, which can be rented in some areas.

In areas of high winds a cedar shingle can be hammered into the ground on the windward side of each tree to protect it during establishment.

To control weeds, cultivation between the rows is recommended for the first few years after planting.

In areas of undependable precipitation a drip irrigation system can be installed at a cost of about $1.70 per tree. These systems use very small amounts of water and are used only for the first two or three years to assure establishment.

Technical assistance for windbreak planting is available from conservation districts, SCS, and state forestry agencies.

Windbreak planted on the contour provides wildlife habitat and in a few years will offer protection from snow and wind, reducing fuel costs.

Duncan R. Warren

Farm Ponds

A farm pond is an attractive improvement to any country unit. Besides its esthetic value it can be used for recreation, livestock water, fish production, wildlife habitat, and fire protection.

A good pond site is one where the largest volume of water can be stored with the least amount of fill in the dam. Look for a place where the valley is narrow at the damsite and the reservoir area is wide and flat.

Size of the watershed above the site is important. It must be large enough to supply the water needed to fill the pond, but if it is too large it may be expensive or impractical to construct a dam and spillway to handle peak flows.

Extent of active erosion on the watershed above the pond is an important consideration. Too much erosion can fill your pond with sediment in a few years. It may be wise to delay building the dam until the watershed is stabilized with vegetation.

You will need to determine the suitability of the soils before building a pond. Soils with a high clay content are generally best, since they are relatively impermeable. Sandy or gravelly soils do not hold water and are unsuited for both dam construction and reservoir area.

It is necessary to clear the pond area of existing vegetation before beginning construction. This eliminates safety hazards and the possibility that the decomposition of plants may cause the dam fill to become unstable.

Depth of water in the pond is important. Shallow ponds produce excessive aquatic vegetation that is detrimental to most uses. Specific water depths are required for various species of fish, especially where live water is not continuously moving through the pond.

Ponds should not be constructed near feedlots, sewage disposal fields, mine dumps or other pollution sources.

If you cannot locate a good site on a natural drainage channel, you might investigate the possibility of an off-channel pond. This type of pond depends on an alternate water source, such as a spring or diversion from a nearby stream.

Off-channel ponds have some advantages over the in-channel type. Water supply can be controlled and sediment problems are usually eliminated.

It's advisable to test the water quality before you build a pond, especially if you intend to stock it with fish. Most fish species have definite tolerance limits to toxic elements, pH levels, and water temperatures.

Cost of building a pond may vary considerably, depending on size of the dam and the complexity of the site. Most small farm ponds are built for around $2,000 to $4,000.

Many states have laws that regulate the use of water and that specify acceptable criteria for dam design. Before you start building a pond, find out what laws apply.

You will need engineering assistance to help you select a proper site and to prepare plans and specifications for the dam.

You can obtain this kind of help from your local soil conservation district, or you may want to employ a private engineer.

Pleasing to the eye, farm ponds can be used for recreation such as swimming, fishing, and boating.

R. L. Kent

Don Baldwin

Fences. Fences serve a number of purposes in the country. They mark property boundaries, control livestock movement, regulate access by people and protect property. They may even add to esthetic values.

The most common fences in use today are the barbed wire fence and woven wire fence.

There are many types of wooden fences but their purpose is now primarily ornamental. While attractive, they are very expensive and much less effective in controlling livestock.

Barbed wire fences can cause injury to horses, especially around corrals or small pastures.

Fences represent a significant investment. A good barbed wire fence will cost about 30¢ a foot or $1,584 per mile installed. Woven wire fences cost nearly twice this amount; by comparison, a two-rail wooden fence costs around $1.65 per foot or about $8,700 per mile.

A barbed wire or smooth wire fence should have at least three strands of 12-1/2-gauge galvanized wire with class 2 zinc coating. Additional strands may be needed to confine small livestock such as sheep, goats, or ponies. Fence posts may be wooden or steel and should be spaced no more than 20 feet apart.

Wooden posts should be at least 3 inches in diameter and six feet long. They should be placed at least 18 inches into the ground. Corner and brace posts should be at least 5 inches in diameter, 8 feet long, and placed at least 3 feet into the ground and anchored.

Steel posts need to be driven into the ground so that the top of the anchor plate is level with the soil surface. They should be long enough to allow for a fence height of at least 42 inches.

Wooden posts made from cedar, juniper, osage orange, catalpa, black locust, or redwood contain natural preservatives. Most other wooden posts should be treated with a commercial preservative.

Standards and specifications for fences are available at local soil conservation districts.

Roadways

The importance of a good access road to the country home is frequently underestimated. If improperly designed, it may cause you considerable inconvenience and be a continuing expense.

If you plan to build a new roadway from a public road to your homestead you need information about soil properties, topography, and drainage patterns. You also need to evaluate how well the proposed roadway will function under the most severe weather conditions expected in your locality.

Soil properties are extremely important. Soils with high clay content become very slippery when wet. Sandy soils frequently lack stability and erode easily.

Most roadways require surfacing with gravel or other suitable material. Gravel should be placed about six inches deep on the surface. The common 10-foot-wide tread width will require about 20 cubic yards of gravel per 100 feet of road.

Gravel costs are variable and depend largely on delivery distance. In most areas gravel is delivered for around $7 to $10 per cubic yard. This means a cost of $150 to $200 per 100 feet of road surface. There will be additional costs for spreading and packing the gravel.

Slope stability is an important consideration, especially in steep country. Roads cut into steep hillsides can fail completely if soils are unstable or have moisture moving through the soil profile.

Try to plan your road to avoid steep grades. If possible keep the roadway grade below 6 percent. Steep roads become extremely hazardous when covered with ice or snow.

Improper drainage is probably responsible for most unpaved roadway problems. The road should be sloped laterally to prevent water from running down the road surface. Small graded ridges called water bars are helpful in getting water out of ruts and into a planned drainage system.

Natural drainageway crossings are especially critical. Even the smallest gully may become a torrent during a high intensity storm. Most drainage crossings require a culvert, bridge, or grade dip to keep the water in its natural channel without damaging the road.

Design bridges and culverts large enough. The carrying capacity of culverts is often reduced by sediment or debris. Be sure they are installed at the proper grade.

New roadways can create serious erosion hazards, especially where cut and fill slopes are left exposed. These areas should be revegetated as soon as practical to stabilize slopes. You may also want to plant trees or shrubs to screen unsightly areas.

If your roadway project is complex you might need an engineer or surveyor to help you with planning and design. Local road departments are also good sources of information.

Advice concerning revegetation can be obtained from soil conservationists, Extension agents, and forestry agencies.

Conservation Measures

The term *conservation measure* applies to a wide range of land improvement practices including most of those discussed above.

Conservation measures are applied in various combinations to protect soil and water resources, improve productivity of the land, and create better environmental values.

On irrigated lands, conservation measures include such practices as land leveling, ditch lining, drainage, pasture establishment, and structures for water control.

In non-irrigated farming regions, where erosion from both wind and water are prevalent, common improvement practices include terrace systems, diversions, grassed waterways, grade stabilization structures, windbreaks, and small dams. A number of conservation practices are designed to improve range and woodlands. These include livestock water developments, brush control, proper grazing management, and various tree-management practices.

Special improvement practices to create benefits for wildlife or establish facilities for recreation are also considered conservation measures.

Several government agencies have programs designed to help the rural landowner with planning, financing, and carrying out land improvement measures relating to conservation.

If you move to the country, get acquainted with your local soil conservation district. It can give you the technical assistance needed to plan and lay out conservation practices. SCS provides most of the technical assistance to districts, and maintains standards and specifications for about 130 conservation practices.

You may be eligible for cost-share assistance on certain improvements through programs administered by the U. S. Agricultural Stabilization and Conservation Service. This agency works closely with SCS and is usually headquartered in the same building.

In some western states similar assistance is available under the Great Plains Conservation Program administered by SCS.

If you need financing to establish land improvement practices, contact your county representative of USDA's Farmers Home Administration. This agency makes low-interest loans for certain rural land improvements.

The U. S. Forest Service, through its state and private forestry program, provides excellent assistance in forest management and related activities. Many state forestry agencies provide similar assistance.

Your Farmstead Buildings Can Be of Simple Materials

By Lee Allen

The type homestead you have will determine the kind of buildings you need or can get by with. They can range from windbreaks to barns.

A suburban estate or rural acreage within organized communities requires that structures be somewhat conventional. They will probably have to meet building codes, and the esthetic value may be important. Since the owner will have a source of income, the cash cost may be a minor consideration.

On the other hand, homesteaders far removed from codes, jobs, powerlines and neighbors will be primarily interested in low cash outlay and the structure's utility. The remote homesteader has much more latitude in design, and often will use different materials. In fact, the design depends on the choice of materials at hand. The labor must be done by the homesteader based on sound building principles, with the emphasis more on permanence than beauty.

The financial situation of many homesteaders will be somewhere between the two examples described. Most designs, then, will be a blending of features to accomplish the most desirable compromise of low cost, convenience, function, permanence, and beauty.

In all cases the structure will be a capital asset, and its effect on the overall value of the estate is important. Even though the structures may be planned for your sole use, consider what value different designs will have on your property value should you sell the property. In some cases the estate's increased value may be enough reason for a construction project, since many homesteads have been created only to be sold at a tidy profit.

Lee Allen is Research Agricultural Engineer, Alaska Agricultural Experiment Station, Palmer.

The Building Site

For buildings to fulfill their purpose, they should be built according to an overall plan. While generalities of good homestead design can be stated, (placing buildings conveniently around the access and residence), differences in individual terrain make planning for each homestead a necessity.

First consideration in locating the buildings will be freedom from surface water drainage. Buildings should be situated on high ground. When they must be located on slopes, drainage ditches can be provided so water doesn't enter the structures. Each animal shed should be located where water can't accumulate, or a mound of some sort will have to be constructed.

In all cases drainage from the roof should run away from the building. Your plan should place buildings away from drainage channels.

Take care to see that contractors or other construction people do not leave a depression around building foundations, or that settling of fill does not result in low spots to collect water adjacent to structures. You also will need to see that future buildings do not block drainageways from your initial ones.

Observe the direction of local winds when locating structures, especially in northern climates where drifting snow may pose a threat to animal housing. Use natural windbreaks, and consider plantings for this purpose. Don't let valuable land blow away or your animals become sick or unproductive. If your site is wooded, cut only those trees that need removal; but don't allow excessive shade on your garden if you expect it to be productive.

Many factors will influence the location and orientation of your buildings. Winter sun shining into animal enclosures will do much to keep them dry and aid the health and productivity of the animals. Possible overflow from nearby streams that may become frozen or flooded should be studied. Many streams overflow in winter when ice restricts the flow in the main channel.

Neighbors or oldtimers in the area may be able to advise you on quirks of nature that may affect the building sites you pick.

Of course, your basic wants and needs will be paramount. Estimate the size garden you need, and the kind and number of animals you wish to care for. Consider every factor you can in creating your homestead design within limitations imposed by the terrain.

Building Materials

Every building has two basic parts; the structural framework and the weather-protective material. Protection from weather generally is provided by a weatherproof surface material and a separate insulating layer. Sometimes all the needed properties are combined in a building material with structural strength, weatherproofing and insulating value.

Depending on the purpose the structure is designed for, and the climatic conditions, more or less of each building component will be combined to provide the degree of environmental control needed. Properties of materials chosen and the design also determine how long the structure will last. For instance, adequate roof overhang to avoid wet walls can compensate for using a building material that deteriorates or rots readily when wet.

Many conventional structures use a wood frame of poles or studs to support the roof. Walls are covered with wood or metal to cut down air exchange, and a non-structural insulating material is attached to the frame to prevent excessive heat loss. Add a waterproof material to the top of the structure to protect against rain and snow.

Many refinements or adaptations are possible but basic functions of the structural parts remain the same.

Often building materials can provide two or more required properties. Logs or sod, for instance, have structural strength, and can be cut to restrict air flow. They have good insulating value in the thicknesses commonly used and provide resistance to deterioration from moisture.

If a stout frame can be made, many materials will provide windproofing and insulation, especially if an adequate roof prevents major damage from moisture. Hay, straw or moss can provide good weather protection and warmth.

Remember, the key to homestead living is to cut cash expenditures. You can do this if you use ingenuity in your own designs and rely on materials gathered locally at little or no cost.

Ventilation

A primary design feature in structures, especially for animal housing, is ventilation. Most animals withstand cold, but not wet and drafty conditions. Animal hair gives good insulation unless it's ruffled by air currents or gets wet. Ventilation removes moisture from animal structures.

Air enters the structure and is slightly warmed by the animals' body heat. This warmed air is able to hold more moisture in the form of water vapor than does cold air, so it absorbs the excess moisture in bedding and from the animals' breath.

Replacing this warm moist air inside the structure by cool, dry outside air provides the drying effect of ventilation. Design of animal structures needs to provide for this essential air exchange. Besides its drying effect, ventilation provides fresh air for breathing and air exchange to carry away undesirable odors and air contaminants.

As the structure is made drier by ventilation, the inside environment becomes a poorer place for the increase and spread of disease germs or organisms that cause sickness in animals. In enclosed buildings, holes or slots are left in the structure to allow for air exchange.

For short periods when wind and temperature are severe, the holes can be closed. Open up the building when the storm is over so excessive moisture, odors and air contaminants don't build up.

Removing moisture from the structure by ventilation reduces a major threat to structures as well as to animals. Rot in wood is caused by microorganisms which can grow in the wood only when moisture is present. The rot stops when the wood dries out, but becomes active if the wood gets wet again.

The most common protection against moisture and wood decay is to place structures on concrete foundations above sources of ground moisture. Another way is to use wood in moist locations which has been treated with a material toxic to growth of rot organisms. Most of these toxic materials resist insects that burrow in wood as well. Many also deter attack by rodents or larger animals.

With treated wood, cheaper designs can be used, since considerable strength is gained by building around treated posts planted in the ground. Buildings that might otherwise last only 4 or 5 years can last 30 to 50 if made with foundations of treated wood.

Sources of Plans

If you want a conventional animal shelter, or storage for crops or machinery, there are several good sources of plans. State universities maintain plan services with plans suited to the climate of that area.

Building supply stores not only sell lumber, but have plans for structures. Plywood, lumber, and concrete trade associations will supply plans showing how to use their products economically and safely.

Metal building manufacturers not only provide plans and

materials, but can probably suggest a reliable contractor in your area to erect the structure if you prefer.

Any farm magazine abounds in sources of commercial agricultural buildings, so buildings large enough for commercial production will not be discussed here.

If you use native materials found on or near your homestead for little or no cost, then you likely will have to create your own designs using the building principles discussed earlier. You can save money by cutting trees to take to the sawmill. Many small mills will saw your timber into rough lumber on a shares basis, allowing you to use more conventional materials and readily available plans at reduced cost.

Don't disregard dry straw, grass or weeds as insulating material. Wood shavings or sawdust are good sources of inexpensive insulation. Most of these materials have about half the insulating value of commercially available insulations, so you will need about twice the normal thickness.

Some materials that you can provide for yourself, like ground sphagnum moss or coarsely ground peaty materials have as good insulating value as commercially available fill-type insulations. Common moss from ponds or lakes can be dried, possibly ground up in a feed grinder, and then packed into your walls to provide good insulation. When packing in pieces of moss or other materials that are not ground up, take care that no voids are left to create cold spots in the walls.

As a homesteader you most likely will shun credit and, conversely, some suppliers of credit will shun your efforts to obtain it. You then will have to live within your present means, and the type buildings you elect to build will depend on your present financial condition. Most homesteaders will want to substitute labor for standard building materials as much as possible.

Basic Sheds

The homesteader will experience a certain pride of ownership in a sound, permanent structure built with little cash outlay. Poles or logs are the most basic and easiest to use building materials, if available. The simplest animal shelters can be made from poles lying horizontally to form three walls, with sod or straw around the structure for windproofing. Poles may be used to form a slightly sloping shed roof with branches, straw and earth on top to shed water.

For northern climates, enclose the shed with a fourth side and a doorway opening. Generally the doorway can remain open to a fenced run. For some animals hang a canvas or other wind protection over the doorway. Don't use a solid wood door as your animals should have free choice of their shed or runway area.

An improvement would be double pole walls so straw or other fill insulation is better contained and protected. Another variation to the basic structure would be to trim the logs so they offer more resistance to air movement, as in log cabin construction. You may be able to find scraps of conventional building materials, like pieces of plywood. Don't hesitate to collect these when available.

Wood can be saved by planting the poles of your structure in the ground. Use a windproofing and insulating material to fill the spaces between the posts supporting your roof. Wire fencing can hold a pack of straw, hay or weeds.

If you have income or savings you intend to use in construction, consider buying polyethylene plastic in 4 or 6 mil thickness. Roll roofing or corrugated sheet metal are fine to use if you can afford them.

Polyethylene film is about the best multi-purpose material. It is vapor proof, windproof, rot proof, and affected by sunshine only over a period of time. When protected from damage by wind and sun it lasts almost indefinitely.

Slab Siding

Another excellent building material available in many areas are slabs, the bark-covered outside portion of logs generally discarded when the interior part is cut up for lumber. Slabs make excellent siding for a pole structure insulated with straw.

Many mills will edge the better slabs at low cost so they are smooth on three sides and have bark only on the one curved side. These three-sided slabs can be used for almost any rough construction project with studs or light rafters. As siding on your house or buildings they have the attractive appearance of log construction. Unsided slabs often are used in roofs to hold up the layer of straw insulation.

Slabs make excellent fencing materials as they offer wind protection for your animals. Remember that slab fences will catch drifting snow, so consider this in designing your pens.

If you can get slabs at a reasonable price, you will have a good supply of a versatile building material and plenty of scraps and odd pieces for kindling your fires.

Sod was the basic building material of grassland homesteaders of the past. Where supplies are available, sod is probably most suitable for stacking around structures for its insulating value. Sod roofs are surprisingly waterproof when built up over shingles of birch bark that would otherwise blow away. Modern sod roofs are best built over polyethylene film.

Stone construction will usually be beyond the energy and talents of the average homesteader, although stone is one of the most permanent building materials. Stonework requires massive foundations, much time for collecting and preparing materials, a good deal of skill, and purchase of cement for the bonding mortar. It is unlikely that stone construction will be as inexpensive as other types, but you may have to give it a try if no other material is available.

Several magazines and catalogs advertise complete metal buildings ready to erect on your lot. These are suited for the subsurban handyman, but are impractical for the homesteader. They are generally designed with the lightest materials possible to meet minimal snow and wind loads.

Experience has shown that a high percentage of these small metal buildings become damaged from weather or use. Doors are often flimsy and do not close properly after a while. The buildings require a better foundation than usually provided, so this is an expense you may fail to consider when ordering.

Sounder structures for homesteaders can be made by purchasing standard building materials—but no decrease in strength or function occurs if your tool storage or other shed is made from rough lumber or slabs.

Building Codes

Despite what many who become involved with them think, building codes are not adopted to harass people trying to build. Building permits and codes are to help you put up buildings that are safe, and adequate for the intended purpose. They also help assure the person who might someday buy your homestead (or you if you are presently buying) that the buildings are indeed safe.

Generally the codes require electrical wiring that will not cause fires, foundations adequate to support the structure, and buildings far enough apart for fire safety. The permits assure that a qualified representative of the granting authority has an opportunity to check over your plan.

Some states maintain authority for water, sewer and other permits. Generally the county exercises building authority outside city limits. Almost all towns have building codes and require permits to build or remodel.

If your homestead is outside the jurisdiction of any building authority, permits are not required. But once you determine that codes apply and permits are needed, be careful to follow the procedures outlined.

Probably there is little chance you can get out of following the codes, should they apply to the area where you build. They will prevent you from putting up substandard or inferior structures.

Some rural residents may want to erect a unique structure, or use new materials or building techniques not accepted by their local code. Then you need a variance from the chief building inspector or other authority. Obtain the required permits in advance.

Insurance

Whether you need fire and liability insurance is a personal matter. Some experts say every homeowner should have these basic types of financial protection. Certainly if you have a large investment in home and buildings, live near a town, and can afford it, fire insurance seems a prudent purchase to most people.

But fire insurance for remote homesteads should be investigated. Since policies have minimum insured values, it may cost as much to insure a bare bones home as an expensive one. Some areas cannot be covered by policies, so the company may not be forced to pay even if your house burns.

When you are outside services of a fire district, the premiums will be much higher. It may also be impossible under some policies to get back an equitable value for your labor when you have built the home yourself, especially if you have a unique design constructed with local materials. In any event, make sure the insurance company will pay the dollar amount you think it should in case your buildings burn.

Liability insurance should also be purchased based on your perceived need and ability to pay. Since this insurance protects the other fellow from things that might be your fault or from accidents he might experience on your property, your individual financial situation will be a factor.

If you feel responsibility to pay for damages that the courts may determine to be your fault, then you may want liability insurance. If you have little equity to lose, and feel that your neighbor and his children should look out for themselves while on your property, then you can probably do without insurance. Remember that in case of some accident to another person, you could lose your property and buildings to pay the cost of the damages.

Advice and Help

Your state university has research experts on agricultural and country living problems. At each university there are Extension specialists whose job is to keep informed on research findings and the latest agricultural practices. They make information available to home agents and county Extension agents whose prime purpose is to educate the public in order to help them solve their problems.

This county Extension or home agent will be close and easy to find, and can answer most questions you have. On special problems, the agents can enlist the help of state specialists or the university research staff.

Through the Cooperative Extension Service, your state university maintains a plan service, where copies of all farm building plans considered appropriate for your state can be purchased for a few dollars. Many U.S. Department of Agriculture and state bulletins can be obtained at no cost, so that you can read up on a variety of subjects.

Banks and government lending agencies offer free counseling, especially if you apply for a loan. Libraries are prime sources of all kinds of information, where you can read—at no cost—copies of the latest how-to-do-it magazines.

Don't forget to talk to neighbors. They probably have more information on local climate and other special conditions than you could find elsewhere. And they may be willing to share this with you.

When you get close to actually building a new structure, sort through all these information materials and settle on a final plan.

Remember that each plan is a compromise of features and prices. Finalize, on paper or in your mind, just what you intend to build before you start. Occasionally you may find a plan just as you want it, but generally the plans of others can be improved for your purposes.

Plans and suggestions for individual structures that follow may be just what you need, but more likely they will suggest ways to combine your materials and talents into the plan just right for you.

Building Barns

Barns are the classic buildings of rural America, for storing crops and sheltering animals. However, a barn need not be of the large size and classic shape most people visualize.

The barn must be sized to your needs and financial situation. You will have to decide which farm functions you want confined to the barn and which should have a separate structure.

Use your barn for storage and animal pens. Generally more than one kind of animal can be kept, if you have separate pens. Use outside pens and crude shelters for pigs and goats while keeping a few chickens and a cow in the barn. These outside animals can be moved into the barn during storms.

You probably should have a separate feed storage room, though small, to keep wild animals as well as your farm animals out of your feed.

Confine chickens in a good pen or separate room. It can be either wood or wire mesh. If predators like weasels or foxes are in the area, your chicken pen needs to be tight enough to keep them out at night. Use large windows in your chicken pen area or a plastic covered wire mesh to admit sunlight.

Space required for chickens can be as little as 2 or 3 square feet per bird, so that 16 to 25 chickens can be kept in a 6 × 8 foot enclosure. Their crowded situation is helped by an outside run with access through a small door.

If you plan to have one cow, a second stall can serve for hay storage. Until you get your workshop built, reserve space for a work area out of the weather. Functions of your barn

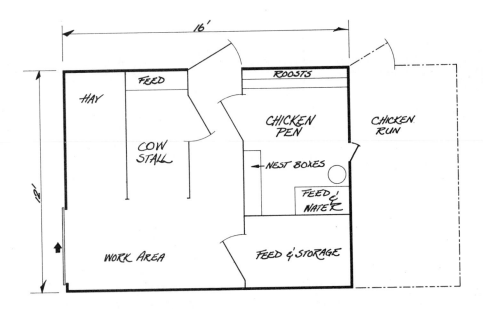

will change from time to time. As you move some activities to their own separate building, you will make space in your barn for other storage or activities.

Build your barn as large as you can afford, consistent with getting the construction done within your limitations of time and materials. It will be your largest, most expensive and most time consuming project, after your home. Thus you will have to consider whether you want to build your barn first or construct a pig pen, goat house, or tool storage at the outset.

Chicken Coop

Chickens need a well-lighted structure, as they are responsive to light conditions and day length. You may provide artificial light in a basement or windowless insulated structure, or you can provide adequate window area for natural lighting. In northern areas your egg production will be better in winter if you provide a few hours of artificial light, but you can still maintain some production without it.

Chickens also have the advantage that in summer they can be turned out to "scratch" for themselves. If left to run in your garden, chickens may destroy small plants. Besides fencing a poultry run where you can throw a great variety of weeds and other food items, you may want to fence your chickens out of the garden. Thus they can have free run to forage your woodlot and fields without doing damage.

Small flocks usually are housed on a floor (which can be earth) covered with a litter of straw, rather than kept in cages as for commercial production. Your chicken coop will then have an area for roosts with a dropping pit beneath, and an area along one wall with built-in nests.

On your floor or suspended from the ceiling you need a feeder for chicken mash and a feeder or box for grit. A waterer or two completes the equipment. You can provide a "scratch" feed of cracked corn, wheat or barley, simply by spreading it on the floor. The chickens will dig in the litter and help keep it stirred up and dry.

Poultry houses for small flocks are not usually heated in winter, but you may want to consider supplemental heat for severe climates. Be careful if you use a space heater, as contents of poultry houses are flammable.

If you keep the concentration of adult chickens at about 2 1/2 square feet per bird in a well insulated house, the chickens will generate enough heat to keep the house above freezing in all but the worst weather. Build the small door to your chicken run so it can be closed.

You will be able to keep a few ducks or geese with your chickens if you choose. These will nest on the floor.

Housing for Pigs

Housing for your pigs can be crude, but provide them a dry place to sleep and protection from drafts. In severe weather, pigs will burrow into their bedding until you can't see them at all. Due to their compact shape they can stand cold, but are not able to bear prolonged hot spells without shade.

To get started you may want to purchase a weaner pig and raise it in a small shed in a fenced pasture or enclosed pig pen. Once you get used to raising your pig to butchering weight of 200 pounds or more, you will know if you want to keep a breeding sow and raise your own litters.

To overwinter the sow, you will want a windproof shed. This is a chance to use imagination and ingenuity to construct a sound building from materials at hand. A low structure 6 × 8

Insulated pig house made of double wall pole construction with insulation of packed straw, moss or weeds.

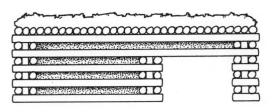

ELEVATION

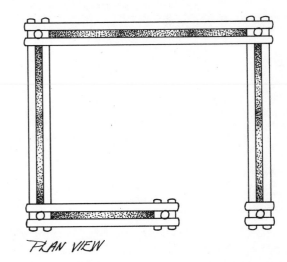

PLAN VIEW

feet with a door 4 feet high and 2 1/2 feet wide is adequate. For milder climates a tight shed that is uninsulated will do.

Goats, Pigeons, Rabbits

Goats can use the same simple housing as pigs, or a larger barn shared with other animals. Your goat shelter should be connected to a run or pasture with a high fence.

The "proper" way to house pigeons is to build a loft with an attached enclosed flyway. By the time you have finished the building, bought feeds and equipment, and paid for probably expensive breeding stock, you will decide that it will be a long time eating squab to get back your investment.

But, if your structure is such that the pigeons can be turned out to forage most of the year, you will have little feed cost and can show a profit more quickly. Pigeons have traditionally used barns or haylofts for shelter and soon learn if there are any feedlots, grain elevators or other easy sources of feed nearby.

With this approach, you can adapt almost any shed or barn to house your birds. They require some nesting spaces and horizontal shelves for walking and resting. Since they are fliers, they will concentrate near the ceiling and need some roosts or landing areas.

Pigeons tend to go back to the home where they were raised. Thus you will want to have a place to keep your breeders so they won't leave. Once the young learn where home is, you should have pigeons from then on.

Rabbits do nicely with minimum care in a shed or barn. If you want them confined, you will probably need a floor in your rabbit shed as they like to dig burrows and will eventually get out. If they have straw or hay for bedding you only have to put the buck in your rabbit pen occasionally, and keep them in feed and water, to have a batch of young rabbits about ready to eat at any time.

Even less care will be required when you turn your rabbits out in the spring, summer and fall. They will provide themselves shelter under buildings and anywhere else they can hide. When fall comes and there is little green grass and garden residue left, you can catch your breeding stock and confine them to the rabbit shed. You catch them by feeding them a little inside the shed.

Rabbits are prolific, so you will not have the expense of keeping many breeders over the winter. When your shed is warm and well bedded you may even get a litter or two in winter to keep you in meat.

Housing Horses

Horse barn plans are available through the university plan service at most cooperative Extension Service offices. If your state university does not handle plan materials of the Midwest Plan Service, you can obtain their *Horse Handbook* #15 by writing to the Midwest Plan Service, 122 Davidson Hall, Iowa State University, Ames, Iowa 50011. The price is $2.

If you have adequate acreage, facilities for putting up hay, some pasture land, and want to keep a horse at minimum cost, then you will not need the complicated and expensive structures for keeping a horse on a small acreage. With free access to a 10 × 12 open front shed that is well bedded, your horse will do fine.

A barn with a 12 × 12 foot space for a box stall or a 5 × 9 space for a tie stall is also adequate horse housing. Under these conditions be prepared to let your life revolve around the horse, as it will take most of your spare time and cash for its care.

Storage Sheds

Storage for tools, hay, machinery and equipment needs only a roof, or roof and walls to prevent deterioration.

Storage for tools and small equipment usually needs a windproof sidewall to prevent entry of dust, rain or blowing snow. Prime needs are adequate shelves, hooks, pegs or compartments to hold the stored items.

Store grains, dry feed in bags, and other food items in areas designed to keep out wild animals, mice and rats. Wood buildings with a few cracks can be improved by nailing metal from tin cans over cracks and holes. Or else use 35-gallon garbage cans for feed containers, or construct mouse-proof feed storage boxes.

In general storages bear in mind the importance of keeping items dry, especially feedstuffs. If your soil is moist, build a floor in your storage and allow air to circulate freely beneath the floor. This air circulation will carry away moisture from the ground before it can enter the storage area and cause problems.

Further Reading:

U. S. Department of Agriculture, *Building with Adobe and Stabilized-Earth Blocks*, Leaflet 535, on sale by Superintendent of Documents, U.S. Government Printing Office, Wash., D.C. 20402. 35¢.

U. S. Department of Agriculture, *Roofing Farm Buildings*, Farmers Bul. No. 2170, on sale by Superintendent of Documents, U. S. Government Printing Office, Washington, D.C. 20402. 45¢.

Water and Waste Disposal, Vital for Your Few Acres

By Stephen Berberich and Elmer Jones

Water is a basic necessity of life. For your small farm or homestead, the success of just about every activity depends on having a safe and unfailing water supply. Careful design and construction of the water system is essential.

Directly related to your water system is a sanitary sewage disposal system. Both should be planned at the same time if possible.

By profiting from current knowledge on rural water and sewage disposal systems, you can have good to excellent water service and sanitary waste disposal at reasonable cost. Private systems can be constructed which will give long satisfactory service without damaging the environment or endangering the health of your family or community. You start by defining your needs.

To estimate water needs, consider the future. Whether you choose to install a new well, reconstruct an old one, tap a spring, or add water to your system from a cistern, reservoir or storage tank, you should plan for any dream projects or goals as well as for your immediate needs.

Research studies have produced this formula for estimating home water needs: The largest water requirement for a single fixture (usually the bathtub or automatic washer) plus 1/4th the requirements for every other fixture (the kitchen sink, the shower, each toilet, etc.) equals your home's peak water demand—those periods when the well and pump must supply water continuously.

The unit of measurement for the formula will be gallons per unit of time, such as gallons per minute, or GPM. After you've

Stephen Berberich is an Information Specialist for Federal Research, Science and Education Administration, Beltsville, Md. He lives on 75 acres near Chesapeake Bay. Elmer Jones is Cooperative Agricultural Engineer, Cooperative Farm Buildings Plan Exchange, Beltsville.

established a reasonable GPM demand, a water source and delivery system can be set up with capacity to meet the demand.

50 Gallons a Person

Many studies show that home water use is 50 gallons a day per person, an average for people in this country. However, the figure is only an average. Many people use less. Some use much more.

Wells which meet demands of the farmstead home will usually also meet the water demands of a small scale farm operation, one with just a few head of livestock, for example. For more elaborate projects, such as automatic stock watering or extensive irrigation, the water source serving the home may have to be re-tooled to reach higher peak demands of the farm. Or new water sources may have to be developed.

By having a private water system, you should know more about well construction and sanitation than city friends who depend on a municipal water system. However, familiarity with your waterworks does not rule out the threat of water-borne diseases. If one factor in the system is most important it is sanitary protection of the source of your water. Contamination of a source can be caused by sewage, animal wastes, or chemical pollution of various kinds.

Newly constructed wells can lead to contamination of ground water, unless precautions are taken. In the process of

Table of water requirements for individual fixtures to help estimate home water needs

Water Uses	Flow Rate in Gallons per Minute (GPM)
Household Uses	
Bathub, or tub-and-shower combination	8.0
Shower only	4.0
Lavatory	2.0
Toilet—flush tank	3.0
Sink, kitchen—including garbage disposal	4.0
Dishwasher	2.0
Laundry sink	6.0
Clothes washer	8.0
Irrigation, Cleaning and Miscellaneous	
Lawn irrigation (per sprinkler)	5.0*
Garden irrigation (per sprinkler)	5.0
Automobile washing	5.0
Tractor and equipment washing	5.0
Flushing driveways and walkways	10.0
Cleaning milking equipment and milk storage tank	8.0
Hose cleaning barn floors, ramps, etc.	10.0

* Some irrigation sprinklers have more water capacity than shown in this table. If the capacity of your sprinkler is known, substitute that figure.

drilling, boring or digging a new well, natural earth barriers to surface and subsurface waters will be disturbed. The well itself can become a low resistance path for contaminants to travel from ground level to below the water table. However, the path can be sealed off with a grout made of neat cement and water.

Your best assurance of the proper installation, materials and location of the new well is to hire a licensed well driller. Authorized drillers know the water-bearing strata in their locality and should follow state health department regulations.

Sometimes ground water serving a farmstead can become contaminated from nearby wells which are old, poorly constructed, and unable to hold out surface drainage.

Remember this model to help avoid contamination problems: Three things are needed for the entry of surface contaminants into the ground water supply, (1) a contaminant, (2) a transmission path, and (3) a transporting medium. Exam-

Shallow wells can become polluted more readily than deep wells. Note that pollution can come from underground sources as well as surface sources. (Adapted from *American Association for Vocational Instructional Materials*).

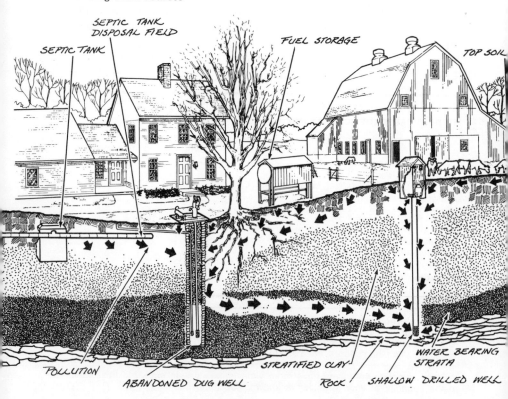

SEPTIC TANK
DISPOSAL FIELD

SEPTIC TANK

FUEL STORAGE

TOP SOIL

POLLUTION

ABANDONED DUG WELL

STRATIFIED CLAY

ROCK

SHALLOW DRILLED WELL

WATER BEARING STRATA

ple: (1) harmful bacteria, (2) a well bore, and (3) surface water or rain.

Rebuilding Old Well

It may be more economical to reconstruct an old well than to install a new one from scratch. Before reconstructing the old well, however, ask the following questions: Are there any obvious contamination sources near the well site? Is there excellent drainage? Will the proposed well structure be an adequate barrier against contaminants?

Also before you begin, measure yield of the old well. If the water is laden with sediment, clean out the well before taking a measurement. You can normally have a rural property with an existing water source evaluated by the county health department.

The old well may have been designed for use before electricity was readily available for pumping. It may be inadequate after you add modern pumps and plumbing. However, old wells often can be made deeper to reach a greater flow of water.

With modern well casings and pumping equipment, a reconstructed old well no longer needs a large diameter. A new casing of five inches or less will meet farmstead water needs. In figuring the cost of renovating the well, or installing a new one, add the cost of closing the old well bore properly.

A typical procedure is to backfill the hole with sand, after the new casing is placed and special well-grade gravel is poured around intake screens at the base of the casing. As the sand is filled around the new casing, water will surge into cracks and crevices of the old walls, carrying sand with it. This improves sanitary protection of the new system.

Spring Water

Springs, or natural flows of water from the ground, can also be developed as a water supply source. If you have a good spring on your property, ask these questions before acting:

Is the spring water good quality? Is the flow adequate? Could a gravity feed be set up from the spring? If not, would the cost of pumps and piping be within reason? And finally, because springs and areas of seeping ground water are frequently flooded, can the water source be protected from contamination?

You should have control over any storage reservoir for spring water. If the spring discharges more water than you can use, set up a diverting device for water to go past the reservoir except when needed. Also consider digging diversion ditches around any protective housing for the spring.

One more point on developing your water source. If you plan a business catering to the public such as a campground or a riding stable, your water system probably will have to comply with the Safe Drinking Water Act by following the latest regulations for well construction and sanitation.

Beyond your source of water, performance of the system is determined by other components, specifically the design, size, and maintenance of the pumps, valves, pipes, and storage facilities.

A widely accepted term for storing water is intermediate storage. It applies only to drinkable water held at normal atmospheric pressure in a storage facility carefully designed and constructed to protect quality of the water. A separate pump (besides the well pump) is needed to distribute water through the system.

With intermediate storage, a well yielding less than one gallon per minute can provide good to excellent water service for a home. A three to five GPM well can provide enough water for about an acre of lawn and garden besides home usage.

Intermediate storage at or below ground level usually provides the best per dollar storage. The water is protected from frost, and in summer comes to the spigots at cool temperatures. On the other hand, water stored in overhead tanks or held under pressure can be over ten times more expensive. The water tower or stanchions are costly. And with pressurized tanks, not much water can be held at reasonable cost.

Fire Fighting

Stored water can serve the farmstead in many different ways. It is available during power failures, or used for fighting fires, or for first aid in emergencies. The local fire chief can help you plan adequate storage to handle fires on the farm. For example, if you have 5,000 gallons available, a rural fire department pumper can fight with 500 gallons a minute for 10 minutes.

Lack of available water is a major problem in rural fire control. Sometimes even if water is at the scene it can't be delivered when friction in the plumbing reduces the water pressure.

Friction loss actually affects more than fire protection. Even with adequate well-pump capacity and intermediate storage, friction loss from poorly designed plumbing can mean the difference between a high performance water system and an inefficient one.

In the design and construction of private water systems,

too much emphasis is placed on quantity and quality of the water and not enough on quality of the service. It is unpleasant, to say the least, when competition for water between fixtures becomes a major part of planning household activities.

How would you change your water system if given the chance? Most farmers who cooperated in a dairy farmstead study answered by saying they would like to be able to take showers and be totally unaware of water use in the kitchen.

Replacing Valves

Sometimes much of the friction loss in older water systems is caused by undersize globe valves, also called compression stop valves. They should be replaced with ball-type valves. Small-diameter pipe can also contribute to friction loss. For underground pipes, use at least 1 1/4-inch pipe. You won't save cash with smaller pipe—the same size trenches must be dug.

Water for farm work is related to labor costs. Even with no hired hands, when you need a certain amount of water, you want it pronto! You don't want to waste time and money waiting for it. Minimize friction loss.

With good planning and management, a large area can usually be sprinkler-irrigated at reasonable cost. (Energy requirements will normally be less than one kilowatt hour per 1,000 gallons of water for a well-designed system.)

If you choose crops which have critical watering periods at different times, a larger area can be irrigated than your

Sometimes pressure loss in older water systems is caused by the friction created in undersize globe valves (left). They should be replaced (right) with ball-type valves. (Adapted from *American Association for Vocational Instructional Materials*).

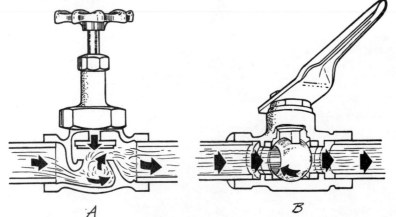

A   *B*

water supply might indicate. A small deep well pump, intermediate storage, and a small pump between storage and sprinklers can do the job.

Water from surface sources such as streams, ponds, lakes, or rainwater stored in cisterns should only be used if ground water is unavailable. Any surface water will be polluted. Contact your local health official about proper treatment and disinfection. Also, tapping surface waters may also involve special water rights problems, so investigate before proceeding.

Creating a farm pond is another way of getting water, especially for non-drinking purposes. You should have full control of the water coming into the pond. If the pond is to be fed from a stream or spring, this may mean excavating for the pond above the level of the source and pumping water up to it. If a watershed drains into your pond, keep the watershed grassed, free of barns and septic systems.

The pond spillway—that part of the pond's banks or dam where excess water exists—should be large enough to handle flooding from heavy rains.

Waste Disposal

As a rural resident, your responsibilities broaden. Besides meeting the needs of your family and the immediate community, you are now a steward of the land and the community in a larger sense. Most pertinent here is that the farm be managed so water is not wasted and the environment not polluted by ill-conceived waste disposal systems.

This versatile farm pond provides beauty, recreation, and limited irrigation.

E. Barker

The 19th century invention of the septic tank-soil disposal system brought the indoor toilet to rural America. A properly functioning, properly located septic system is still a very efficient way of disposing of sewage. Today, it remains the most common type of private sewage system.

The septic system consists of two parts, the septic tank and the soil disposal area, also called the drainage or distribution field.

The septic tank is where sewage solids separate from liquids and where bacteria begin to decompose the solid material. However, the soil disposal area performs most of the sewage treatment. Starting at the outlet pipe of the septic tank, a system, usually consisting of 4-inch corrugated plastic pipe, extends through trenches containing gravel buried beneath the ground's surface.

Discharge from the tank is a gray, somewhat odorous liquid, referred to as septic tank effluent, which carries suspended solids. The effluent goes through the drainage system and passes through holes in the piping. Soil bacteria and fungi then decompose the solids into inert matter. The result should be a clear, bacteria-free and odor-free effluent. But stop—that is only how the system *should* work.

Although the septic system is an old idea, many systems have been poorly designed and constructed. A large portion of those in operation today simply do not work properly, and are serious threats to health and environmental quality.

Clogging Is Villain

Research indicates that with proper design and management, modern septic systems can have nearly infinite life. The research shows that most premature failures are due to clogging of the soil disposal area.

Breakdowns can also result from poor septic tank performance, high soil moisture during construction of the system, failure to properly evaluate soil drainage, or overloading the system. In short, septic system failures are caused by lack of foresight or by neglect.

The inlet to the septic tank should be nonfouling and designed to cause a minimum of turbulence in the tank as sewage enters. From the intake, heavy sewage particles settle into a layer of sludge at the bottom of the tank, and lighter-than-water substances, such as grease and fat, float to the top forming a layer of scum.

Between the scum and sludge layers, there is a zone of clear liquid. The outlet, located at the upper level of the liquid

149

zone, should permit only a minimum of solids to exit onto the soil disposal area.

However, especially with single compartment septic tanks, turbulence from incoming sewage, gas bubbles rising from the sludge, and the like may allow too much suspended solid material to leave the tank. The soil disposal area then becomes clogged.

Many tanks have two compartments, the second serving as a settling chamber. These tanks can reduce the amount of suspended solids discharged into the soil disposal area by 25 to 30 percent. For any given design, the larger the tank volume, the more suspended solids will be removed.

Until you are familiar with how often your septic tank needs pumping out, have the system inspected once a year. Part of the scum and sludge can not be eaten by bacteria and must eventually be cleaned out.

Because the septic action of the tank depends on bacteria, nothing should be flushed into the tank which will kill or retard growth of the bacteria. They will tolerate moderate amounts of soaps, detergents, disinfectants and similar household products. But don't overdo it.

The septic tank should be located on the downhill side of the water source and at least 50 feet away. Size and location of the soil disposal area depends on the terrain and capacity of the soil to absorb sewage liquid, the effluent.

Heavy impervious soil, a high ground water level, or insufficient land area can create problems for the soil disposal area. When soil around the disposal area becomes saturated with moisture, for any reason, decomposition of solid particles in the effluent slows down. The system clogs. Effluent can back up in the septic tank. Ground water can be harmed.

The problem may seem a physical one, but it is really a matter of the wrong chemistry. Organic compounds in sewage

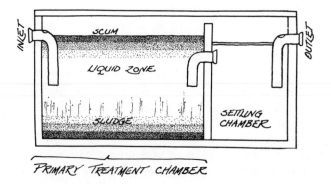

PRIMARY TREATMENT CHAMBER

or septic tank effluent need lots of oxygen to decompose properly in the soil. When the soil is loaded with moisture, not enough oxygen is present and the system stalls.

New ways have been developed to keep the purifying action going under difficult circumstances, and still avoid ground water spoilage.

Pressurizing

One way is to pressurize the system. A time-controlled pump sends effluent into the disposal system for only brief periods at a time. Corrugated drainage pipe is not used. Instead, the pump sends effluent through 1- to 1-1/2-inch diameter plastic pipe which has small holes every 30 to 42 inches.

For sandy soils, the pump should be set to eject effluent into the pipes four or five times a day. For heavier soils just once a day may be enough to prevent clogging.

Another way is to alternate between two or more drainage fields by using diverter valves. Alternating offers the advantage of giving the soil of each disposal system a rest. If one system clogs, it can immediately be shut off without stopping operation of the entire septic system.

For many soil types, one disposal system can be used for six months to over a year, before switching to a fresh system—one with enough oxygen in the soil to properly decompose sewage.

Serial Distribution

A third way, the serial distribution system, has proved very effective on soils where there is no risk of polluting the ground water. A series of trench sections are separated by dams or earthen barriers. As effluent flows into the system, the first trench section is forced to fill completely before the liquid can go to the next section.

Flooded sections serve as contact chambers, where bacteria decompose suspended solids. The biological activity continues to improve effluent quality in each succeeding section.

This system is often used on hillsides, but it is equally effective on flat land.

If the serial system should fail, only the first trench usually needs to be replaced. The series of trenches can also be alternated.

The pressure distribution system could be used for subsurface irrigation of shrubs, trees and lawn. Normal amounts of effluent entering the soil from a septic system far exceed what landscape plants would require.

However, if you are hooked into an expensive town water system, consider making the soil distribution system larger, with separate laterals and two-, three-, or four-way diverter valves to accommodate the irrigation. Plant nutrients in sewage effluent make such a system worth considering.

Recycling, Composting

No discussion of waste disposal on the farm would be complete without touching on garbage disposal. Two things here—the compost pile and waste recycling.

With some industrial resources becoming scarce or expensive, recycling will likely get more popular. On your farmstead, you may consider having separate disposal containers for paper, glass, metals such as aluminum and steel, or other categories. In some regions good return rates on these items already make the effort worthwhile.

If you pay for trash collecting, recycling will reduce that expense. If you burn wood for heat, paper is probably more valuable for fuel.

Composting is an ancient art, and a modern science. On the farmstead it can be used to provide fertilizer for the garden or it may be an integral part of raising crops, depending on how much organic material is obtainable.

Leaves, hay, sawdust, crop residues, weeds, grass clippings, kitchen wastes, manure and other organic materials are layered in piles and allowed to ferment. With the aid of oxygen, provided by turning the piles occasionally, the material breaks down quickly into an excellent soil conditioner or fertilizer.

A small compost pile can keep a vegetable patch in top condition. If a lot of composting material is shredded, piled and turned with power equipment, orchards and field crops can also benefit from the additional humus and fertility that compost gives to the soil.

Further Reading:

Planning for an Individual Water System, American Association for Vocational Instructional Materials, Engineering Center, Athens, Ga. 30602. 1973. $6.95.

U. S. Department of Agriculture, *Treating Farmstead and Rural Home Water Systems,* Farmers Bulletin No. 2248, on sale by Superintendent of Documents, U. S. Government Printing Office, Washington, D.C. 20402. 35¢.

U. S. Department of Agriculture, *Water Supply Sources for the Farmstead and Rural Home,* Farmers Bulletin No. 2237, on sale by Superintendent of Documents, U. S. Government Printing Office, Washington, D. C. 20402. 35¢.

Power Sources, Equipment for Life on a Few Acres

By Wesley Gunkel and David Ross

Modern farming and our style of living require the use of energy in much greater quantities than in the past. While returning to nature on a few acres may reduce dependence on energy and modern equipment for some, it is hard to escape their use. Power is needed to perform the many tasks found in maintaining and using a few acres.

The few acre site may open new energy sources to the owner. Natural energy from wind, water, the sun, wood or coal may be locally utilized. Electricity, natural gas, fuel oil and coal may be supplied by commercial companies.

Equipment such as tractors and tractor-operated implements, household appliances, and power tools use many sources. Family living requires surprisingly large quantities of energy. More than 20 percent of all the energy in the United States is consumed in the home. Over half of this energy is used for heating.

Both your life style and the nature of any operations on your few acres will determine energy requirements. The location of your place will determine the most likely energy sources. Most few acre operations will be near electricity or a petroleum fuel source. However, the electrical service may be inadequate for large electrical motors, particularly 3-phase motors. Accessibility for fuel deliveries may be poor during some months, and larger storage facilities may be needed.

Electricity obtained from a central station is fairly dependable and reasonably priced as a rule. Electrical generators on your few acres are ideal for standby operation. If generators are your only source of electricity, some form of energy storage is needed; direct current (d.c.) and storage batteries are used.

Wesley Gunkel is Professor of Agricultural Engineering, Cornell University, Ithaca, N.Y. David Ross is Extension Agricultural Engineer, University of Maryland, College Park.

Average wattage and amount of energy used per year for selected home appliances

Appliance	Wattage	Annual Energy Consumption (kw.hrs.)	Equivalent Gallons of Gasoline
Air-Conditioner (room)	1,500	2,000	54.6
Can Opener	150	5	.14
Clock	2	17	.47
Clothes Dryer	4,500	1,000	27.0
Clothes Washer (automatic)	500	100	2.7
Coffeemaker	900	106	3.0
Dishwasher	1,200	340	9.4
Electric Blanket	200	180	5.0
Fan (attic)	350	270	7.4
Food Freezer (15 cu ft)	350	1,200	32.8
Food Mixer	125	10	.3
Food Waste Disposal	500	30	.8
Frying Pan	1,200	240	6.6
Hair Dryer	500	15	.4
Hot Plate (2 burner)	1,250	100	2.7
Iron (hand)	1,000	150	4.1
Lights	1,000	800	22.0
Radio (solid state)	5	20	.6
Range	13,250	1,550	42.3
Refrigerator (frost-free 12 cu ft)	600	1,200	33.0
Sewing Machine	75	10	.3
Television (black & white)	160	400	11.0
Television (color)	350	540	14.8
Toaster	1,200	40	1.1
Vacuum Cleaner	630	45	1.2
Water Heater	4,500	4,500	123.9
Water Pump (shallow)	500	231	6.4

These systems are satisfactory for electric lights but have limited capacity for large power demands.

Other sources of commercial energy that may be available at your site are natural or LP gas, gasoline, fuel oil or coal. All of these fuels can be used for both power and heat. Gasoline and diesel fuel are used regularly for tractors. LP gas is normally used for heating, but can be used in cars, trucks and tractors fitted with special LP gas carburetors.

In select locations, water and wind energy may be useful for performing some operations. Work on utilization of wind energy has come from many sectors and may result in equipment directly coupled to a windmill, or in wind-generated electricity.

In 1850, one percent of the total energy consumed in the United States was supplied by wind. Since then more than 6 million small windmills of less than 1 horsepower each have been built. These windmills pumped water, generated electricity, and performed other similar tasks. Over 150,000 are still in use.

The amount of power available is directly proportional to the cube of the wind velocity, which illustrates the importance of wind velocity. Doubling the wind velocity means 2 x 2 x 2 or eight times the power. As a rule of thumb, continuous winds over 8 miles an hour average are needed to operate a wind-powered electrical generator.

Suitable site characteristics include high annual wind speed, no tall obstructions upwind for some distance depending upon the height of the windmill, top of a smooth well-rounded hill, open plain or shoreline, or mountain gap that produces a wind funneling. Consult the windmill manufacturers for specific information.

Waterpower is another potential source of natural energy available at a few farm sites. In the past, waterpower provided the energy for grinding flour, sawing wood, and generating electricity. Today waterpower is used primarily for generating electricity in large central hydroelectric plants. Very few of the original small scale systems remain. Interest has developed in harnessing water in rivers and streams for limited supplemental energy.

Wood, Solar Energy

Wood is a source of energy available at many sites. While only a small fraction of America's fuel needs are now supplied by wood, it can be used advantageously if a sufficient, low cost supply is available. Fireplaces have been installed in many homes for decorative purposes. Few are used for primary or supplementary heating, as the efficiency is a low 10 percent. Improvements have been made in wood-burning equipment to increase the combustion efficiency of fireplaces and stoves.

Wood can be obtained from your own well-maintained woodlot or from forest land nearby. State and local regulations will govern wood collection from public lands where a permit may be needed.

Solar energy has been used in recent years to heat and cool buildings, dry agricultural products, power irrigation pumps, generate electricity, heat water, and for other purposes. Solar cells are used to generate electricity from solar energy. Costs of solar cells and associated equipment are expected to come down with time.

Passive solar systems make use of the building design to capture and store heat. Windows or structural components which absorb heat energy are examples. Passive systems are relatively inexpensive and use few or no moving parts. They are designed into the structure and not added onto it later.

Active solar systems use large heat-collecting panels, pumps or fans and storage facilities. The equipment is relatively expensive, requires space, and must be maintained. Except in moderate temperatures, the solar system cannot economically provide all the heating needed. A backup conventional heating system will provide 50 to 70 percent.

Putting finishing touch on heat collectors (bottom). Solar heat can move directly from collectors into the house or be stored in crushed rock beneath house, as shown in diagram. Working drawings of solar house are available from Extension Agricultural Engineers, Cooperative Extension Service at state land grant universities. Ask for *Plan 7220, Solar House.*

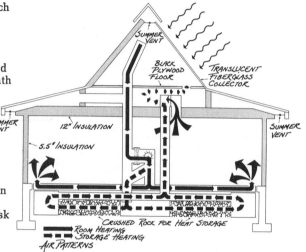

Popular magazines carry articles about build-it-yourself systems and how they perform. A well designed build-it-yourself system will be economically feasible before one purchased and installed by a contractor; however, the lay builder must be capable of doing the job.

Two rules of thumb on sizing the solar heating system are: 1) the collector area is equal to one-fourth to one-third of the house floor area, depending on the site and geographical location, and 2) a liquid heat storage system needs 1.5 to 2 gallons of fluid for each square foot of collector, while an air heat storage system needs one-half to three-fourths of a cubic foot of rock per square foot of collector.

In addition, the house should be insulated to standards equal to electric heat. Often the many possible conservation practices give sufficient savings in an existing house so that the additional expense of putting in a solar system does not give an economic return.

The fastest payoff comes in heating household water. No structural insulation is needed and comparatively small collectors—30 to 100 square feet—are used. The storage tank is connected to the household hot water heater by a heat exchanger. Fluid in the collector does not mix with the household water, so there is no contamination.

Insulating the attic of a house.

Field Machinery

A wide range of tractors and implements are available from local farm machinery dealers. Most of the larger tractors and implements sold are manufactured in the U.S., while many of the smaller tractors—particularly diesels—are imported.

Small compact tractors, frequently called lawn or garden tractors, are also manufactured in the U.S. and are sold by either farm machinery dealers or other specialized stores. Garden tractors are designed primarily for light, estate duty and are not intended for continuous heavy service.

Consider product reliability, equipment warranty, dealer reputation, and availability of spare parts when you purchase machinery.

A successful farm business requires careful management of land, labor and capital. In most cases, machinery costs are high and exceeded only by land costs. Thus it is important to manage machinery properly. This includes planning the use of machinery for timely and productive operation, selecting proper types and sizes, proper care to maintain performance and reliability, replacing obsolete or wornout machinery at the right time, and deciding if a custom operator should be hired or a machine leased.

Most operations involve several different crops with specific tillage, planting, pest control and harvesting requirements. Ideally each crop should have its own set of specialized implements to produce maximum yields. More equipment in turn means higher overhead costs. Lack of adequate equipment can delay getting crops planted or harvested on time, reducing yields and product quality.

Using larger machines reduces labor costs since they complete the job faster. But while large tractors can cover more acreage than smaller ones, they also have higher overhead costs. Smaller tractors have less capacity and may cause delays in key field operations, resulting in a lower crop yield.

By carefully analyzing the work to be done in the available time, you can select the right size.

Farm tractors and equipment are commonly measured in terms of horsepower or machinery width. Machines should be selected on the basis of their capability to perform a given job within a certain time. This rate of performance is called field capacity, and is measured in either acres per hour or tons per hour. Some harvesting and processing machines are measured in bushels per hour but this is an inaccurate measure due to variations in grain moisture and densities.

Field capacity is based on actual time spent in the field.

If it requires 5 days, working 8 hours a day, to plow an 80-acre field, the field capacity is 2 acres per hour. Field capacity = 80 acres divided by 40 hours = 2 acres/hour.

Field efficiency is the ratio of the effective capacity of a machine to the theoretical capacity. It measures the relative productivity of a machine under field conditions. Operating the machine at less than full width, lost time spent filling hoppers, cleaning and adjusting machines in the field, all reduce field efficiency.

Some of the time lost in doing field work cannot be eliminated. Other lost time can be substantially reduced by careful planning and good management.

Servicing Machines

Keeping farm machinery in top mechanical condition is one of the best ways to improve field efficiency. Machines should be serviced regularly and adjusted correctly. Neglecting this can cause expensive repairs or complete overhauls.

An operator's manual provided with all new machines

Making adjustments on machinery may give an opportunity for a relaxing work break (below). Indian girl gets tractor lesson (right) in Alaska. (Photos selected and captions written by Yearbook Editor.)

Charles O'Rear

George Robinson

gives proper instructions for servicing, repairing and adjusting. Manuals for older equipment can usually be obtained from the manufacturers. Taking time to read the manual before using the equipment will eliminate future problems and save you money.

Replace farm machines when they become unreliable for completing the job on time. It is difficult to keep a wornout machine repaired once it starts having excessive breakdowns. When purchasing a replacement, consider a used machine. Frequently machines in good mechanical condition have been traded in because larger equipment was needed. If these fit your needs they are a good buy.

Leasing machinery or hiring custom operators are alternatives to owning farm equipment. In some cases custom operators can complete the work faster and cheaper than you can. This is especially true when you have only a few acres and specialized machines are needed.

When considering hiring a custom operator, talk to other people who have used his services. Waiting for a custom operator to arrive can be expensive if the crops are not planted or harvested at the optimum time. Timeliness is important when you compare leasing equipment, owning equipment, or hiring a custom operator.

Comparing Costs

Costs of owning and operating machines should be determined before comparing viable alternatives. Machinery costs involve both fixed and operating costs. Fixed or overhead costs include machinery depreciation, insurance, interest, shelter, and taxes. Operating costs include fuel, labor, lubrication, maintenance, and repairs. Fixed costs generally depend upon how long a machine is owned while operating costs relate to its use.

Depreciation is a loss in the value of a machine caused by age, wear, and obsolescence. Frequently this is the largest machinery cost. There are several different ways to calculate depreciation. Internal Revenue Service methods can be used to calculate this cost.

Premiums paid for machinery insurance are included in the total costs of owning farm machines. Frequently only the more expensive equipment is insured while a calculated risk is taken on less expensive machines. Typical insurance costs run about 0.3 percent of the original list price.

Interest on borrowed capital or cash used to purchase farm equipment is a direct expense. Prevailing interest rates are used to calculate this machinery cost.

Storing machinery under shelter is a good management practice in most of the U.S. Studies on expensive tractors, combines and other machinery have generally shown that storing these machines adds one or two years to useful life. Typical shelter costs are less than one percent of the machine's original cost.

Taxes paid on machines are included in the cost of owning them. Typical annual charge for taxes are one to two percent of the machine's value at the beginning of the year.

Typical fuel and lubricant needs amount to between 20 and 30 percent of total machine cost. The ratio of fuel and lubricant cost to total machinery costs increases as annual use of the machine increases.

Fuel use can be reduced by matching tractor horsepower to machinery size. Other fuel-saving tips include eliminating unnecessary tillage operations, combining operations so the tractor can operate at full load, and throttling back and shifting to a higher gear when pulling a light load.

Labor to operate farm machinery is an operating cost and should be included in total machinery cost. Either hourly wages paid for hired labor or a fair value for your own time should be used to determine the cost. This cost is particularly important when comparing custom machinery costs with owning or leasing machinery.

Maintenance and repairs are operating costs since the amount spent for repairs is proportional to machine use. While some repair costs are caused by deterioration, the majority are due to use of the machine.

Repair costs are almost impossible to predict because of the many factors involved. Excessive repair costs can almost always be traced to excessive speed, overloading, poor daily maintenance, abuse of equipment, and ignoring first signs of problems.

Repairs should be made to maintain the farm machine's reliability and keep it performing at top capacity.

Mowers, Lawn Tractors

Small equipment will meet the needs of some activities on a few acres. Large lawn areas and small gardens require powered equipment but not the large equipment used for field operations.

Riding lawnmowers will handle the grass-cutting chores. A garden tractor offers more versatility to do mowing, snow-blowing, plowing, roto-tilling and many other tasks. The bigger the job and the more tasks you have to do, the greater the horsepower and special features the tractor will need. Lawn tractors are light duty while a garden tractor is often heavier duty and more capable of handling attachments that plow, disk, or till the soil. Some manufacturers distinguish between the two types.

These tractors are commonly powered by air-cooled single-cylinder gasoline engines ranging from 7 to 14 horsepower. Above 14 horsepower, a two-cylinder air-cooled or four-cylinder water-cooled engine will be used.

Electric-powered compact tractors are available and operate with minimum noise and vibration. The batteries are recharged by plugging into a 115 volt convenience outlet. Electric-powered tractors do have some drawbacks, including relatively short operating time between battery recharges and generally greater expense.

Features such as front, center or rear power takeoff (PTO), standard Category O three-point hitch for attaching and raising and lowering rear mounted equipment, hydrostatic or automotive type transmissions, lug-type or lawn tires and wheel weights should be considered. The PTO supplies power to attached equipment. Small lawn tractors have no standard PTO location, speed, or type.

While manufacturers have made some efforts to standardize attachments, the best bet is to purchase the same brand tractor and attachments. Some attachments are not available for the complete size range of compact tractors since the smaller tractors do not have enough horsepower to operate them satisfactorily.

Small agricultural tractors have standard three-point hitches and are designed for continuous duty.

Available tractor attachments

Aerifier

Cultivator

Disk Harrow

Duster

Earth Auger

Electric Generator

Fertilizer Spreader

Flail Mower

Forklift

Front-end Loader

Front-mounted Scraper Blade

Garden Cart

Hydraulic Wheel Scraper

Insect Fogger

Landscape Rake

Land Leveler

Lawn Mower (Reel, Rotary-sickle)

Lawn Roller

Lift Boom

Plow

Powersaw

Rotary Broom

Rotary Plow

Rotary Tiller

Seeders

Snowblower

Sprayer

Stone Picker

Tool Bar

Tractor Cab

Transplanter

Trencher

Turf Sweeper

Water Pump

Wheel Rake

Small 20 to 45 horsepower agricultural tractors are available from local farm machinery dealers. These tractors are specifically designed for continuous duty field work and cost more to buy than compact garden tractors. They should be equipped with a three-point standardized Category I hitch for attaching the various implements.

Three-point hitches provide near parallel lift of the attachments, give better depth control of tillage implements, and provide weight transfer for heavy plowing. Standardizing the hitch lets you attach any brand of Category I implement to any Category I tractor.

Options available usually include a choice of either diesel or gasoline engines. Diesel engines cost more to buy initially but require less maintenance and use a lower priced fuel. Consider the total number of hours you plan to run the tractor each year before deciding on the economics of a diesel versus gasoline tractor.

Bear in mind that many of these small agricultural tractors —particularly the diesels—are imported. Before purchasing, make sure your dealer has adequate repair parts or that they are available from some central location in the U.S. Spare parts shipped from abroad take longer and can delay getting your tractor repaired and back into use.

Small Tools

A wide selection of hand and power tools are available from local garden centers, hardware stores, and mail order houses. Select good quality tools as they will give better service and with proper care last longer than less expensive equipment. Name brand tools usually are more reliable but you should shop comparatively, noting the best features and quality of each.

After buying power tools and equipment, read and follow the printed instructions. More tools are damaged and broken by improper use than are worn out through a lifetime of service.

Store hand tools and equipment in an indoor area. A small combination farm shop and tool storage is ideal. Small cutting tools should be cleaned and sharpened before storing. A light film of oil or grease spread thinly on hoes, saws, shovels, shears, trowels and similar tools will provide rust protection.

Larger equipment that cannot be stored under cover should be protected from the weather. Canvas or plastic sheets make excellent covers; tie or tape them securely to the machine. Winterize engines that you need to store for long periods.

Consider operator safety when you either purchase equipment or operate it in the field and on the road. The Occupational Safety and Health Act requires safe and healthful working conditions for everyone. You are not only responsible for your own safety when operating machinery, but also responsible for other people working for you or using your equipment.

Machinery is potentially dangerous. Treat it with respect. Farm machinery designers have built in many safety features, but it is impossible to protect an operator who disregards all safety precautions. Practicing good judgment and common sense is the best protection.

Safety precautions you should follow include: (1) keep equipment in good operating condition, (2) don't operate machinery on steep sidehills or near ravines, (3) use a large enough tractor to handle the load—a tractor can pull a heavier load up a hill than its brakes can control coming down, (4) don't turn at high speeds—centrifugal force can cause side tipping, and (5) don't rush to finish a job and become careless.

Machines are built to help you and they will do an excellent job if handled carefully and treated properly.

How to Make
the Most of It

Part Four

John Messina

Five Years on Five Acres— Tips to Go About It Right

By Arthur Johnson

Many of us have a creative need to be productive in our spare time. For some, this means woodworking, cooking, painting, or writing. For others, the need can be filled by part-time farming. All of these should be considered hobbies and cost money to pursue, but other people may be willing to buy some of the products of our creativity.

I've lived for five years on a few acres such as you probably envisioned when you picked up this book. At different times I've raised beef, sheep, pigs, chickens, ducks, geese, turkeys, vegetables, strawberries, and fruit trees. I've tried various ideas to find the best, and also read and compared the experiences of others in similar situations.

Although living and producing on five acres is not the same as on 500, many techniques and methods can be adapted for small-scale operations.

Keep in mind some basic guidelines when starting to farm on a few acres. The first is, be cautious. Begin by producing on a very small scale, perhaps only enough to satisfy your own needs. Learn what problems you will encounter: diseases, insects, climate, and the requirements of quality and timeliness. These problems are likely to become more severe as production scale increases.

I heard of a man who bought 100 ducklings from a farmer thinking he could sell dressed ducks for a large profit. When he accidentally killed one of the ducks, he hurried back to the farmer to ask how to pluck and clean ducks. Clearly, that man did not realize what he was in for.

An important factor is whether you can produce your product profitably with acceptable quality, and what the local market is for that product. Some simple record-keeping will be

Arthur Johnson is Assistant Professor, Agricultural
Engineering Department, University of Maryland,
College Park.

necessary or you may end up subsidizing the food budgets of your customers.

Pleasing the Clientele

Pleasing your customers so they will return requires you to sell the best of your product and guarantee satisfaction. One man sold a pig to a customer and accidentally broke its shoulder when loading it on the truck to be hauled to the butcher. He had to replace the pork shoulder with one from his own pork to satisfy the customer.

Knowing your market is probably the most important factor in developing a profitable enterprise. Determining where or to whom to sell your produce takes some work, but is very important. Markets may be already available, as a local farmers' market, or you may have to create your own. This requires some selling on your part to get started, such as advertising in a local paper, and some system to keep in touch with past customers—especially if they are satisfied. These customers make the best advertisement for you. Be sure they are completely satisfied if you wish to sell to them again.

There are economies of size. Small-scale farms cannot compete against large farms if they sell to the same markets. It's a good idea to market directly to the consumer, charging a fair price above wholesale and below retail. With food items, there is a considerable spread between farm market value and store shelf price. You and the customer can both do well by a direct transaction.

Selling directly to the consumer requires you to produce consumer-oriented crops, not field crops such as wheat. Strawberries, vegetables, fruits, honey, eggs, and popular meat animals are among the most marketable items. Cater to needs of the local populace, and grow items which are not abundantly available in your local area.

A good source of information on all matters is your local Cooperative Extension Agent. Many of them are becoming more aware of the needs of small-scale farmers. Your state department of agriculture may also be of service.

Production methods may vary considerably depending on the crop and farm size. With your main source of income not derived from the farm, you can experiment with new ideas. Mistakes may not be quite as costly as on large farms.

Regardless of your small-scale farm's size, a considerable investment is still required. Rearing 10 feeder pigs to slaughter calls for more than two tons of feed, plus cost of the pigs, plus cost of water, labor, buildings, and other equipment. Out-of-pocket costs are at least $700. An investment for dwarf fruit trees might be $10 per tree and five years to production. In general, the more profitable the crop, the larger the initial investment required.

Techniques

Decide on appropriate production methods. You can use large-scale production methods or something completely different. An example of the difference is buying a tractor, mower, rake, and baler to produce hay vs using a lawn mower to chop grass to make silage. Both methods allow your animals to be fed during winter. Another example is the use of a large tractor and cultivators vs rotary tiller vs herbicides vs mulch to control weeds in a strawberry plot.

I prefer a balanced farm with little waste. This means several crops, both animal and vegetable. Pig waste can be used to help grow strawberries and vegetables; waste vegetables can be used to help feed pigs. If pig waste is used, a fall application is recommended.

Ground unsuitable to vegetable growing can be used as pasture for sheep, goats, or cows, and fruit trees can be grown in the same area. Ducks, geese, and chickens help keep fertility of the area high. All of these can be marketable crops in your area.

In all of this, you must have tools and equipment. To reduce your investment, watch announcements of farm auctions and sales. Used tools can often be bought for a small fraction of their price when new.

Remember there are laws and regulations governing production and sale of food items, and these vary from location to location. Often the sale of processed food is prohibited unless the processing was done in inspected facilities. Examples are butchering and cutting of meat and poultry, and handling of milk.

Pesticides, herbicides, other sprays, and feed additives must be used within registered restrictions (crops, application methods, withholding periods). Beehive inspections are required. Use of sewage sludge for fertilizer may be restricted while use of composted sludge may not be.

It is wise to know the regulations that apply in your area, especially if you deal directly with the public.

Small-scale farming can be a rewarding hobby. Often it requires most of your free time and does not allow you to leave home for extended periods of time. But for those who like it, every day is a vacation.

Pigs and Pumpkins

By Nancy P. Weiss

New England farms conjure up images of Robert Frost poems with bending birches and stone walls between neighbors. Some farms in Connecticut closely resemble the popular impression. My husband and I own an 87-acre farm in Pomfret, a small town in the northeast corner of Connecticut. Built in 1790, the farm hugs a hillside and is reached by traveling up a long dirt driveway lined with maple trees.

Come Spring, the maple trees are tapped. While sap buckets are picturesque, at our farm they have been replaced with plastic tubing which conducts the sap down the hill to collection vats. Maple syrup is but one of several sources of either revenue or crop for family use which is produced on the farm.

Our choice to live on the farm was made deliberately and at considerable cost in many respects. Jim's jobs take him to Hartford each day, a distance one way of 49 miles. In Hartford he works as a stockbroker for a large investment firm and serves as the state legislator for seven rural towns. The legislative district, based on population, is geographically one of the largest in the state.

Our personal experiences on the farm have helped put Jim in closer touch with the problems of many of his constituents, who are either commercial or part-time farmers. Through farm ownership, we have worked with various USDA agencies and participated in a number of programs. The insight gained through this first-hand experience is far more educational than learning of the work of the agencies through committee hearings or constituent contact.

I work for the Cooperative Extension Service as Assistant Director of the 4-H Program in Connecticut. Our lifestyle requires that I commute at least 20 miles each day to the University of Connecticut campus in Storrs where my office is located. Although I was brought up in a rural area, my family did not farm. My work with 4-H has provided me with experiences and an appreciation of farm life that is reflected in our choice to live on a farm.

Some days the juxtaposition between our professions and our farm is amusing and quite perplexing to non-farm people.

Last year during the legislative campaign, I was often forced to call a halt to planning meetings or sign-painting parties while the volunteers helped me chase our wandering pigs

Nancy P. Weiss is Assistant Director, 4-H and Youth, Cooperative Extension Service, The University of Connecticut, Storrs.

back into their pen. Poor fencing was the problem, and highly intelligent pigs made quick use of our inability to take the time to properly fence them. News reporters laughed in astonishment when we dashed out the door of the partially restored farmhouse to call the hogs back to their pen.

A tea party was disrupted one hot summer day by the exhaust from a faulty diesel tractor stuck in the hayfield.

Early fall mornings saw us up at dawn to spray the pumpkin patch, our first effort at commercial growing, only to find that the harvest, nearly the entire crop, had rotted on the vine from heavy rainfall for several weeks. Our small, peppery Scotty dog killed all of the chickens one morning—all that is, that were left following an attack by a ravenous fox. Finally, a pet billy goat died from eating nightshade after two weeks of hand feeding him and $61 in veterinary bills.

Bean Smash

The disappointments have been great, so we have tended to stress the successes. A bumper crop of green beans brought income from a roadside stand which more than paid for the investment of seed, fertilizer and plowing. An added boon was the sight of a freezer full of home-grown produce, which easily meets our family requirement for vegetables, provided we don't become too sick of green beans, broccoli and snow peas.

The land, long depleted by intensive growing of corn and poor soil conservation practices, is being slowly brought back into good condition. An abandoned apple orchard bore much fruit after several Saturday mornings of pruning. The woodland, once scorned as useless, provides wood for three fireplaces and a woodstove.

Most of all, there is a great sense of satisfaction. In restoring our house and a small half-house, used for many years by the farm help, we discovered beautiful hand-done stenciling, a wainscoted "keeping room"—a central room of the house, wide pine board floors, and a long-abandoned bake oven.

Although the investment in money has been larger than we had hoped, much of ourselves has been put into the farm in the year and a half we have owned it.

As we pry ourselves out of our cars after long days and long commutes from work, we both feel a surge of energy that comes straight from the land.

Several winter evenings while we sat snowbound in the house, we spent our time planning better and more efficient ways to use the buildings and improve the land.

Orchards Can Be Profitable, Good Management Is the Key

By Harold Fogle and Miklos Faust

Orcharding is more intensive than many other cropping systems. Under ideal situations, good livings are made on some fruit farms of about 20 acres.

However, orcharding requires a location suitable for fruit trees, and heavy initial investment in land preparation, equipment, nursery trees and in planting and caring for the young trees with little return for three to seven years.

Heavy demands are placed on the operator to get maintenance operations (pest control, weed control, pruning, fertilization, irrigation) done on a timely basis, to find a profitable outlet for the fruit, to have adequate labor available to thin and harvest the fruit at the proper times, and to keep abreast of new varieties, control practices, planting systems and marketing practices.

About 20 acres is a minimum orchard unit if no other enterprises supplement the orchard income. The requirement for, and investment in, machinery and supplies will be about the same whether 2 or 20 acres of fruit are grown. If the machinery can be adapted to other enterprises, can be shared with other orchardists or rented, or if custom services are readily available, a smaller orchard unit may be practical.

The decision to operate a small orchard should be based on careful analysis of local costs and expected returns from the orchard as a complement to other enterprises. Thought must be given to minimize conflicts in timing of harvesting and maintenance operations.

As with other perennial crops, certain areas of the United States are more subject to risks such as low temperature damage to trees, inadequate fruit set due to spring frosts, damage

Harold Fogle is Research Horticulturist and Miklos Faust is Laboratory Chief, Fruit Laboratory, Plant Genetics and Germplasm Institute, Beltsville Agricultural Research Center, Beltsville, Md.

Considerations before establishing an orchard

- Need for training of operator
- Availability of advice
- Types of fruits grown in the area
- Rainfall distribution, needs for irrigation.
- Winter temperatures
- Temperatures during blossoming period of various fruits
- Marketing possibilities
- Availability of labor for harvesting and other operations
- Variety mix for labor distribution
- Variety mix for orderly marketing
- Variety mix for pollination requirements

- Machinery needs
- Need for storage facilities
- Possible rental of machinery or custom operations
- Site: air drainage
- Site: soil drainage
- Soil preparation and erosion control
- Soil preparation: nutritional requirements of fruit types
- Purchasing, storage and handling of pesticides
- Human factor — tolerance to pick-your-own operation, pesticide regulations, etc.

to developing fruits and trees by hail, wind, and rain, inadequate moisture or too cool temperatures for growing large fruit or fruit which matures at the proper time.

The Great Plains areas generally are poorly adapted to fruit growing because of low temperatures and lack of sufficient summer rainfall. Most northern states are subject to occasional winter freezes and spring frosts which reduce the chances of annual crops. In the Deep South, certain fruits and

Susceptibility of tree fruit crops to environmental hazards

Hazard	Apricot	Apple	Cherry (sweet)
Spring frosts	xxx	x	xx
Winter freezes	xx	x	x
Insufficient chilling	x		xxx
Brown rot	xx		x
Rain cracking	x		xxx
Bird damage	xx		xxx
Tree borers	xx		xx
Hail	xx	xx	xx
Excessively high temperature			x
Insufficient moisture	x	x	xx
Poor soil drainage	xxx	x	xx
Bacterial diseases	xx	x	xx

x = slightly susceptible   xx = moderately susceptible   xxx = highly susceptible

172

specific varieties of other fruits may not receive sufficient winter-chilling to blossom and put forth leaves properly.

Any prospective orchardist should first look for successful orchards in the immediate area. If there are none, do not consider an orchard unless you can determine that you have an exceptionally good site. For example, in northern areas a site on the windward side of a large lake may escape low temperature extremes while this moderating effect is not generally felt in the immediate surroundings.

Several years of records of minimum temperatures as close to the proposed site as possible should be observed. If winter temperatures often drop to –22° F (–30° C), no fruit should be grown in the area. If temperatures fall below –13° F (–25° C) in some years, probably peaches, plums, nectarines, and apricots will be thinned and may suffer tree damage. A site which seldom has temperature below –4° F (–20° C) should be sought.

Marketing possibilities may determine what kind of fruit can be grown profitably in an orchard. In an established fruit-growing area, several options may be available. Commercial wholesale outlets for fresh fruit marketing, canning, freezing, or other processing uses may be readily available. Close to large population centers, an attractive fruit stand or pick-your-own and rent-a-tree orchards may be feasible.

You should carefully analyze the total requirements of one or more such operations before deciding on the appropriate outlet. For each type of operation, rather specific requirements in spaces, facilities and temperament are needed.

ctarine	Peach	Pear	Plum	Prune	Sour cherry
X	XX	XX	XXX	X	XX
X	XX	X	XXX	X	X
X	XX		X		XX
X	XX		X	XXX	X
X	X		X		X
X	X	X	XX	X	XX
X	XX		XX	XX	X
X	XX	XX	XX	XX	XX
		X		X	X
X	X	X	X	X	X
X	XX	XX	XXX	X	XX
X	X	XXX	XX	X	X

Another critical decision which must be made before planting is whether to grow only one fruit type or a mix of several types. For fruit stand or pick-your-own operations, usually a number of fruit types as well as several vegetable and berry types would be indicated. Pick-your-own would require several varieties of one or more fruit types. Some growers might prefer a concentrated harvesting season of only one type for wholesaling.

Risk of a complete crop failure is reduced as you grow more types of fruit because of differences in hardiness and in blossoming and harvesting periods.

Apples and some plums bloom later and are hardier to winter freezes than other fruit types. You might be limited to these fruits in areas subject to early frosts.

Peaches, nectarines, cherries, and pears blossom several days earlier, varying with location. Apricots and Japanese plums bloom even earlier and require sites with greater freedom from frost.

If other factors seem favorable for growing fruit, the critical decision of specific sites for the fruit plantings must be faced. Commercial experience in the local area should be followed in selecting sites for the various types.

Even though an eastern site might seem relatively frost-free, do not plant nectarines in preference to apples, because the Brown Rot fungus damages nectarines in high-rainfall areas. Draw on the experience of the nearest Experiment Station and the advice of your county Extension agent in making the decisions.

Soil type usually should be a sandy loam. Clay soils often are subject to poor water drainage. All fruit types are susceptible to "wet feet."

If steep slopes are involved to get the necessary air drainage for frost control, you need to control erosion with contour planting and maintenance of sod.

Previous use of the site is important. Crops such as tomatoes, potatoes, and melons favor buildup of harmful nematodes and wilt fungi. Newly cleared forest land may be infested with root-rotting organisms which also attack some fruit species.

Select an Orchard System

A basic decision must be made regarding the orchard system to be used. Standard-sized trees result in fewer trees in a given acreage but have the disadvantages of requiring harvesting from ladders, and more powerful equipment for spraying, pruning, and thinning. And they produce fruit several years later.

Apples are available on fully dwarfing stocks and a range of semidwarfing rootstocks as well as "spur-type" trees. The stone fruits also are available on *Prunus* spp. rootstocks and pears on quince rootstocks for dwarfing, but these are less satisfactory than the apple stocks. The investment in trees multiplies as smaller trees are planted closer together.

Trees on the dwarfing stocks grow differently in different areas, so follow local recommendations for planting distances. Poorer soils produce smaller trees and thus more trees can be grown per acre.

In general, dwarf trees require better soil and better air drainage than standard trees, and they produce fruit earlier.

Most peach and nectarine varieties are self-fertile and may be planted in solid blocks. In the other fruit types, however, a pollenizer variety is usually needed. Most fruit nursery catalogues indicate satisfactory pollenizers for specific varieties.

A rule of thumb for providing pollenizers is every third tree in every third row, which exposes one side of each main variety tree to a pollenizer tree. For ease of handling, it may be more feasible to interspace two to four rows of the main variety with one or two of the pollenizer variety. Usually both varieties can be commercially acceptable and can mature at the same time—or at different times if a spread of maturity is preferred.

With peaches, it is now possible to have good adapted varieties ripening at weekly or even semi-weekly intervals for almost four months in areas suitable for this fruit. The ripening period of the other fruit types is shorter but a sequence of good varieties is possible.

Give thought to the ripening sequences if several fruit types are grown. Apricots and cherries have distinct ripening seasons earlier than the other fruits. However, pear ripening coincides with that of later peach or early apple varieties. Hence, you should decide which fruit crop will be ripening in each period, unless supplemental labor is available for harvesting.

Since there will be no return until the trees reach bearing age, there is a temptation to intercrop with vegetables or other crops. Generally this is not advisable, since requirements of crops planted between the rows are quite different from those of the trees.

Machinery used for the trees does not adapt to the other crops, and purchasing specialized equipment for the vegetables for 3 to 5 years' use is too expensive. Spray materials necessary to control pests on the non-bearing trees may not be permissible for use on the vegetables.

Thus, neither crop may receive the care it needs at the appropriate time, or one crop will be neglected. If both fruits and vegetables are to be grown, they should be on separate sites.

Invest Adequately

Do not consider fruit growing unless you provide for adequate equipment. A tractor for spraying, mowing, and cultivation is required. A rotary mower or disk will be needed to control weeds in the middles. If chemical control of weeds in the tree row will be used, a 3-point hitch drum-type sprayer is needed.

Commercially marketable fruit cannot be grown without the protection of timely sprays with effective spray materials. In fruit areas, custom spraying may be available but you should be assured of its availability before planning the protection solely through custom applications. Low volume speed-sprayers are less expensive and more effective than high volume ones, if properly used.

A tractor fork-lift and harvesting wagon probably will be needed also. Arrangements are necessary for harvesting buckets, boxes or bins, and ladders, if standard trees are used.

Cold storage facilities are almost a necessity if a fruit stand is planned or if fruit will be stored into the winter.

The use of dwarf trees and close planting distances may require staking or wire trellises, and will call for smaller equipment.

Should you consider buying or leasing a small established orchard or a farm with a small orchard? Many factors must be considered and expert advice sought.

Some questions should be answered in an analysis of the prospects. Is the orchard in good condition and, if so, are the trees young enough to remain productive? Are the varieties commercially desirable? Is the mix of types and varieties suitable to the operation planned?

The production and profit history of the orchard should be determined if possible. Are the necessary outlets available for the type operation planned?

If the orchard is not in good condition, what is the reason? Are there indications that the site or soil conditions are not suitable?

Have the trees been neglected and, if so, can the orchard be renovated or must it be replaced? If replacement seems practical, are there replant problems such as arsenic in the soil from previous sprays, high nematode populations, root-rotting organisms, viruses, or nutrient deficiency indications?

Determine whether storage facilities are available or will be needed.

Unless you can reasonably expect a profit from the management you propose for the orchard site, don't buy it. In most cases, it will be better to invest your labor in establishing an orchard suited to your own situation.

Planting and Management

Soil preparation of the orchard site should be planned well in advance. If high populations of nematodes or root-rotting organisms are indicated by previous cropping or by soil tests, fumigation may be advisable. Fumigation is expensive but should reduce tree loss and give weed control the first year as a bonus. An alternative for peaches is use of resistant rootstocks if rootknot nematodes are high.

Certain rootstocks, such as Stockton Morello for cherries and peach-almond hybrids for peaches and plums, will alleviate excessive moisture or high alkaline conditions.

A green-manure crop the year previous to planting is a good idea. If soil pH is lower than 6 to 6.5, liming is required to supply calcium as well as to correct pH. (pH indicates the soil's relative acidity or alkalinity).

Be careful in selecting varieties to plant. Check on recommendations of your county Extension agent and of experiment stations with conditions similar to yours. Time spent with nursery catalogues or nurserymen specializing with fruit types in your general area will help you select adapted varieties.

Varieties differ widely in their adaptation, so don't make a large planting of an untried variety. Test a few trees first, if the variety is new to your area.

A prospective grower can propagate his own trees. But this involves growing rootstocks, finding a "clean" source of budwood, making the propagation, maintaining the nursery, and waiting usually two years to have trees comparable to those of commercial nurseries. Instead, select a reputable nursery and order medium-sized trees of adapted varieties. Order not only the variety you want but the rootstock that seems appropriate for your conditions.

Many of the catalogues give useful information on planting distances, number of trees needed per acre for different planting distances, effective pollenizers, planting directions, and training and pruning advice.

Number of trees per acre and years required to reach full production for various tree fruit types

Fruit type	Density	Number of trees per acre	Years to reach full production
Apricot		75-200	5-7
Apple	Low	Less than 100	15-20
	Moderate	100-200	9-15
	High	200-500	6-9
	Ultra high	More than 500	4-7
Cherry (sweet)		75-150	9-15
Nectarine & Peach		100-200	4-6
Pear	Moderate	100-200	6-8
	High	200-500	4-6
Plum & Prune		100-250	6-8
Sour Cherry		100-200	5-7

Follow the customary planting time and procedure for your area. Where winters are mild, fall planting gives the roots time to become well established before growth starts. Have the nursery store your trees until you can plant, unless you have facilities to store them properly.

Avoid drying out the roots during planting. Usually it is best to plant them directly from a barrel of water into the soil. Water them in if the soil is dry. Adding some water will ensure better contact of roots and soil.

The training system starts with planting. The planting distance determines how much space a tree can occupy without undue shading.

Detailed pruning instructions are available in the 1977 Yearbook of Agriculture and in numerous bulletins. Usually the newly-planted trees are topped at 2 to 3 feet and 2 to 4 main leaders are selected, if available, for standard trees. More specialized pruning is necessary for high density trees. In either case, keep pruning to a minimum necessary to develop the main leaders and avoid overlapping limbs.

Well-exposed wood which is not part of the permanent tree structure can be saved for early fruit production.

Use summer pruning as much as possible to prevent delay in fruiting.

Fruit harvesting should be timed to place the highest quality product possible in the hands of the consumer. For successful marketing, appearance of the fruit is very important. Production of high quality fruit requires picking at proper maturity. There are many maturity indices but development of the undercolor and softening of the flesh are most reliable.

The proper stage depends somewhat on the marketing outlet to be used and local advice on this should be sought. For example, fruit to be shipped to distant markets undergoes some ripening in transit and handling and, therefore, cannot be picked fully tree-ripened.

Apples may be stored for varying periods in refrigerated storage at about 31° to 34° F (0° to 2° C), depending on the variety. Pears usually are stored a few weeks to four months depending on variety. The other fruits usually lose quality after storage for two weeks or less. Modified atmosphere storage can extend the period of storage. (The atmosphere is modified to 2% oxygen, about 3 to 5% carbon dioxide, with the remainder nitrogen).

Too much emphasis cannot be placed on seeking all the advice available in your county Extension office, nearby experiment stations, commercial and amateur growers in the area, and reputable nurserymen. Helpful handbooks are available in most fruit-producing areas for the adapted fruits.

Fruit growing can be an expensive and even disastrous hobby, unless proper sites are chosen and the needed skills are learned. At the same time, careful operators need not be overwhelmed by the complexities of running an orchard.

Further Reading:

U. S. Department of Agriculture, *Establishing and Managing Young Apple Orchards,* Farmers Bul. No. 1897, on sale by Superintendent of Documents, U. S. Government Printing Office, Washington, D. C. 20402. 1976. 35¢.

U. S. Department of Agriculture, *Growing Cherries East of the Rocky Mountains,* Farmers Bul. No. 2185, on sale by Superintendent of Documents, U. S. Government Printing Office, Washington D. C. 20402. 1977. 50¢.

U. S. Department of Agriculture, *High-Density Apple Orchards— Planning, Training, and Pruning,* Agric. Handbook No. 458, on sale by Superintendent of Documents, U. S. Government Printing Office, Washington, D. C. 20402. 1975. 85¢.

U. S. Department of Agriculture, *Growing Apricots for Home Use,* H&G Bul. No. 204, on sale by Superintendent of Documents, U. S. Government Printing Office, Washington, D. C. 20402. 1973. 70¢.

U. S. Department of Agriculture, *Growing Nectarines,* Agric. Inf. Bul. No. 379, on sale by Superintendent of Documents, U. S. Government Printing Office, Washington, D. C. 20402. 1975. 40¢.

U. S. Department of Agriculture, 1977 Yearbook of Agriculture, *Gardening for Food and Fun,* references on page 259, on sale by Superintendent of Documents, U. S. Government Printing Office, Washington, D. C. 20402. $6.50.

Grapes are Versatile Fruits and Have Market Potential

By J. R. McGrew

The grape is a versatile fruit. For dessert use, there is a wide range of flavors—fruity, spicy, muscat, and of textures—crisp, melting, juicy. There now is the same range of colors in seedless varieties adapted to most parts of the United States that is available in seeded varieties.

Fruits may easily be processed into jelly or breakfast juice for use throughout the year. With some additional effort, wine from your own vines can enhance your meals.

Grapes can be marketed for all the above purposes (except as wine which is strictly controlled by State and Federal laws) and may be a profitable, though sometimes risky, source of income.

Grape growing has been tried at some time during the last 400 years in almost every area of this country. Grapes are produced commercially where the climate is favorable, diseases are successfully controlled, and vines survive well enough to pay back the costs of a vineyard.

In established grape areas the competition for markets is greater than in areas where more care and skill are required or where the risk of failure is greater from frosts or rots due to untimely rains. Modern pesticides and use of varieties adapted to particular regions have extended the areas where grapes may be profitable.

If you can accept an occasional crop failure and plant those varieties best suited to your locality, rather than attempting to grow the "premium" varieties of the better commercial areas, grapes can be grown almost anywhere provided there is a frost-free season of 140 days or more.

There are too many varieties, climates and judgments of quality to cover this subject in detail here. A general listing of types of grapes will help you narrow down the possible choices.

J. R. McGrew is a Research Plant Pathologist with the Fruit Laboratory, Plant Genetics and Germplasm Institute, Science and Education Administration, Beltsville, Md.

At that stage you should contact your County Agricultural Agent or the Extension specialist for fruit crops from your State Agricultural Extension Station for suggestions and recommendations.

Vinifera (European or California types). Among these varieties are found the highest quality grapes for table, raisin and wine use. Unfortunately, they are very susceptible to diseases and some are damaged both by winter cold and by warm spells in the winter.

A few small plantings have been made outside the recognized West Coast vinifera areas but they are considered too risky to recommend for general commercial plantings.

American. These are the hardy, disease-tolerant varieties. Most are derived in some degree from the wild American fox grape (*Vitis labrusca*). Well known examples are Concord, Delaware and Niagara. Several modern varieties in this group are seedless and well suited for table use.

Hybrids include the so called "French hybrids" developed in France over the past 80 years primarily for wine. They are more neutral in flavor than most American varieties and a few are well adapted for table use.

The hybrids are crosses between European grape varieties and various native American species. They were selected for the fruit quality of their European ancestors and the disease and insect tolerance of their American ancestors.

Both traits vary widely, from nearly wild with excellent resistance to pests, to high quality with modest resistance. As a group, hybrids are much easier to protect against the many pests than are the vinifera varieties.

Ripening of hybrids differs widely and there are varieties suited to most growing seasons. Also, in this group are several of the more recently introduced wine and table grapes originated in the U.S.

Southeast Grapes. Pierce's Disease of grapes is widespread in the Southeast U.S. It limits the survival of vinifera and most of the American and hybrid varieties. Muscadines (Scuppernong types) are tolerant as are a few of the older American varieties and some recent American varieties originated in Florida.

Whether rootstocks are necessary and if so, which one to use, is especially confusing to the beginner. For most U.S. vineyards only the *vinifera* varieties may need to be grafted.

A rootstock enables you to grow a variety under conditions where the own-rooted (non-grafted) vine might fail. Reasons for failure include *phylloxera* root-louse damage (on heavy soils) and nematode damage (on lighter soils, especially when replanting a vineyard).

Certain rootstock varieties may increase vigor, permit good growth on lime soils, reduce damage from droughty or wet soils, or tolerate certain soil-borne diseases.

Rootstocks are advisable only when one of these particular situations has been found repeatedly in the area. The rootstock variety chosen is of the group that best corrects the situation.

Before planting grapes you must know what is to be done with the crop. A quarter-acre (100 vines) can produce a ton of fruit (50 to 60 bushels). This amount would make 100 to 200 gallons of juice or wine.

The date of harvest of a variety can vary a week or two from year to year. When the fruit is ready or if the weather turns wet, picking cannot be put off, and once the fruit is picked it must be used promptly. A ton or more of grapes with no visible market can lead to a state of panic.

Except in areas where the grape supply is at or near saturation, there are several ways to market the crop. Successful marketing depends on the correct choice of varieties made when planting the vineyard.

Cost of establishing a vineyard is generally estimated at around $3,000 per acre exclusive of actual land cost. This includes vines, trellis posts, wire, pesticides and labor for the approximately three years before the vines begin to produce.

Labor requirements for weed control, pruning and training of vines, and pest control is high. Many of the specific chores must be done at the right time or major (and expensive) problems can develop.

Successful grape production always depends on keeping ahead of the problems. You must take preventive steps against weeds, insects, and disease. Once any of the pests build up in a vineyard, control is more costly and risk of financial loss greater.

Recognizing and staying ahead of the problems extends to all phases of grape growing. For example, weak posts should be replaced promptly or the entire row may go over.

Other problems that must be considered are rabbits that can prune off young vines, deer that seem to prefer tender grape shoots to almost anything else, and birds that flock to the vineyard before the grapes are ripe enough for you to enjoy them. The cost of fences or netting may need to be added to the estimates for growing grapes.

Site Selection

To be productive, grapevines require full sunlight. Nearby trees, even when they do not shade the vines, compete for moisture and provide birds with a perch from which to invade the vineyard. Wet soils are not good for grapes.

At least three feet of medium to heavy soil or five feet of sand should be available. Shallower soils reduce vigor and size of crop.

Grapes are adapted to a wide variety of soil types. The low fertility of light or poor soils can be corrected with fertilizers. Deep, rich soils can result in overly vigorous vines that have poor clusters, mature later, and are of lower quality for wine.

The extremely steep hillside vineyards above the Rhine are picturesque and the slope does increase the sunlight and warmth reaching the vines, but only the high prices received for these special grapes justify the labor.

Less extreme slopes may be practical in areas where air drainage gives protection against frosts after the tender vine shoots have begun to grow. In most areas of this country, nearly level to rolling ground is more practical for grapes. Avoid frost pockets, those areas where cold air tends to collect at the bottom of slopes.

Spacing of vines is determined as much by the equipment to be used as by the varieties grown. Ten to 12 feet between rows is recommended so that tractors and sprayers can travel between the vines.

A closer spacing is possible if you use smaller equipment, but the actual yield per acre does not increase in proportion to the number of vines.

In humid areas where diseases are prevalent, close spacing of rows slows the drying of leaves and fruit and makes disease control more difficult.

Spacing between vines within each row is determined by the vigor of the variety.

You want the vine foliage to at least meet between the vines so that maximum use is made of sunlight. Some overlap of foliage is usual and expected.

For vigorous vines 9 to 10 feet will usually fill the trellis with foliage.

For some of the less vigorous hybrids, a spacing of 6 feet between vines is preferred.

Fruit yield is based on foliage exposed to the sun, so as long as the trellis is filled, optimum yield can be expected. Increasing the number of vines does not increase the yield.

Size of Crop. One factor in grape growing that cannot be ignored, and cannot be overly stressed, is that a vineyard can produce only a given sized crop without loss of fruit quality or damage to the vines.

Sunshine ripens the grapes. Under ideal conditions, a long growing season with no cloudy days or rain, a one-acre vineyard (regardless of the number of vines) can produce about 15 tons of properly ripened grapes per year. When the growing season is shortened, or if the available sunshine is reduced by clouds or rain, the maximum yield may be only a quarter as large.

If you permit the vines to bear too much fruit, sugar will be lower, vines may be damaged, and fewer fruitful buds produced so that during the following year the yield will be much smaller.

Varieties and the number of vines to plant for one's own use are determined by your goals. For juice, 10 Concord vines set at 8 to 10 feet apart will usually provide 50 quarts of grape juice each year.

For table use, each vine will produce 5 to 15 pounds of fruit. One or two vines each of 10 varieties chosen to ripen over a period of 1½ to 2 months will usually produce enough for lavish home consumption and gift baskets for friends.

For wine, a quarter acre is the maximum size for a home winemaker operating under the Federal permit (BATF 1541) limit of 200 gallons per year. The size can be scaled down to suit your goals.

Varieties grown will depend on the type of wine desired. Some varieties are acceptable for either table or wine use, and the wine grapes do make interestingly different unfermented juice or jelly.

Sale of Grapes

Successful marketing depends on planting the right varieties for both your climate and the intended end-use of your customers. The quantities you can expect to sell to each customer vary greatly.

Roadside stands are a popular marketing method and grapes make an excellent addition to other fruits and vegetables in attracting customers. Three classes of grapes can be marketed:

· Table Grapes. Top quality, well-ripened clusters are required. Individual sales are small, ranging from 1 to 5 pounds and some sort of container is required to prevent damage to fruit. Price per pound is high.

The great range of varieties can be used advantageously. Early to late ripening plus cold-stored clusters of some tougher varieties can further extend the season. Possible flavors include Concord, muscats, and neutral grapes. Seedless varieties will be especially popular.

- Juice and Jelly Grapes. The flavor generally expected by a customer is the Concord-type of fruity fox grape. Quantities for each sale may range from a half to a couple of bushels. Smaller clusters and a trace of imperfect berries are acceptable. Some not quite ripe berries with their higher pectin content are even desirable for the jelly maker.

The neutral or muscat-flavored grapes make interesting juice and jelly. The offer of a sample of the finished product is a good way to convince a prospective customer.

- Wine Grapes. Home winemaking is a popular hobby and there appears to be a good market among those who do not have space or time to grow their own.

The quantities you can expect to sell to each customer range from 1 or 2 bushels (enough for a 5 gallon carboy) to several hundred pounds.

The wine quality from each variety is determined by proper ripeness of the grapes. While perfect fruit is desirable, a few damaged berries can be removed when the customer prepares fruit for fermenting.

Sugar levels of 20 to 22 percent are desirable and lower levels may mean lower prices.

The price received for good quality wine grapes sold by the bushel can be attractively high. The better varieties of hybrids and—if you can grow them successfully—of vinifera varieties are in demand by home winemakers.

Most of the State Agricultural Experiment Stations issue bulletins that recommend preferred varieties and suggest others for trial plantings. Small commercial wineries found in most areas of the country can give further suggestions on varieties that have done best for them.

Pick-Your-Own

When quantities of fruit such as these are purchased, the option of customer picking may be of mutual interest to both grower and customer.

The grower can eliminate the cost of harvesting, except for some supervision in the vineyard, when the customer is willing to harvest the grapes. The customer reduces his cash outlay and is able to do some trimming-out of defective fruit while picking.

If the grower decides to specialize in wine grapes, there are several ways to make the marketing more attractive to home winemakers.

After the fruit is in the basket, the home winemaker is faced with the problem of crushing and pressing white grapes or stemming and crushing the red grapes. Several growers, who may themselves be home winemakers, have purchased small commercial stemmer-crushers ($200 and up) and presses ($150 and up) and either lend or rent these to their customers.

This has been carried even further by some growers who offer (for a price) instructions in wine making, guidance and equipment for sugar and acid determination and suggestions for best handling of each batch of wine.

One successful operation that started this way now contracts with other growers to supply additional grapes, sells home winemaker supplies, and has a bonded commercial winery.

Customers among the home winemakers are found through advertisements in newspapers or by convincing a reporter from the local paper that your vineyard operation is worthy of a feature article. Ripe grapes, back to the farm, do-it-yourself-wine are all catchy subjects of local interest.

Several State Agricultural Experiment Stations issue annual lists of "Pick-your-own" fruit and vegetable growers that receive wide distribution. Local wine appreciation groups or clubs usually include some home winemakers.

Once you find good customers, keep them informed through your own mailing list of which grapes are available and the expected harvest dates. Telephone calls may be essential to assure that the crop is picked when ready.

Pressing grapes can be fun for a youngster.

Kevin Hayes

Other Markets. You may be able to sell grapes to supermarkets, local grocery stores, farmers' markets, or to stores that cater to home winemakers. These potential outlets should be questioned before planting to determine what varieties and quantities they might purchase. Good fruit quality is required for continued sales.

You may be able to sell your entire crop to a commercial winery. Prior arrangements—even a contract—on varieties, acceptable sugar levels and price should be made before planting. The price per ton received will be lower, but so will be the cost of marketing.

Because grapes are expensive to grow and take several years to reach maturity, consider several strategies in planning a vineyard.

If you are fortunate enough to know exactly which varieties your market requires and that these varieties are well adapted to your area, your worry will be whether you can provide the essential attention at the times required.

But in many areas the choice of varieties is not clear-cut and there is an element of experimentation involved. Then you should be more conservative in vineyard size and be willing to try only 5 to 10 vines each of several varieties.

It is possible to go too far in experimenting. One or two vines each of many varieties may enable you to select certain kinds that do grow well for further planting. But orderly marketing of small batches of fruit is difficult and quantities of fruit produced are not sufficient for the home winemaker to evaluate their wine potential.

It is better to start with a small vineyard of 10 to 100 vines to which you can give proper attention than to plant a couple of acres without considering the costs, labor, and unexpected problems that may lead to a disappointing failure.

The prospect of a good harvest of grapes is exciting and potentially profitable, but it does not "just happen." There is the continuing series of chores and always the prospect of a crop failure from frost, diseases, hail or birds. The grape grower must be willing to accept such risks.

Many State Agricultural Experiment Stations publish bulletins and guides to grape growing and a few have information on wine making. These are usually available through your County Agricultural Extension Agent or may be requested (or purchased) from the publication office of other States.

Two to Five Acres of Berries Can Sweeten Your Income

By Gene Galletta, Arlen Draper, and Richard Funt

Some 15 to 30 acres of intensively cultivated land is usually thought of as the necessary farm size to make labor-saving equipment pay for itself. However, several of the berry crops (namely strawberries, blueberries, blackberries and raspberries) offer a unique opportunity for a small or average sized family (of four) to make a good supplemental income on two to five bearing acres.

Other berry crops such as cranberries, gooseberries, currants, or elderberries, demand very specialized culture (cranberries), or have limited usage (elderberries) or limited popularity (gooseberries and currants) to make their culture profitable on a small scale.

Berry crops need a lot of labor, but a family can usually manage all but the harvesting on a small acreage. Berry culture requires considerable knowledge of the crops and their care, and a large initial investment per acre. However, berries offer a high return per dollar invested because of a generally low supply and high demand in many regions of the United States.

The advent of direct farm to consumer marketing (U-Pick or Pick-Your-Own) reduces harvest labor cost and provides the consumer with high quality fruit. Future demand for berry crops promises to be high because of their appeal as sources of dietary enrichment and their varied uses.

Strawberries

The garden strawberry (*Fragaria x ananassa*) is a perennial herbaceous plant in the Rose family. The strawberry plant has a short thickened stem (called a crown) which has a growing

Gene Galletta and Arlen Draper are Small Fruit Breeders, Fruit Laboratory, Science and Education Administration, Beltsville, Md. Richard Funt is an Assistant Professor and Extension Pomologist, Department of Horticulture, University of Maryland, College Park.

point at the upper end of the crown and forms roots at the crown's base. The leaves are borne along the crown on long petioles (leafstalks) arranged in spiral fashion around the crown.

At the juncture of each petiole and the crown, a bud is borne which may grow into one of three structures according to the environmental stimuli the plant receives:

- A runner or specialized elongated stem, which normally forms daughter plantlets at every other node, if the temperatures are warm and the daylength long
- A branch crown or new stem under the same temperature and daylength conditions under which runners are formed (some strawberries are runnerless forms and produce many or multiple branch crowns)
- Flower stalks (scapes) bearing 1 to 15 flowers, usually under short days and low temperatures ("June Bearers"). However, some strawberries flower under long days ("everbearers")

The strawberry fruit is the ripened receptacle bearing many small "seeds" (achenes). The seeds form following pollination of female parts of the flower, which are collectively arranged on the fleshy receptable (modified stem).

Strawberry plant roots are usually confined to the upper 6 to 12 inches of soil. As such they are subject to soil moisture fluctuations and severe weed competition.

Strawberry roots are vulnerable to nematode attack and several root-rotting fungi, notably red stele (Phytophthora), Verticillium, and Rhizoctonia. Frequently soil is chemically fumigated to kill these pests and many weed seeds before planting. Several newer strawberry cultivars are resistant to some of these soil fungi.

Strawberry plants grow on a wide variety of soils and tolerate wide ranges in soil acidity and composition. Generally the lighter soils (sandy) are chosen for early maturity. Strawberry soils should be reasonably high in organic matter and well-drained. Strawberry roots will not tolerate poorly drained soils.

Control of weeds by mechanical and/or chemical means is essential to successful strawberry culture.

One acre inch of water per week during the growing season is considered optimum, as is a slightly acid soil reaction (pH 5.5 to 6.0).

Strawberries grow in almost any climate, but it is critical to select cultivars adapted to your region and tolerant to its pests. Varieties developed at a certain latitude are generally adapted to within 3 to 5 degrees of the same latitude at the same elevation. A 1,000-foot increase in elevation reduces the mean temperature 3° F. A variety is adapted to a region if it grows and produces well in response to the region's prevailing environment and pests.

Climatic factors which influence strawberry adaptation are low winter temperatures (remember that strawberries are frequently covered by snow or straw in areas with cold winters), high summer temperatures, amount of winter chilling, length of growing season and days.

Important strawberry pests are viruses, nematodes, mites, and fungi and insects that attack the root, leaf, crown or fruit. Most insect pests and many of the fungal pests can be controlled with chemical sprays.

Water must be added to supplement natural rainfall and to protect flowers and green fruit against spring frosts in many locations.

Blueberries

Cultivated blueberries (*Vaccinium* species) are woody perennial bushes of the Heath family. The commonly grown types are tall multiple-caned (stemmed) plants with shallow and fibrous matted root systems.

It takes 3 to 4 years for the plant to grow to mature size (they are usually pruned to 5 to 10 feet in height and 3 to 8 feet in width and depth). The single leaves are arranged in spiral fashion along new shoots which grow from each stout cane. There may be several flushes of new growth each season.

The buds between each leaf and the stem become one of two types: *leaf buds* which will expand into a leafy shoot in the following season; or *flower buds* which will bear a cluster of 2 to 12 flowers the following spring. Each flower, if pollinated, will develop into a fleshy, many-seeded berry. A mature bush bears literally thousands of blueberry fruits.

Blueberries generally grow well on moist, but reasonably well-drained, acid soils with a high organic matter content. Mineral soils can often be modified for blueberry culture by adding organic matter and/or soil acidifiers. Certain types of blueberries may be grown on acid mineral soils without soil amendments.

Blueberry roots are very fine, usually grow in the upper foot of soil, and are easily damaged by excessive fertilizer salts and standing water. Hence blueberry plants are frequently planted in low-lying, quite acid, sandy areas (optimum pH 4.5) on raised beds. Special attention must be given to weed and water control in blueberry fields.

Cultivated blueberries cannot usually be grown successfully where winter temperatures are lower than −20° F, where the growing season is less than 160 days, and where there is less than 1,000 hours of temperatures under 45° during winter. Rabbiteye and southern species hybrid blueberries require lower amounts of winter chilling, and succeed in the Gulf Coast and Southern U.S. areas.

Specific cultivars are adapted to different regions. (See article by Galletta and Draper in the 1977 Yearbook of Agriculture, p. 279-283, for suitable varieties and cultural practices).

Principal blueberry pests are the stunt and red ring spot viruses, two fungal stem cankers, a fungal root rot, a number of leaf and stem fungi, a bud mite, several fruit-rotting fungi and worms, and a number of chewing, sucking, and tying insects. These are controlled by resistant varieties and chemical pesticides.

Bramble Plants

Blackberries and raspberries (bramble plants) are woody, multiple-stemmed, usually thorny plants in the genus *Rubus* of the Rose family. Roots live for many years in brambles, but the stems usually live for only two seasons (biennial), vegetating in the first year and bearing fruit in the spring of the second season. Exceptions to this growth habit are certain raspberry varieties (everbearers) which fruit in the fall on first-year canes, and in the next spring on the same canes.

The canes may be either erect, semi-trailing or trailing (the latter two need to be trellised), depending on the particular species of *Rubus* involved. The stout canes grow and branch, forming laterals during the first growing season. During the next season buds on the laterals grow and produce short shoots which bear both leaves and flowers.

Like the strawberry, the female pistils are aggregated around a raised receptacle (or stem axis called the torus in blackberries) in each flower. However, each pistil ovary develops into a fleshy drupelet bearing a single, hard, internal seed upon pollination. The drupelets adhere together, separating from the receptacle in raspberries, and including the central stem axis, when harvested, in blackberries.

Raspberries and blackberries will tolerate almost any soil type as long as it is well-drained to a depth of 3 to 4 feet. Both crops grow well on soils having good organic matter content and water-holding capacity and a slightly acid soil reaction (pH 5.5 to 6.8). Soil grubs and certain weeds should be controlled before planting.

Some bramble plants spread so the row width must be controlled in subsequent years.

Raspberries are particularly hardy in areas with cold winters, withstanding temperatures of —35° F for red and —25° for black (and purple) fruited types. However, raspberries are frequently injured in milder areas having fluctuating winter temperatures.

Blackberries may be injured at —15° F, while thornless blackberries, boysenberries and youngberries may be injured by 0° temperatures.

Generally, raspberries are not too successful south of USDA minimum temperature zone 6 (—10° to 0° F) nor are blackberries north of that zone. (See USDA Farmers Bulletin 2160, p. 8, 1975 revision, for map of plant hardiness zones. A map also is carried inside the front cover of the 1972 Yearbook of Agriculture).

Raspberries are especially sensitive to soil nematodes, viruses, and a number of insect and fungus troubles. Blackberries are troubled by the bacterial diseases crown and cane gall, the fungus diseases anthracnose, leaf and cane spot, orange rust, double blossom and a considerable array of insects. Resistant varieties and pesticides are the usual means of control.

Berry Culture

To successfully grow strawberries, blueberries, and brambles, you need to understand some general principles. All these crops grow on acid to alkaline soils, pH 5.5 to above 7.0, except blueberries which require an acid soil in the range of pH 4.0 to 5.5. These crops grow best on well-drained soils that have a good supply of water. For top yields they should be irrigated in dry periods, and in strawberries spring-frost control by irrigation is desirable in most areas.

They should all be fertilized with a complete fertilizer, such as 10-10-10, at the rate of 50 to 75 pounds of actual nitrogen per acre varied to the soil's fertility. This recommendation is for established mature plants.

Young and new transplants can be easily damaged by fertilizer. Apply lower rates initially and increase the rates up to the full amount on mature plants of blueberries and brambles. Rabbiteye blueberries probably should not be fertilized the first year.

Distribute fertilizer over the entire root zone just before times of greatest growth and fruiting.

Prune the woody crops to remove dead and injured canes and to balance and distribute the fruit load. Most brambles fruit the second year on canes produced the previous season. Fall-fruiting red raspberries can be cut back in late winter and will fruit the following fall on new canes.

Training small fruits depends upon the plant's growth habit. Trailing blackberries and red raspberries are tied to a wire trellis for support; erect blackberries and black raspberries need not be.

Erect blackberries are planted by covering pencil-size root pieces with soil and permitting suckers to form a hedgerow, no wider than 24 inches. Red raspberries also sucker freely and row width must be narrowed. Trailing and semi-erect types (such as the thornless varieties Smoothstem, Thornfree, Black Satin, and Dirksen Thornless) and black raspberries sucker very little and are propagated from cane tips that root in contact with the soil.

Planting density varies with the particular kind of berry. With black raspberries, blueberries, trailing and semi-erect blackberries, and strawberries grown in hill culture, the number of plants set is the same as the desired number to be fruited. Red raspberries, erect blackberries and strawberries grown in the matted row are set with fewer than the desired number of plants and allowed to multiply.

Weeds, insects, and diseases must be controlled by a combination of cultural practices and pesticides. Soil fumigation before planting is expensive but pays well in controlling most weeds, soil insects and nematodes, and some disease organisms.

Blackberry plants before (left) and after pruning.

Planting and Planning Guide for Strawberries, Blueberries and Brambles

Type	Planting distances		Planting stock	Years from planting to economic return	No. of bearing years	Average mature yield (lbs)
	Between rows (ft)	In the rows (ft)				
Strawberries						
matted rows	3.5-4	1.5-2	1-yr runners (virus free)	1	2-3	1/row foot
hill	4-5	0.5-1	1-yr runners (virus free)	1	2-3	1.5/plant
Blueberries						
Highbush	8-10	4-5	2-yr plants	3	25+	6-8/plant
Rabbiteye	10-12	6-8	2-yr plants	4	30+	12-15/plant
Brambles						
Blackberries						
Erect	8-12	2-6	root pieces	3	10-12	1.5/row foot
Trailing	8-12	6-8	rooted cane tips	2	5-10	9/plant
Raspberries						
Red	8-10	2-4	1-yr suckers (virus free)	3	10-12	1.5/row foot
Black	8-10	3-4	rooted cane tips (virus free)	3	3-4	1.5/plant

Fruits of these crops develop from flowers pollinated usually by local bee and insect populations. It is a sound practice to grow at least two rabbiteye blueberry cultivars in each ripening season for cross pollination. The other small fruits are self fruitful, but can benefit from cross pollination by more uniform fruit maturation and increased berry size.

A straw mulch for winter protection is beneficial for strawberries in certain areas, and plastic mulch is used in some areas for moisture conservation and fruit protection. Blueberries and bambles also benefit from mulching, particularly on soils not optimally suited for these crops. Incorporating peat moss in the planting hole improves establishment and growth.

Economic Aspects

Small fruit production involves a high initial investment for plants and machinery. Estimates of the total cost per acre during the first year for strawberries, raspberries, thornless blackberries and blueberries were $1,100, $900, $3,200 and $3,400 respectively using 1975 and 1978 input prices.

Strawberries begin to produce heavily in the year after planting while brambles and blueberries start producing heavily in the third to fifth year. However, returns per acre and per dollar invested over several years can be the highest. Thus you need to regard such enterprises as a long term commitment. In

Costs, labor and rates of return per dollar invested per acre over the life of the crop for several small fruit enterprises in Maryland. (Based on 1975 input prices for all except blueberry which is on 1978 input prices)

Crops (years grown)	Cost Per Acre		Labor/A/Yr Bearing year(s) hours	Rate of Return	
	First year	Total life		Hired Harvested %	Consumer Harvested %
Strawberry (4)	$1,100	$ 6,700	143	52	65
Thornless blackberry (12)	3,200	32,000	183	39	57
Blueberry (12)	3,400	52,400	180	30	41
Raspberry Red (12)	900	14,600	116	25	26
Black (12)	900	12,200	116	13	16

planning for this commitment there are a number of considerations.

Labor requirements vary for each month of the year. The jobs, number of hours and time of year will vary with species, geographic region, plant density and yield.

Hand harvest labor may represent 60 to 80 percent of total costs over the life of the crop. Generally 8 to 12 persons per acre per day are required to harvest the ripe fruit.

Pick-your-own harvesting, where consumers pick the fruit, provides an excellent means of reducing harvest labor costs, selling a perishable crop quickly, and receiving payment immediately. Pick-your-own harvesting is most advantageous within 25 to 50 miles of a large metropolitan area.

Virus-free, disease-resistant, hardy plants are worth the cost because healthy, vigorous plants provide the highest yields. Trellis materials should be of good and lasting quality. Select chemicals on local Extension recommendations. Take advantage of soil and tissue testing services provided by some state agencies.

Equipment purchases can be most challenging in matching price and specific uses. For example, a tractor and sprayer are two indispensable implements. A small high-volume sprayer costs less than a low-volume sprayer. However, the high-volume sprayer will require more water and labor per application. The low-volume sprayer will require a larger tractor than the high-volume sprayer. Hence, if time is the limiting factor (as in weekend farming), you may need the more expensive equipment.

Tractors must be matched to the PTO horsepower required by implements. Emphasize these points to a dealer when you

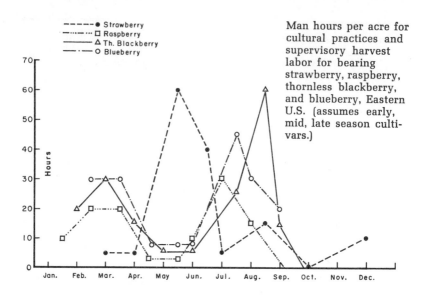

Man hours per acre for cultural practices and supervisory harvest labor for bearing strawberry, raspberry, thornless blackberry, and blueberry, Eastern U.S. (assumes early, mid, late season cultivars.)

purchase equipment. A reputable dealer can furnish parts and repairs in a short period of time.

Renting, loaning or doing custom work for other growers is a way to reduce the total cost of operating equipment. However, many operations are timely and vital to success, and waiting more than a few days for a sprayer can lead to a crop failure.

Selecting a Crop

First, consider the amount of labor available within the family. Remember that teenagers will not be able to assist during school hours. Weather conditions and shorter days re-

Picking strawberries is easier with a carrier that holds several baskets.

strict working time in winter. Berry crops require a large amount of hand labor. If labor is limited, note which crops have high labor requirements during the same period of the year, and choose just one of them.

Next consider only a small acreage to start. This will reduce the financial risk as well as provide an oportunity to learn

Suggested Berry Crops and Regions of Adaptation

Hardiness Zone*	Approx. Range of ave. minimum temperature (°F)	Suggested Berry Crops	
10	30 to 40	Strawberries	(Winter producing varieties developed in Calif., Fla.)
9	20 to 30	Strawberries	(Early Spring vars. developed in Fla., La., Calif.)
		Blueberries	(Short Chilling Rabbiteye in upper half, low chilling highbush throughout zone)
		Blackberries	(low chilling trailing types)
8	10 to 20	Strawberries	(Mid Spring vars. developed in N. C., Ore., Wash.)
		Blueberries	(Rabbiteye vars., canker-resistant highbush in Carolinas and Va., and Northern highbush in Pacific Coast States)
		Blackberries	(Erect and trailing, but different varieties are adapted to eastern and western U.S.)
		Raspberries	(Certain reds and blacks in cooler areas of zone)
7	0 to 10	Strawberries	(Late Spring vars. developed in Md., Ark., N.J., Wash., Ore.)
		Blueberries	(Highbush types)
		Blackberries	(Thorny and thornless erect and trailing types)
		Raspberries	(Reds and blacks in cooler areas of zone)
6	−10 to 0	Strawberries	(Late Spring vars. developed in Md., N.J., N.Y., Ill., Ark., Canada)
		Blueberries	(Highbush)
		Blackberries	(Erect types and some thornless vars. if protected)
		Raspberries	(Reds and blacks developed from Minn. to the East and in British Col.)
5	−20 to −10	Strawberries	(Late Spring vars. developed in N.Y., Mich., Wisc., Minn., Canada)
		Blueberries	(Highbush in warmer areas)
		Blackberries	(Hardiest erect types)
		Raspberries	(Most of the hardy red and black fruited types)
4	−30 to −20	Strawberries	(Early Summer vars. devel. in Minn., Wyo., Alaska)
		Raspberries	(Hardiest reds)

* See inside front cover of 1972 Yearbook for map of hardiness zones.

the biological, technological and economic requirements before expanding. Plan ahead for expansion, rotation and location of crops, particularly if you have a pick-your-own operation.

The initial cost of land, plants and equipment will generally require borrowed capital if personal assets are limited. You should anticipate a rate of return greater than the interest rate you pay the bank.

If you have limited cash and labor to invest, then a pick-your-own operation having strawberries, and/or red and black raspberries, would be a logical choice. If cash is not limited, then choose strawberries, raspberries and thornless blackberries. If ample cash and labor are available, you might grow strawberries, raspberries, thornless blackberries and blueberries.

Since the species of berries are adapted to different regions, and varieties within a type of berry are adapted to smaller areas within a region, consult your local Extension service or agricultural experiment station for particular berry variety recommendations.

In land use planning, set aside sufficient additional acreage for new plantings as present fruiting areas decline in vigor and productivity. The marketing season may be profitably extended in most areas if compatible vegetable, floral, nursery, and tree fruit crops are grown near the berry plantings.

Further Reading:

Funt, R. C., *An Economic Comparison of Several Small Fruit Production and Harvest Systems in Maryland, 1975*, Md. Agr. Expt. Sta. Misc. Publ. 922. 1977.

Galletta, G. J. & A. D. Draper, *Just About Any Home Garden Can Produce Blueberries*, 1977 Yearbook of Agriculture, 279-283, on sale by Superintendent of Documents, U. S. Government Printing Office, Washington, D. C. 20402. $6.50.

Hill R. C., Jr., J. D. Utzinger & E. J. Stang, *Strawberries Like Full Sun —And A Good Deal of Attention*, 1977 Yearbook of Agriculture, 265-271, on sale by Superintendent of Documents, U. S. Government Printing Office, Washington, D. C. 20402. $6.50.

Tomkins, J. P., *Cane and Bush Fruits are the Berries*, 1977 Yearbook of Agriculture, 272-278, on sale by Superintendent of Documents, U. S. Government Printing Office, Washington, D. C. 20402. $6.50.

U. S. Department of Agriculture, *Growing Raspberries*, Farmers Bul. No. 2165, on sale by Superintendent of Documents, U. S. Government Printing Office, Washington, D. C. 20402. 1974. 35¢.

U. S. Department of Agriculture, *Thornless Blackberries for the Home Garden*, H&G Bul. No. 207, on sale by Superintendent of Documents, U. S. Government Printing Office, Washington, D. C. 20402. 1975. 35¢.

Vegetables Are Appealing If You Don't Mind the Work

By Allan Stoner and Norman Smith

Growing vegetables is often an excellent way to make the best use of a few acres. Using all or part of your land to grow vegetables may be appealing regardless of whether you intend to derive pleasure or profit from the property. But keep in mind that while most gardeners can grow vegetables for fun, growing them as a money-making venture is a serious business requiring many talents and skills.

If it is important to earn income from the land, raising vegetables may seem just the thing since most of them yield a large return per unit of area. On the other hand, vegetables require a lot of work and should not be tried unless adequate labor is available and you are willing and able to put forth the effort to properly manage the planting during the growing season.

Before deciding to grow vegetables on your few acres, determine suitability of the land for vegetable production. Most vegetables grow best in a slightly acid, fertile, well-drained soil. However, soil acidity and fertility can be altered by adding fertilizer and lime, while poorly drained soils can be tiled to improve drainage.

Without exception, vegetables grow best in full sunlight. Easy access to an abundant supply of high quality water is also important or even essential in some areas.

Length of the frost-free growing season, and the temperature throughout the season, may have a bearing on the feasibility of growing vegetables.

If you intend to make a profit growing vegeables, one of your first concerns should be how to market your produce. First, a distinction should be made between crops grown for fresh market and those grown for processing.

Allan Stoner is a Research Horticulturist, U.S. Department of Agriculture. Norman Smith is Cumberland County Agricultural Agent, Bridgeton, N.J.

Most of the processing crops are grown on large acreages so that they can be machine-harvested. A small-scale farmer using small machines or hand labor finds it very difficult to compete with these large mechanized farms. Ten years ago you could grow a few acres of tomatoes for a local processor and make $100 to $200 per acre, but rarely is this possible today.

In contrast, vegetables for fresh consumption often can be grown by a small-scale farmer-grower or by a larger farmer using mostly hired laborers.

Some of the most popular vegetables for fresh consumption include sweet corn, tomatoes, peppers, summer squash (yellow and green), cucumbers for slicing or pickling, snap beans, lima beans, eggplant, acorn and butternut squash, okra, muskmelons and watermelons. Also popular are the salad crops including cabbage (green, red and savoy), lettuce (iceberg, leaf,

Greenhouse tomatoes grown in peat vermiculite mix in troughs (left) may be profitable for home retail sales if fuel costs can be held down. Plastic tunnels (below) protect tomatoes from wind outdoors, and help produce earlier crops.

Vincent Abbatiello

Boston, bibb, romaine, escarole and endive), collards, kale, parsley, beets, and carrots.

Marketing fresh vegetables can be a large or small venture depending on your crops and method of selling. If your farm is a long way from the consumer you may choose to sell through a broker-shipper who buys your crop and sells it to a chain.

For successfully selling directly to a chain store, you have to be a large grower or located in an area with many smaller growers so that you can pool a large volume of mixed vegetables for sale to a buyer. Your returns from this type of selling will be about 25% of what you see the vegetable selling for in your local grocery.

Selling direct to the consumer gives you the opportunity to acquire a larger share of his dollar and is likely to be simpler for many small farmers. You can do this with a pick-your-own operation, a roadside stand, local deliveries to smaller stores or other roadside stands, or by organizing a city retail market.

Large chain stores are not likely to buy much local produce, so don't plan on a large store next to your farm as a customer. Their marketing techniques are not geared to working with small producers since they handle large quantities in trailer-load lots from production areas where they can obtain a dependable supply of a known quantity and quality at very competitive prices.

Roadside Sales

A successful roadside stand can be developed if you are in an area where people like and are willing to pay for fresh vegetables which are often of higher quality than they can buy in their local stores.

The most popular vegetables for roadside sales are sweet corn, tomatoes, and salad crops. An important factor in their popularity is that they are so highly perishable. Roadside marketing makes it possible to reduce the time interval between harvesting and consumption, thus making it possible for the consumer to have the highest possible quality.

Your crop mix for roadside marketing can also include peppers, cucumbers, yellow and green squash, new potatoes, cabbage, muskmelons, watermelons, beets, carrots and cauliflower. Strawberries also fit into roadside marketing as do pumpkins grown for the Halloween trade, Indian corn, and some winter squash.

When developing a pick-your-own operation, give thought to what vegetables are best suited to this system. Since many inexperienced people will be harvesting, crops that are easily harvested and do not require difficult decisions on their being at the proper stage for harvesting are best adapted to pick-your-own. Tomatoes, snap beans and cucumbers are examples.

On the other hand, some experience is required to be able to determine when sweet corn is at the proper stage of maturity for harvest. Therefore, unknowledgeable people harvesting corn can cause significant damage to a planting and still not end up with high quality produce.

Once you have decided to grow vegetables and how you will market them, consider what crops will grow well in your area, the potential demand, the likely monetary return from each, and what specific varieties are best adapted to your area.

If you plan to grow several vegetables, a comprehensive plan for using the available space and the planting sequence should be developed before putting in the first seed. Also decide what might be grown in future years so a crop rotation system can be developed.

If space is limited, you will probably want to grow crops that will give the largest return per unit of area. Examples are tomatoes, peppers, summer squash, and cucumbers. On the other hand, sweet corn, pumpkins and winter squash need a great deal of space and may not be good crops to grow.

Unending Harvest

Sound planning before the time to plant can result in a continuous harvest throughout the season. One way to use space effectively is to make a sequential planting of crops with different temperature requirements and/or different rates of maturing.

For example, an early crop of lettuce, cabbage or other cool season vegetable can be followed by snap beans or summer squash. These in turn can be followed by another planting of the cool season crops, such as cabbage and lettuce.

Continuous harvests of a single crop can be achieved by successive plantings of the same variety at a regular interval, such as every two weeks. The same result may be achieved by simultaneously planting several varieties of a crop that are known to mature at different rates.

Use cultural practices that will let you minimize the amount of labor required. These might include the use of various mulches, herbicides, or drip irrigation.

Mulches can significantly lessen weed problems and reduce the need for other weed control methods. They also can result in warmer soil, thus earlier crop maturity, conservation of moisture, and reduced disease.

Suggested plant spacing, number of seeds or plants required, and average yield of the common vegetables

Vegetable	Spacing (Inches)		Plants or seed per 100 feet	Average yield expected 100 feet
	Rows	Plants		
Asparagus	36-48	18	66 plants or 1 oz.	30 lb.
Beans, snap bush	24-36	3-4	½ lb	120 lb.
Beans, snap pole	36-48	4-6	½ lb	150 lb.
Beans, lima bush	30-36	3-4	½ lb	25 lb. shelled
Beans, lima pole	36-48	12-18	¼ lb.	50 lb. shelled
Beets	15-24	2	1 oz.	150 lb.
Broccoli	24-36	14-24	50-60 plts or ¼ oz.	100 lb.
Brussels Sprouts	24-36	14-24	50-60 plts or ¼ oz.	75 lb.
Cabbage	24-36	14-24	50-60 plts or ¼ oz.	150 lb.
Cabbage, Chinese	18-30	8-12	60-70 plts or ¼ oz.	80 heads
Carrots	15-24	2	½ oz.	100 lb.
Cauliflower	24-36	14-24	50-60 plts or ¼ oz.	100 lb.
Celery	30-36	6	200 plants	180 stalks
Collards & Kale	18-36	8-16	¼ oz.	100 lb.
Corn, sweet	24-36	12-18	3-4 oz.	10 doz.
Cucumbers	48-72	24-48	½ oz.	120 lb.
Eggplant	24-36	18-24	50 plts or ⅛ oz.	100 lb.
Kohlrabi	15-24	4-6	½ oz.	75 lb.
Lettuce, head	18-24	6-10	¼ oz.	100 heads
Lettuce, leaf	15-18	2-3	¼ oz.	50 lb.
Muskmelon (cantaloup)	60-96	24-36	50 plts or ½ oz.	100 fruits
Okra	36-42	12-24	2 oz.	100 lb.
Onions	15-24	3-4	400-600 sets or 1 oz.	100 lb.
Parsley	15-24	6-8	¼ oz.	30 lb.
Parsnips	18-30	3-4	½ oz.	100 lb.
Peas, English	18-36	1	1 lb.	20 lb.
Peas, Southern	24-36	4-6	½ lb.	40 lb.
Peppers	24-36	18-24	50 plts or ⅛ oz.	60 lb.
Potatoes, Irish	30-36	10-15	6-10 lb. of seed tubers	100 lb.
Potatoes, sweet	36-48	12-16	75-100 plts	100 lb.
Pumpkins	60-96	36-48	½ oz.	100 lb.
Radishes	14-24	1	1 oz.	100 bunches
Spinach	14-24	3-4	1 oz.	40-50 lb.
Squash, summer	36-60	18-36	1 oz.	150 lb.
Squash, winter	60-96	24-48	½ oz.	100 lb.
Tomatoes	24-48	18-36	50 plts or ⅛ oz.	100 lb.
Turnip, greens	14-24	2-3	½ oz.	50-100 lb.
Turnip, roots	14-24	2-3	½ oz.	50-100 lb.
Watermelon	72-96	36-72	1 oz.	40 fruits

Organic materials such as straw and leaves or specially manufactured polyethelene or paper products can be used as mulches.

Recent research shows that aluminum foil or aluminum-coated paper provide the usual benefits of other mulches, plus they may act as insect repellants.

Herbicides, or chemical weed killers, may be valuable to a small-scale operation. However, no one herbicide is effective or legally approved for use on all vegetables. Thus care must be taken to use only those chemicals approved for weed control for a given crop.

Early set on Harris hybrid Red Pak tomatoes growing on black mulch film (right). Squash growing through foil-covered rows helps control virus more effectively (below), without need for aphid control sprays.

Vincent Abbatiello

Using a material not labeled for a crop is unlawful. This can damage the crop and pose a possible problem to the people consuming it. Strict adherence to instructions on the container label is essential.

Drip or trickle irrigation is a watering technique that has evolved in recent years. It involves use of plastic pipe or hose laid very near a row of plants. The pipe or hose has minute holes which allow water to gradually seep out and wet the area immediately around the plants.

This provides a continuous supply of water to the crop and delivers the water to where it is of greatest use. Drip irrigation also greatly reduces the amount of water required to grow a crop—which is particularly valuable in arid areas.

The drip system is especially effective when combined with use of paper or polyethylene mulch. The irrigation hose is centered under the mulch along the plant row. Thus the mulch reduces water loss due to evaporation from the soil surface.

Plant Pests

Keep a sharp eye out for insects and diseases or you may lose all or part of the crop in a short time, or suffer from a poorer quality crop.

Information to help you identify these pests and describing control methods can be obtained from textbooks, USDA and State Agricultural Experiment Station publications and per-

Drip irrigation system on onions will wet an area 2 to 3 feet wide below the soil surface when placed near rows.

Vincent Abbatiello

sonnel, county Extension agents, and agricultural supply agents. Many farmers will visit with you about their experiences and can be a tremendous help on local crop requirements.

The type equipment you need to grow vegetables on a few acres depends in large part on the size of operation you envision. Basically you need some means of preparing the land for planting, of tilling or cultivating during the season, and of applying fertilizer and controlling pests. It may also be desirable to have mechanical equipment available for seeding and/or transplanting. A wide variety of equipment is available for all sizes of farming operations.

Modern farm equipment is extremely expensive, and should not be purchased if you can hire the job done for less. For a small acreage you may find it advantageous to hire another farmer to do your plowing and soil tillage and then buy a new or used small farm tractor for the row crop work of cultivating and spraying.

Most small tractors designed for mowing lawns are not powerful enough to plow satisfactorily, and do not have enough clearance to permit driving over a partially or fully grown crop row for spraying, cultivating, etc. At least a 30 horsepower tractor is needed to plow 9 inches deep to turn over the soil properly and make a good seedbed.

A small heated and ventilated greenhouse is essential if you want to grow transplants for earlier planting. You can advance your harvest season by several weeks if you raise your own transplants of tomatoes, peppers, melons, cabbage and lettuce. It takes about six to eight weeks to grow transplants of peppers, tomatoes, cabbage and lettuce. Melon plants are ready for field planting in three weeks.

An inexpensive greenhouse can be built by covering a pipe frame with polyethelene.

The foregoing gives general information on requirements for growing vegetables for your gratification or for developing a small venture for profit, either on your own or in combination with neighboring farmers.

At the other extreme let's discuss a highly organized group of experienced vegetable growers who strive to obtain maximum production and superior quality of their vegetables through their own efforts and with the benefit of expert advice by Extension Service and their State University. They are favored by good soils and water resources, plus a long growing season. But their ultimate returns are determined by an auction market where prices vary daily according to supply and demand.

In this highly competitive situation each grower must produce higher than average yields to cover expenses and make a satisfactory return on his investment. If yields are reduced by mistakes or unfavorable weather, or prices are low at harvest time, the grower may incur serious losses.

This example of how vegetable culture on small-scale farms can be successful is illustrated by the Vineland area of Southern New Jersey where there are several hundred small vegetable farms ranging from 4 to 60 acres. Gross income from these farms is from $1,500 to $5,000 per acre, depending on crops grown, yields, and fluctuating prices.

These farms are concentrated in southern New Jersey for three key reasons:

- The climate, fresh well water, and soil type are suitable for a planting season that extends from late February to August and a harvest season from April through November
- A cooperative auction marketing system was developed in the 1930's known as the Vineland Produce Auction Cooperative. There the farmers sell most of their fresh vegetables daily to buyers for shipment to the over 100 million consumers in the Northeastern United States and Canada that are within 24 hours driving distance by truck
- The farmers use very intensive production technology which has been developed through the support of the Cooperative Extension Service with Rutgers University-Cook College, the U. S. Department of Agriculture, and Cumberland County

New back-up technology is developed each year by the Cook College staff and the local Extension agents in cooperation with farmers. A book of annually revised recommendations is given to growers before each season begins. The back-up agribusiness group in the area includes a number of seed, fertilizer, lime, pesticide, farm equipment and packaging suppliers, repairmen, brokers and truckers to buy, sell and deliver the fresh produce to city stores.

Credit is provided by many local banks who are farmer-oriented and by the Production Credit and the Federal Land Bank and by the Farmers Home Administration. The pay-back record of these New Jersey farmers is excellent.

Fresh vegetable sales from small farms at the Vineland Cooperative Produce Auction have progressed from $1 million in 1960 to $3.5 million in 1966 to $20.25 million in 1977. Another $5 to $10 million was sold in 1977 to surrounding markets by direct sales or to roadside stand buyers. This growth is an excellent indicator of the successful small farmer enterprises which exist in the South Jersey area.

On the average the farmer receives about 25% to 30% of the consumer's dollar for vegetables sold through the auction market. The other 70% pays for brokerage fees, assembly, hauling, prepackaging, refrigeration and storage, warehouse ownership and operation, selling to stores, and finally selling to consumers.

At the Vineland Produce Auction a buyer can load a trailer with as many as 40 different vegetables on a single day. The mixed trailer load of high quality fresh vegetables is paramount to the success of vegetable production in this area with the buyers' ability to send it where needed by the consumer within 24 hours.

An individual vegetable farmer has little control over the price he receives, except as affected by quality. At an auction market the buyers bid competitively to obtain what they need that day to satisfy their store customers. If the local supply harvested by the farmers is more than needed by the buyers, prices are low. If market demand is high and the supply moderate, prices are bid higher. If an item is in high demand and scarce, prices go even higher.

A major factor in price determination is the supply of the vegetables in the whole country. For example, when farmers in South Jersey are harvesting lettuce in June and October, the supply and quality in the Far West affects the price the farmer receives in New Jersey. Pick-your-own, roadside or other means of selling directly to consumers are not as greatly affected by nationwide supply and demand.

Besides the knowledge and skills required to grow vegetables and care for machinery, personnel management skills are needed to work with labor in order to keep workers productive and happy.

Business skills are required to sell, to borrow money (short and long term), and for record keeping for income tax, social security, unemployment insurance, workman compensation, budgeting and analysis.

This job is often done with the help of the farm wife or another family member.

Pesticides

All New Jersey farmers who use restricted pesticides must pass a test by the State Department of Environmental Protection and be licensed by the state to use pesticides on their farm. Other states have similar licensing procedures.

Without this license, farmers cannot purchase the effective pesticides needed to produce quality vegetables. Low quality, diseased and insect-infested produce would be rejected by the buyers and consumers.

Vegetable growing is the main activity of many of New Jersey's small vegetable farmers. It can be a full-time, year-round job which provides sufficient income for their family.

The busiest season is from March through November. During summer the work day may being at 6 a.m. and end at dark. This usually goes on seven days a week since vegetables need harvesting every day on a farm where a variety of vegetables are grown. A very limited time is taken for church and social activities in the growing season.

During winter the small farmer is busy maintaining and repairing machinery, expanding or repairing buildings, constructing or maintaining greenhouses and growing plants for the next season.

Labor in the small acreage vegetable farm includes family members, local teenagers, and skilled and semi-skilled labor brought in from other areas. About three-quarters of the work on a small vegetable farm is hand harvesting, bunching, and packaging the vegetable for market. The rest of the time is spent in production and maintenance.

To sum up, the use of all or part of your small acreage for producing vegetables can be enjoyable and profitable. Even though growing vegetables takes a lot of work and is beset with a host of possible problems, the potential for high return per acre makes vegetable culture particularly suitable for small acreages.

The beginner will soon discover there is no right or wrong way to grow and market vegetables.. He must learn to anticipate new situations and change his methods to deal with them.

Further Reading:

U. S. Department of Agriculture, *Home Garden Vegetables,* Part 2 of 1977 Yearbook of Agriculture, pp. 102-244, on sale by Superintendent of Documents, U. S. Government Printing Office, Washington, D. C. 20402. $6.50.

Can Ornamental Plants Turn a Profit for You?

By Elizabeth Scholtz and Frederick McGourty, Jr.

Can you make money raising ornamental plants in your spare time? Yes, if you're a fairly experienced home gardener, are willing to work hard and to forego a financial return for at least a year or two. But don't expect windfall harvests of either money or plants, although a few hobbyists eventually go into business full time and are quite successful.

The main pleasure is not financial. It's the opportunity to work from a few to many hours a day with living green things, toward the specific goal of attracting a following of customers by providing the plants they seek—as well as occasional specialties they will be excited to discover. There can be a modest return, enough to help offset the cost of a college education or to provide a retirement cushion.

An informal survey of part-time growers was recently conducted by the authors, and we think the results will be of interest to people wondering whether to take the plunge.

The growers we questioned are, to a person, instinctive gardeners and their satisfaction is primarily in the doing, not in monetary gain. All live in rural or semi-rural areas except one, who is a suburbanite. All except one are married, with the spouse working full time in another profession but with time to help in the raising and—at least to some extent—the care of plants.

Each grower relies on a modest greenhouse for starting his plants and is thus affected by heating costs. Such expenses are of less concern in Atlanta, where it costs roughly $350 a year to provide a 12 x 20 foot lean-to type greenhouse with all services —heat, water and electricity—winter temperatures being held at 60° to 65° F. Costs necessarily would be more in New England or other winter-cool climates.

Elizabeth Scholtz is Director, Brooklyn Botanic Garden, Brooklyn, N.Y. Frederick McGourty, Jr., is Editor of *Plants & Gardens* and Associate Taxonomist at the Garden.

Heating temperatures from wood stoves cannot be as closely controlled as those from oil burners or natural gas units. But they do at least allow the spare-time grower to carry over certain mild-climate plants, such as scented geraniums, which can tolerate extremes of heat but not freezing temperatures.

One Connecticut grower with a lean-to greenhouse that doubles as a pottery workshop burns three cords of wood a winter (for night heat only, at a cost locally of about $150). The radiant heat of the sun provides adequate daytime heating, except on cloudy days. Under such conditions rugged sorts of subtropical perennials and cuttings are kept in fairly good condition.

Because of high heating costs, some semi-commercial growers in the North who depend on an oil unit choose not to start up their small greenhouses until late February. They then concentrate on popular annuals, most of which can be ready for sale in May, after thinning, transplanting and hardening off in cold frames.

In the case of a few slow-growing annuals or plants treated as such, these Northerners may buy small ones from nearby wholesalers and grow them on. This method is favored with fibrous begonias and geraniums, both of which are excellent sellers in May if in flower. Good timing is essential because a

On clear days the sun provides adequate heat for this lean-to greenhouse in which owners offer herbs and ceramic pots for sale.

211

late start to minimize heating costs may result in flowerless plants at peak sales time.

Colorful Annuals

The raising and sale of colorful annuals has much to recommend it because the grower's season is fairly short, starting in the North in late February and ending about July 1. During this period the grower cannot be away from home overnight unless someone is on hand to maintain the heat and tend the plants.

Because the spare-time grower works on a small volume and margin, it is usually not feasible to hire outside help. Even if it were, some growers report, the bookkeeping involved may rule against it.

Standard annuals, the ones customers know best, are usually dependable sellers provided they are in bloom at selling time. These include marigolds, petunias, salvia and zinnias. Begonias and impatiens, though more costly to raise, always seem the first to sell out.

There is a smaller market for hardy perennials, but it should not be ignored as an adjunct to sales, particularly if they are kinds brought into bloom during the peak selling season. These include columbine, bleeding-heart, delphinium, certain phloxes and bearded iris.

An important consideration is that hardy perennials can be grown exclusively in cold frames and do not require a heated greenhouse, though the latter can spur early spring growth or serve as a propagation aid.

One Massachusetts grower, who does not have a large enough retail market to justify extensive raising of hardy perennials, concentrates on growing them for resale to garden centers and small nurseries. With cold frames and an acre of land for field growth, overhead costs are relatively small. The main expense is his time and labor, especially at plant division time.

The growers surveyed stressed that it is essential to raise plants customers want, which are not necessarily personal favorites of the grower. Other advice consisted of starting on a small scale, first getting to know both plants and their market potential, and reinvesting all profits in the business during a building-up period.

Growing too many kinds of plants can be disastrous, the only beneficiary being the grower's own garden. It is better to sell out of a plant quickly than to have many leftovers, although the grower can sometimes dispose of leftovers toward the end of the season by contracting to plant other people's gardens. It does no harm to have a small landscape-consultant business on the side.

"Corner a Market"

Spare-time growing of house plants for sale can be moderately profitable, particularly in mild-winter parts of the country. A Georgia housewife, for example, who finds it most practical to be a small wholesaler, gives this advice:

"If you want to make money at it, you must corner a market and be prepared to produce at the most inconvenient time. Learn to grow for the demand of the season. The best example I know is a man who learned to grow piggyback plants (*Tolmiea menziesii*) so well that he supplied the local market exclusively.

"Go to retailers, ask them about their needs and the prices they would like, and try to provide for that. Don't attempt to compete with Florida growers of foliage plants unless you can raise them better or more cheaply. Instead, concentrate on plants that the big wholesalers have overlooked or disdained because they require special care, but for which there is a market. Plants such as African violets are always in demand.

"Small mail-order businesses are feasible for certain kinds of plants. One woman here has a substantial volume selling hoyas by mail, along with plant accessories. Be careful to advertise at first in free papers and magazines or small local newspapers. The cost of advertising in large newspapers is phenomenal."

Regardless of where you live, there are practical matters to settle before engaging in a part-time business selling plants.

First, find out what the local zoning regulations are and what permits are needed. Many of the greenhouses of semi-commercial growers were already on their sites as hobby units, but their uses may be stipulated by law.

Building permits may be necessary for lath houses or other structures. Parking availability and even the size of signs are considerations, too. In case you plan to use restricted pesticides you will need an applicator's license (see your county Extension agent).

A sales tax number obtainable from the state will allow you to buy wholesale.

It's also a good idea to have a bookkeeper if you're not handy at accounting. Bear in mind the tax advantages: it is possible to depreciate cost of a greenhouse and also to take tax deductions for a business phone, office space and furniture, utilities and other items of expense related to the business.

Dried Flower Arrangements

By Dorothy Kuder Smith

Do you remember Grandmother's winter bouquet stuck in a fancy vase on the upright piano, a lovely fringed scarf spread out under all? The bouquet, collected by Grandmother herself, usually consisted of bittersweet, painted milkweed pods, and pampas grass.

Like other Early Americans, Grandmother used dried flowers to conceal odors of cooking and poor plumbing. She prepared jars of potpourri (dried flower petals mixed with spices) and placed them about the house to give off fragrance when the lids were lifted.

Since our ancestors' time we have discovered untold varieties of plant materials, not only for potpourri and winter bouquets, but for various designs for year-round enjoyment and use.

Today entire families are becoming increasingly interested in growing and collecting plant material for preserving and designing, not only for home decoration, but for profit.

You need only develop a "seeing eye" to spot the many materials nature offers. Besides those found in your own garden, a stroll along a country road or through field and wood will reveal many blossoms, foliages, seed pods, and grasses. With a little knowledge of preserving and designing, they can be put to profitable use.

(Avoid collecting materials from parks, arboretums, and public places. Do not collect materials on the conservation list in your state, unless grown on your own property. Avoid poison ivy with its clusters of grayish berries and its bright red 3-lobed foliage in autumn).

Plant materials, when preserved, remain in good condition for years. In fact, dry garlands and wreaths of blue delphinium and lotus blossoms, in good color and form, were found in opened Egyptian tombs dating back to 1700 B.C.

Four simple methods of preserving are:

Air Drying. Strip foliage from fresh cut flowers. Tie loosely in small bundles, and hang them upside down in a warm, darkened room until dry, 1 to 2 weeks. Store in cardboard boxes. Protect from mice, which attack seeds.

Sturdy flowers, such as zinnias, marigolds, celosia, cockscomb, strawflowers, and herbs and mints dry well using this

Dorothy Kuder Smith teaches flower arranging in the Department of Recreation, Montgomery County, Maryland.

method. Most Williamsburg type dried arrangements are made of air-dried flowers.

- Glycerinizing. Smash the bottom 2 to 3 inches of stems of fresh cut magnolia and other broad-leafed evergreens, and place immediately in a solution, 4 to 5 inches deep, of 2 parts hot water to 1 to 2 parts glycerine (or anti-freeze). Add more solution as it is absorbed. Let remain 2 to 3 weeks or until leaves have become supple and golden brown.

    Glycerinize deciduous branches as oak, dogwood, and beech in mid-summer before sap stops flowing. Materials thus treated remain supple and useful for years. They are excellent alone in large arrangements, or combined with fresh flowers.

- Pressing. Press flowers and foliage between sheets of porous paper (newspaper or a telephone directory) weighted down. Drying may require several weeks. A quick method is to press with a warm iron until all moisture is removed.

    These methods produce a flat effect, unsuitable for arrangements, but useful for pressed flower pictures, laminated lampshades, stationery and other flat designs.

- Burying in Silica Gel. This is a commercial drying agent in which flowers dry quickly, retaining beautifully their form

To make a wreath, cover a wire foundation with bulky fabric to a thickness of 1/2 inch. Then wire dried pods, cones, and fruits from field or garden to the wreath form.

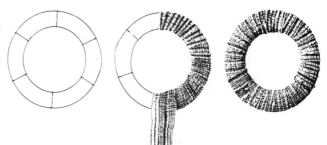

and color. The most delicate flowers are best preserved this way—roses, sweet peas, daffodils and iris.

Flowers dried by this method may absorb moisture from the air, so it is advisable to use them only in the dry winter months, or in arrangements under airtight glass or clear plastic domes.

Detailed directions for using silica gel are included with its purchase. A new and very quick method consists of drying flowers with silica gel in a microwave oven.

Other drying media are fine salt-free sand or borax powder mixed half and half with white cornmeal. Lacking these, fine dry sawdust or "Kitty Litter" can be used, but results may be disappointing.

Many materials need no special attention. Merely cut strong stalks such as corn, grains, cotton, okra, dock, sumac, goldenrod, mullein, teasel, milkweed, rose hips, bittersweet, cattails, and Queen-Annes-lace and stand upright in tall containers until ready to use. These make excellent big arrangements, alone or in combinations with fresh flowers.

Seed pods—such as tulip, daffodil, poppy, martynia, wisteria, and trumpet vine—can be clipped from their stalks and spread out to dry. Magnolia pods, sweetgum balls, sycamore buttons, acorns, nuts, small gourds, and cones should also be dried thoroughly to prevent mildew. The above mentioned are excellent for binding into a dry wreath.

When collecting materials, choose only those in prime condition. To obtain desired color gradations, teasels, Queen-Annes-lace, and many pods and grasses can be bleached and made more effective by soaking for a short time in a solution of household bleach. (Approximately 1/2 cup to 1 gallon water, or stronger if necessary).

Dirty material, if not too fragile, can be washed and dried thoroughly. Clean open cones by brushing vigorously between the scales with a bottle brush or a narrow paint brush.

Plastic Spray

A coat of clear plastic from a push-button can enhances the appearance of cones, acorns, nuts, and all smooth surfaces and reduces the brittleness of baby's breath, goldenrod, and other delicate materials. Made-up designs are improved by a light spray of plastic every six months.

Whether for fun or profit, the basic information given is necessary. Details on creating the designs you wish for yourself, or for sale, are readily available.

Flower-arranging classes in Montgomery County, Maryland, present annually a showing of wreaths, swags, kissing balls, corsages, pressed flower pictures, and arrangements, made mostly of preserved materials. During the exhibition the students freely give information regarding creation of these items.

Almost every community sponsors such shows, workshops, and classes in flower designing in which the public is urged to participate. Libraries, arboretums, garden clubs, and garden magazines also offer valuable information.

By taking advantage of what's offered, almost everyone—including children—can master the technique and find themselves happily involved in a profitable venture.

Since many children have "nothing to do", they might become interested in collecting and preparing material as discussed above. With a piece of ground of their own, they can grow such decorative items as Indian and strawberry corn, gourds, martynia, okra pods, and kaffir corn.

There are great demands for such items in autumn, when children could establish their own roadside stands. Bundles of corn stalks, wild rose hips, bittersweet, and other beauties from the roadside are great for attracting customers.

In Montgomery County, Maryland, where interest is high in this field, many people are selling their products to friends and associates, either from their homes or where they work. Some find outlets through gift shops and department stores. Many have developed a profitable business as professional decorators.

Garden centers, supermarkets and department stores, as a service to customers, may be willing to set up an area for sale of your products.

Plant materials vary from region to region. Those mentioned in this chapter are found in the Mid-Atlantic area. Equally interesting materials in other areas can be subjected to the same methods of preservation described here.

However you use the foregoing information, whether for pleasure or for profit, keeping close to the earth does not cost much, and the generous wealth of nature belongs to everyone.

Further Reading:

A Guide by Plant Family to Foliage Preservation, Arnoldiana Magazine 37(b), 289-304, 1977. Arnold Aboretum, Jamaica Plain, Mass. 02130. $2.00.

Dried Flower Designs, Handbook No. 76, Brooklyn Botanic Garden, Brooklyn, N. Y. 11225. $1.75.

Drying Flowers in a Microwave Oven, American Horticulturist Magazine 56(5), 34-35, 1977.

Making a Mint With Herbs Is Not All That Difficult

By James A. Duke

Herbs are easy to grow, and increasingly easier to sell. Cautious growers can supplement their incomes selling herbs, or grow a variety of herbs for home use including cooking and herb teas, or decoration.

With a 6- to 7-month growing season, you can grow several perennial herbs that sell well at summer garden stands.

My son made vacation spending money by selling herbs at an urban farmer's market. Thyme and chives were his biggest sellers. He started both from seed in pots and marketpacks in our small greenhouse. The chives in 4 x 6 inch peat market-packs sold for over a dollar, planted (at least 5 marketpacks) from a 35¢ seed package.

Other herbs, including thyme, can be subdivided readily by cutting or root divisions. Chives, thyme, and other herbs are ready sellers, weekend after weekend.

There is a big demand both for hanging pots and for herbs. Put the two together and you should have a money earner. Balm, corsican mint, oregano, peppermint, rosemary, savory, spearmint, and thyme have good hanging possibilities.

Perennials more than annuals tend to drape themselves over the edges of pots, making them especially attractive for macrame plant hangers. Annuals like basil are less attractive to the macrame buyer, but still attractive to the adventurous cook. All can sell.

Some buyers are more drawn to the decorative piece of art (pot plus plant) than to pot or plant alone. Some people will buy a hanging rosemary to look at, not to use. My wife, a botanist, spent more money last year on herbs she looks at than herbs she uses.

James A. Duke is with the Medicinal Plant Resources Laboratory, Plant Genetics and Germplasm Institute, Beltsville, Md.

Hanging Pots

Hanging pots need to be watered and drained. Pots with holes in the bottom will drain into saucers also held in the hanger. A good mixture of potting soil and vermiculite, pre-mixed or mixed by the experienced at home, is adequate.

Hanging pots can be started with seeds, but you gain time by starting with rooted cuttings or plant subdivisions. Seed and plant dealers advertise in many horticultural journals.

Like ground plants, hanging pots require water, light, and protection from frost. Many herbs in hanging baskets can be grown in winter in a small greenhouse.

Eventually fertilization will help, since constant clipping of herbs will eventually deplete the soil of nutrients. Bonemeal, well-rotted compost, weed-free manure, or a light sprinkling of 5-10-5 commercial fertilizer can be spread near but not at the base of the plant, perhaps to a radius equalling the plant's height.

Since most of my personal experiences have centered on mints and ginseng, I will emphasize them. Mints offer their magic touch to herbal tea, while ginseng attracts much attention these days.

We drink a lot of herb tea at my house, and I prefer a blend of several herbs to a single herb. Thyme, oregano, and savory are too overbearing to be used alone, but they can enhance other interesting dominants like lemon balm, peppermint, and spearmint.

Easy-Grow Mints

Easy mints to grow for home consumption include apple mint, bergamot, lemon balm, lemon thyme, orangemint, peppermint, rosemary, sage, and spearmint. These are perennials, that come back from the roots in spring.

If you plant mints, keep in mind that many are difficult to eliminate once established. Plan before you plant perennials. Whether you plant seed or plants, determine where you want your plants to reside permanently.

To prevent herbs from spreading, set them out in sunken tins or pots, old tires, in the holes in cinder blocks, or be prepared to do battle with some of the more aggressive spreaders.

Some mints are handsome enough in their own right to be commingled with the ornamental garden. Anise-hyssop, bergamot, hyssop, and several sages (such as pineapple sage) have colorful flowers. Coleus is added to herb borders for its ornamental foliage.

Colorful foliage "varieties" have been developed in many mints. Leaves that are silver-, yellow-, and white-margined or variegated occur in the balm, sage, and thyme species. The purple foliage in various basils, bugles and sages adds to ornamental borders. At the outer edge of the ornamental bed, creeping varieties of mint or thyme can add aroma as well as color to your walkway.

You may place favorite herbs strategically in the ornamental bed, or in a formal "knot" garden.

Land Needs

The *Primer for Herb Growing* (issued by the Herb Society of America) indicates that an 8 x 10 foot plot of land will supply enough herbs for a family of four. The hobbyist can get by with an 8 x 12 foot greenhouse. A perennial grower can get by with a half-acre lot and no greenhouse.

There are true and false tales of ginseng growers making more than $10,000 per half-acre (21,780 square feet). The true tales relate to cautious and serious growers.

It is quite possible to lose money on ginseng. Good 3-year ginseng planting stock can cost more than a dollar per root. With good luck, and a simulated northern habitat, some will produce seed the first year.

To a smart marketer the 20 seeds per plant may repay cost of the root in the first year. But even the cautious grower may lose all the roots to adverse conditions. Chipmunks, groundhogs and other animals may be nuisances in the ginseng garden, devouring the roots and/or tops.

The cautious ginseng or mint grower can make some money on a small-scale farm. The best ginseng grower can survive on a few well-managed acres. The run-of-the-mill mint farmer will make less, unless the facts and/or fiction spread about ginseng could be equally spread about mints. What would happen, for example, if the world believed that bergamot, not ginseng, improved vitality and intellectual acuity and made old men young again?

Once land is prepared, the small herb grower has no special equipment need, but may want drying equipment. The small herb farmer will probably harvest by hand.

We string up our herbs in a dry area out of the sunshine. Herbs dried in the sun tend to lose much of their flavor and color.

Bigger growers of mints or ginseng may need drying racks that can provide ventilation and heat to prevent mildew during humid periods.

If the importance of a plant can be judged by the number of inquiries I get about it, ginseng is the most important herb, followed closely by the medicinal herb, goldenseal. As an herb grower, I would start with these if money were my main objective. I now have a small patch of ginseng intercropped with goldenseal, hoping that the bitterness of the goldenseal will discourage rodents from frequenting my ginseng patch.

Culinary Mints
If herbal tea were the main objective, I would start with a selection of mints. Culinary mints I have grown include anise-hyssop, apple mint, balm, basil, bergamot, bugle, catmint, catnip, clary, horehound, hyssop, lavender, lemon basil, lemon thyme, marjoram, mint, mother-of-thyme, orangemint, oregano, peppermint, rosemary, sage, savory, spearmint, and thyme.

Trying to start ginseng from seeds can be very frustrating to the beginner. Even if you have purchased seeds from reliable dealers, you may have to wait 18 months for the seeds to germinate. Seeds that dried out before you plant them may never germinate. Most mint seeds, on the other hand, tend to germinate readily.

It's cheaper of course to buy seeds than plants, but time also has a value. From planting seeds to harvesting may take five to seven years with ginseng, and at least five months with most of the mints.

Those fortunate enough to have a greenhouse or a cold frame can start mint seedlings to transplant later to permanent sites.

Ginseng does poorly or dies in the ordinary greenhouse. It requires a shady situation, with highly organic forest floor soils recommended. Seeds or roots should be planted 12 to 18 inches apart in beds separated by walkways to permit weeding, mulching (with fallen leaves) and harvesting. Well-drained soils cleared of extraneous roots are recommended (See Farmer's Bulletin No. 2201).

Only two mints, peppermint and spearmint, are grown on a large scale in the United States. Planting and soil requirements for these are discussed in Mint Farming (Agr. Info. Bul. 212). Many mints which do well in full sun can be grown in partial shade. Anise-hyssop, apple mint, balm, basil, bergamot, bugle, curly mint, lemon thyme, peppermint and spearmint have survived well in situations where they get only 4 to 6 hours of full midday sun.

Mints will grow well in most garden soils, but the higher the clay content the more likely the mints will suffer from

221

waterlogging or disease during excessively wet seasons. As a rule, the brighter and drier the habitat, the higher the aromatic qualities of mints. Even so, shade-grown mints are quite satisfactory for fresh herb tea.

Insect Repellers

Mints do not attract most pests. In fact many are insect repellents.

Mint species may contain repellents or insecticides, such as camphor, carvacrol, citral, citronellal, eugenol, furfural, linalool, menthol, and thymol, and fungicides such as furfural, menthol, salicylic acid, and thymol.

Diseases can wipe out mint monocultures, especially in heavy soils. Intercropped with ornamentals or vegetables, mints are less likely to suffer epidemics.

Rodents, though not fond of mints, can be pests in ginseng plantations. Catnip is said to repel rodents but I cannot vouch for this. I have planted some ginseng in empty tins, tops and bottoms of which have been removed. These tin collars may discourage subterranean rodents from eating the $1 roots. Buried screens such as those used for tulip bulbs might work as well or better.

With wild ginseng being considered as an endangered species, its harvest may be outlawed. This would improve the outlook for cultivated ginseng. There is no proposal to place cultivated ginseng on the endangered species list. Meanwhile the practical harvester will dig ginseng in autumn when seeds are mature, and plant these seeds, or sell them to someone who will plant them.

Mints, like ginseng roots, should be harvested only when they can be properly dried. Mint leaves can be stripped off plants in the field, leaving the roots and stems intact, or the plants can be cut and tied into bundles.

For herb tea during the summer, I prefer fresh-picked mints. As winter approaches, I bag up the leaves and dry them to keep the family in tea over winter. For home consumption, both ginseng and mints can be frozen.

The small grower needs to develop a rather personal market. Our herbs are sold at a farmer's market on Saturday. Usually an enterprising herb grower can arrange to sell his wares at local hardware stores, craft shows, roadside stands, specialty shops, seed stores, or supermarkets.

Dried herbs, seeds, even living plants can be sold locally or via mail order. Some herb farmers package their wares as tea bags or even potpourris for market.

222

Scientific names of herbs

Anise-hyssop: *Agastache foeniculum*
Apple mint: *Mentha rotundifolia*
Balm: *Melissa officinalis*
Basil: *Ocimum basilicum*
Bergamot: *Monarda didyma*
Bugle: *Ajuga reptans*
Catmint: *Nepeta mussini*
Catnip: *Nepeta cataria*
Chives: *Allium schoenoprasum*
Clary: *Salvia sclarea*
Coleus: *Coleus blumei*
Corsican mint: *Mentha requienii*
Curly mint: *Mentha spicata* var. *crispata*
Ginseng: *Panax quinquefolius*
Goldenseal: *Hydrastis canadenis*
Horehound: *Marrubium vulgare*
Hyssop: *Hyssopus officinalis*
Lavender: *Lavandula vera*

Lemon basil: *Ocimum citriodorum*
Lemon thyme: *Thymus vulgaris* var. *citriodorus*
Mint: the mint family *(Lamiaceae)* or any of several mints of the genus *Mentha*
Mother-of-thyme: *Thymus serpyllum*
Orangemint: *Mentha piperita* var. *citrata*
Oregano: *Origanum vulgare*
Peppermint: *Mentha piperita*
Pineapple sage: *Salvia rutilans*
Rosemary: *Rosmarinus officinalis*
Sage: *Salvia officinalis*
Savory: *Satureja spp.*
Spearmint: *Mentha spicata*
Summer savory: *Satureja hortensis*
Thyme: *Thymus vulgaris*
Winter savory: *Satureja montana*

Of course, no money can be made without a market. You should gage your market before planting your seed. I believe the markets for herbs, herb teas, and especially ginseng roots will increase in coming years.

An attractive hanging pot with a healthy herb can bring $5 or better today, and probably more tomorrow. If you prepare them yourself, 90 percent of that can be profit. Selling seven a day would put your profits over $10,000 per year. That may seem improbable, but it's not impossible.

Further Reading:

Herb Society of America, *A Primer for Herb Growing,* Herb Society of America, Boston, Mass. 1971. 50¢.

U. S. Department of Agriculture, *Mint Farming,* Agr. Inf. Bul. No. 212, GPA Publications Division, Washington, D. C. 20250. 1963. Free.

U. S. Department of Agriculture, *Growing Ginseng,* Farmers Bul. No. 2201, on sale by Superintendent of Documents, U. S. Government Printing Office, Washington, D. C. 20402. 1973. 35¢.

U. S. Department of Agriculture, *Herbs for Flavor, Fragrances, Fun: In Gardens, Pots, In Shade, In Sun,* 1977 Yearbook of Agriculture, 217-223, on sale by Superintendent of Documents, U. S. Government Printing Office, Washington, D. C. 20402. $6.50.

How You Can Grow
Food Organically

By Wesley Judkins and Floyd Smith

Organic matter is extremely important for improving the physical condition and productivity of the soil. It makes plowing and cultivating easier. It also increases the nutrient reserve and water-holding capacity of sandy or clay type soils.

The gardener derives several benefits by mulching with organic matter. It reduces erosion caused by runoff of rain or irrigation, increases infiltration of water into the soil, and conserves this moisture by reducing evaporation. Organic matter helps to suppress weed growth.

Some good organic materials to use as mulch are leaves, lawn clippings, fresh sawdust, fine wood shavings, pine needles, chopped straw, ground corncobs, shredded tobacco or sugar cane stems, peanut hulls, or cottonseed hulls. These materials do not add important amounts of nutrients or have a significant effect on the pH (relative acidity) of the soil.

The dead vegetable plants in your garden should be chopped down and left on the ground as a protective mulch during winter. This trash mulch reduces erosion and improves organic matter content of the soil when the garden is prepared for planting in spring.

Unmulched areas in gardens and fields, not occupied by growing crops, should be planted to green manure crops such as rye, ryegrass, millet, sorghum, or crimson clover. They will reduce leaching of nutrients and increase organic matter for the next crop as they are worked into the soil.

Organic waste materials such as leaves, manure from livestock and poultry, treated sewage sludge, and the organic portion of urban trash collections can be used as fertilizer, mulch, or compost.

Wesley Judkins is Professor Emeritus of Horticulture,
Virginia Polytechnic Institute & State University,
Blacksburg. Floyd Smith is Research Entomologist
Collaborator, Florist & Nursery Crops Laboratory,
Agricultural Research Center, Beltsville, Md.

224

Some cities accumulate leaves in huge piles during fall collection periods. After several months of composting, the material is available at little or no cost to gardeners. This is a practical way to reduce environmental pollution and supply organic material for gardens and farms.

The solid portion of sewage may be effectively salvaged and used as fertilizer. Composted sewage sludge has a composition of about 5 percent nitrogen and 2 percent phosphoric acid, and is an excellent organic fertilizer.

Insects, Disease

Time of planting is important in avoiding losses by diseases and pests in certain regions. Since seed corn maggots destroy early plantings of beans and corn, you should delay planting until the soil warms. Early maturing varieties of sweet corn can avoid the worst earworm problem. Delay plantings of summer squash to avoid early season activity and resultant damage by the squash vine borer.

During recent years, plant breeders have made tremendous contributions to agriculture by developing new varieties resistant to diseases. These allow large yields of high quality crops to be produced without the use of chemical sprays.

Selling organically-grown produce.

George Robinson

When planning for vegetable production in a home garden or commercial enterprise, consult your county Extension agent or seed catalog for information on disease-resistant varieties. Excellent new introductions are available each year. Comparable insect-resistant varieties have not been developed.

Some vegetable crops are highly subject to damage by pests or disease organisms. Others are relatively pest-free. The beginner should first plant only trouble-free crops, later trying the more difficult ones after gaining experience.

Attack by cutworms can be prevented by placing a simple collar of stiff paper (cut from a drinking cup or milk carton) around newly set tomato, cabbage, and pepper plants—and even sweet corn. The collar should extend about 1 inch into the soil and 2 inches above ground.

Slugs that emerge at night from hiding places in wall crevices, loose mulch, piles of plant stakes or trash, can be trapped under pieces of board, shingles or flat stones laid in the garden. Lift them each day and destroy the slugs.

Slugs are attracted to shallow vessels partially filled with beer into which they crawl and expire. Slug baits moistened with a teaspoon of beer will be twice as effective.

An aluminum foil mulch around low growing plants reflects the ultraviolet rays from the sky and repels flying insects (including aphids, leafhoppers, thrips, Mexican bean beetles, and cucumber beetles) from landing on the plants. Summer squash, Chinese cabbage, lettuce, and peppers have been protected from virus infection transmitted by aphid feeding. Beans and cucurbits have been protected from chewing and sucking insects.

Black polyethylene mulches, used extensively by commercial fruit and vegetable growers, help to control weeds, conserve moisture and prevent leaching of fertility in the garden. They also keep the produce from resting on the soil, thus reducing rot infection from soil contact.

Blacklight traps are frequently advertised for control of insect pests in gardens and on farms. Although great numbers of moths and other insects are attracted to individual black lights and captured in the attached traps or killed on electric grids, there is little or no reduction of the pest insects that attack vegetables.

Sometimes insect pests in the vicinity of the trap will be greater than normal. Insects attracted to the light may not enter the trap, but linger to lay their eggs in the vicinity. Likewise, certain bait traps—as for the Japanese beetle—may increase the infestation in the trap's vicinity.

Routine inspection and handpicking of tomato hornworms on a small planting of a dozen or so tomato plants is highly effective and less time consuming than preparing and applying a spray. In some years, hornworms may not appear at all. Handpicking can also eliminate small infestations of squash bugs, Mexican bean beetles, and potato beetles.

Interplanting

A recent calendar for home gardeners lists a number of plants that should be placed among your vegetables to deter cabbage worms, Mexican bean beetles, Colorado potato beetles, Japanese beetles, borers and tomato hornworms.

Experiments by research entomologists at State and Federal Experiment Stations have shown no beneficial results from such interplantings except for the reduction of one type of nematode by marigold roots.

Moreover, these experiments showed that Mexican bean beetles and Colorado potato beetles found their respective host plants in mixed plantings. Onions and garlic supported thriving populations of onion thrips and mites, and had no measurable repelling effect on cabbage worms, bean beetles, cucumber beetles and aphids that infested their respective interplanted host plants.

Few gardeners ever see the most efficient parasites and predators at work among the pests on their plants. Examples of beneficial insects are: the yellow and black banded thrips; the tiny Orius plant bug; syrphid fly larvae; aphid lions—the ugly looking larvae of the delicate lacewing flies; and larvae and adults of our native ladybird beetles that suck the juices from plant-feeding thrips, spider mites, aphids, young caterpillars, and leafhoppers.

Often during periods of cool damp weather, epidemics of insect disease caused by bacteria, fungi, or viruses will suddenly destroy thriving populations of pests—especially aphids cabbage worms, cabbage loopers and other caterpillars.

Until recently the Mexican bean beetle has defoliated beans, lima beans, and soybeans over wide areas without the depressing effect of parasites or predators. A tiny wasp was recently introduced from India that lays 10 or more eggs in each bean beetle larva, and soon the larva turns black and dies.

This microscopic parasite disperses for 10 miles or more in search of bean beetle infestations.

The parasite does not survive our winters, for lack of food. But if reintroduced each season from laboratory cultures, it has the potential for reducing the bean beetle to a minor pest.

Each Orius bug destroys 20 or more flower thrips per day. He and his fellows are responsible for reducing high spring populations of this insect to low levels for the remainder of the season.

Our native ladybird beetles, that come into our gardens in late spring, lay their orange-yellow eggs among aphids on vegetables where each alligator-like larva sucks the juices from 10 to 20 aphids per day for a total of 300 or more during its growth period. Thus, thriving aphid colonies developing in early spring virtually disappear for the summer and do not reappear until autumn when temperatures are lower and the ladybird beetles less active.

Praying Mantis

In contrast to the highly efficient parasites and predators discussed above, much attention is given to less effective techniques. Some amateur ecologists urge you to buy praying mantis egg masses and pints of ladybird beetles and release them in your garden for season-long insect control.

You should realize that the praying mantis hatch from the egg masses in late spring. The tiny mantids (mantis) scramble for safety—usually into dense shrubbery—to avoid being eaten by their brothers and sisters. Of the hundreds that hatch in the spring, only a few survive until fall and they are usually found in the shrub border; rarely on the more exposed vegetables where you need them.

One authority has stated that "the chief benefit to be derived from the purchase of mantid (mantis) egg masses is the feeling of virtue in believing that you have established a highly beneficial insect which will protect the neighborhood by destroying many harmful garden pests.

"Of the hundreds of young mantids that come tumbling out of a case, perchance a few will survive. With avid appetites and rapacious front legs they capture many insects; including their brothers and sisters, and harmful insects as well as beneficial insects.

"Nevertheless, the mantid is a handsome insect that is interesting to have around. So let us continue to protect it and encourage others to do the same, but do not depend upon it to rid our garden of all noxious pests."

The ladybird beetles you buy are collected from their hibernating quarters in California canyons and shipped to you. When you release them in your garden they usually disperse to other areas just as they disperse from their hibernating quarters in canyons or woodlands—often for several miles—

in search of cultivated fields. Few, if any, remain for long in your garden.

Ladybird beetles found in your garden are local, naturally occurring beetles which migrate from hibernating sources in early spring.

The information set forth in this chapter will aid you in producing an abundance of many but not all kinds of vegetables in most years, without resorting to use of chemical fertilizers or sprays. In some years with poor crops it will be necessary to accept foods of lower quality. Experience will enable you to select and grow only the more reliable, trouble-free vegetables.

Further Reading:

Biological Control of Plant Pests, Handbook No. 34, Brooklyn Botanic Garden, Brooklyn, N. Y. 11225. $1.75.

Encyclopedia of Organic Gardening, The, Rodale Books, Inc., Emmaus, Pa. 18049. 1971. $14.95.

How To Grow Vegetables and Fruits by the Organic Method, Rodale Books, Inc., Emmaus, Pa. 18049, 1971. $13.95

Judkins, Wesley P., *Organic Gardening—Think Mulch.* 1977 Yearbook of Agriculture, pp 78-83, on sale by Superintendent of Documents, U. S. Government Printing Office, Washington, D. C. 20402. $6.50.

Natural Gardening, Handbook No. 77, Brooklyn Botanic Garden, Brooklyn, N. Y. 11225. $1.75.

Starting Up a Kennel to Board, Groom Dogs

By Dennis A. Hartman

A dog grooming and boarding kennel can be a lucrative supplemental or full time business which requires little acreage.

Aside from a love of animals and willingness in an emergency to be on call at any time through the day or night, you must also have a thorough knowledge of dogs, their management and grooming, and a keen business sense of responsibility. The experience of having worked for a kennel operator will also be valuable.

To be skilled in grooming, attend a professional grooming school. It is important to learn grooming from a qualified instructor so that you may pick up correct procedures rather than learn some good and some bad techniques from an unqualified person. There are over 120 different breeds of dogs which vary considerably in hair coat, size, and temperament, thus complicating the task of learning to groom.

Gross income of a boarding and grooming kennel will vary, depending on size and efficiency of management. Boarding and grooming fees also differ within a locality as well as between areas of the United States. Fees are less outside a metropolitan area. On the other hand, building space and labor also cost less. Check building, operation, and boarding fees of your area to gain an idea of net income possibilities.

As a prospective kennel operator, you should know the negative aspects. Several years may be required to build up a clientele. Someone must be present at all times. Since holidays are busy periods, it is difficult to be away then. Customers often break appointments. Cleaning pens and runs is hard work. Some personal attention must be given each dog. Dogs may bite the operator. Dogs may come in very dirty or loaded with parasites. Some dogs may never be picked up.

Dennis A. Hartman is Professor of Dairy Science, Virginia Polytechnic Institute and State University, Blacksburg, and a member of the National 4-H Dog Committee with numerous publications on dogs to his credit.

Before building a kennel, check with the zoning authorities in your town or county. Some areas have sanitation, noise and nuisance laws. You may be required to purchase a special license.

The Animal and Plant Health Inspection Service, a branch of the U.S. Department of Agriculture, and a similar service provided by your state have the responsibility of inspecting kennels twice yearly.

Right Location

Having your kennel available to sizable population areas is most important. It's best to locate on a main access route so customers may leave and pick up their pets without undue travel. The kennel area should be landscaped attractively and have adequate parking space with a driveway. Buildings must be kept in good repair and well painted.

Major clients of a boarding and grooming kennel are families in the middle age group. The dog is sometimes a substitute for children, or for children who have left home.

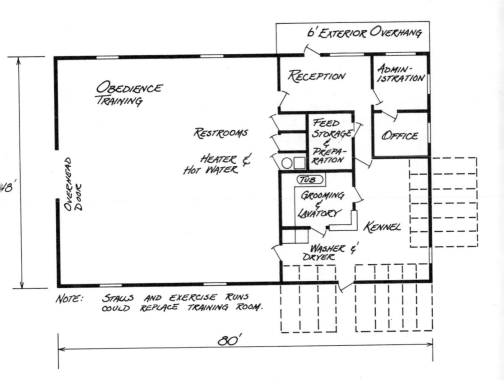

Customers, therefore, insist upon clean and well managed kennels in which their pet will receive the same loving care received at home.

Time spent in planning your kennel will pay big dividends. Most kennels are begun on a small scale and enlarged as the business grows. So design your plans for easy extension. Minimize the labor involved in feed handling, feeding, and cleaning of pens.

Many dog food companies have plans for kennels and pamphlets for feeding and managing dogs, free for the asking. You will also get many ideas from visiting kennel operators in your state who are at a distance and not competitors.

Design of your kennel may be dictated by the investment you are prepared to make. In some instances it may be economical to rebuild existing buildings. However, this can often be costly. Adding insulation, drains, new flooring, and many other necessary changes can be more expensive than new construction.

Concrete block construction is suggested. Kennels made of wood are susceptible to chewing and cannot be sanitized as completely as those made with concrete blocks. Thus savings in low construction costs are not necessarily the best for a long-term operation. Building cost estimates will not be given here because of the many variables. Get several construction estimates before building.

Put your kennel on a high, dry spot which permits drainage in several directions. Design of the kennel will depend upon personal preferences. However the work area should provide rodent-proof storage space, space to prepare food, sinks, and a refrigerator and freezer if fresh meats are to be fed. Dog stalls should be arranged for easy access through the central service area.

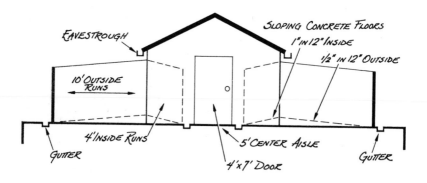

Concrete floors are recommended as they may be easily cleaned and sanitized. Vinyl tile on the floor of the reception and grooming area is useful as you may need to damp mop several times daily to control dog hair.

Insulation Important

Roof style is a matter of choice. However, insulation is important as it will reduce the cost of heating in cold climates and prevent excessive heat from the sun in warm areas. Ceiling insulation will also absorb noise.

Have as much window space as possible to maximize natural lighting and ventilation. Provide plenty of ventilation to reduce odors and prevent moisture buildup inside. Screen the windows to keep out mosquitoes and flies.

The heating system is often placed in the work center. Forced warm air heating systems using gas, oil or electricity are most suitable, but avoid strong air movements creating drafts. Remember that heat rises and the comfort zone of workers may differ from that of dogs as the dogs' body mass is much closer to the floor. If a trap or guillotine door system is used to allow the dogs access to the run outside, devise a rubber flap to eliminate drafts around edges of the door.

The stall floor should slope from the outer wall to the front of the stall to a gutter at the rate of half inch per foot, thus allowing good drainage when cleaning and sanitizing. The exercise run adjoining on the outside should slope away from the building at this same rate to a six-inch gutter drain for easy cleaning.

Sixty pounds of pressure in a 1-1/4-inch hose is best for washing down exercise pens and inside cleaning. If runs are hosed down three times a day, the total time for cleaning may be reduced by about a third.

Septic tanks with bacteria-destroying enzymes are satisfactory. However, they may need to be pumped more often than for normal homeowner use. A sewage lagoon is better for a large operation. Before deciding on either, check with the proper town and country authorities.

Exercise Runs

Exercise runs six by ten feet are adequate for large dogs, and four by ten foot runs for small and medium dogs. Large dogs may injure their tails and hind quarters if the runs are too narrow.

Concrete flooring, metal posts, and chain link fencing

are suggested. Asphalt paving may be suitable in some states. However, in hot weather asphalt can become soft and stick to the dogs' feet.

A six-inch curb between runs will be helpful in cleaning, and also prevent cross infection among dogs by stopping wash water from crossing over. To reduce noise, have a two- to three-foot wall, with fencing above, every six or eight runs. Exercise fencing should be eight or ten feet high, although some dogs can climb fences of this height. An overhang may be needed to prevent climbers from escaping.

Fencing of one-inch mesh is recommended for small dogs so they can't get their muzzles through to fight other dogs or wear down their face hair. One-and-a-half to two-inch mesh fencing may be used for large breeds.

A perimeter fence four feet high protects against dogs escaping, and stops neighborhood dogs from being a nuisance. This fencing may be of a less expensive type than chain link. However, it must be small enough in mesh to prevent small dogs from squeezing through. Two- by four-inch mesh is suitable. The gate in the perimeter fence should be of steel tubing with an automatic latching device.

Although you, the owner, may be required to be on call in an emergency on a 24-hour basis, it is important that you have a relief system built into your program. You can't survive too well if you must always jump up from the dining room table when a car arrives at the kennel to drop off or to pick up a dog.

Your goal as the operator of a boarding and grooming kennel should be to return each dog in better condition than when it came in. This means weight, health, and freedom from external parasites. Trimming the nails and improving the coat by brushing are extras that will encourage business.

The Contract

In the boarding contract include the following points: date received, date to be picked up, customer's name and address, emergency telephone, breed, sex, color, age and name of dog, your right to secure veterinary attention if necessary with payment by owner, and to dispose of the dog should it not be claimed and fees paid.

Have a veterinarian on call at all times to treat any dogs that get sick. Your vet will also be able to advise you on sanitation practices.

It is recommended that you require all dogs you board to have an up-to-date certificate of vaccination for rabies, distemper, hepatitis, and leptospirosis. This gives you protection against passing these diseases in your kennel. It also provides business to vets in your area. These vets will realize the importance of the policy to you as well as the economic return to them, and therefore will support your business.

After you build your facility, carry out an active advertising program. An ad in the yellow pages of your telephone directory will likely be your best method. Also important are radio spots on stations designed especially for middle-aged people. Run these spots several weeks before major holidays and during the summer vacation season. Advertise, too in the pet section of your local newspapers.

A well written and illustrated brochure made available to veterinarians to hand out to potential customers is another method.

"Word of mouth" recommendations passed from a customer to friends will ultimately be your best advertisement, however. Thus, providing good service is vital for continued business.

Further Reading:

American Kennel Club, *The Complete Dog Book*, Doubleday and Company. $14.95.

Dog World Magazine, 10060 West Roosevelt Road, Westchester, Ill. 60153. Published monthly. At pet shops and large newsstands $1.25. Yearly subscription $12.00.

Small Marketer's Aide #71, Small Business Administration, Washington, D. C. 20005. Free.

Year-Round Gardening With a Greenhouse

By Conrad Link and David Ross

The greenhouse is a specialized structure designed for growing plants year-round. A clear or translucent cover permits sunlight to enter, which heats the greenhouse during the day. When excessive sun heating occurs, ventilation is needed. During cold nights and much of cold days, a heating system is required to maintain the desired temperature.

After the initial investment in land and the greenhouse structure, the main expense will be for heat and labor. If the owner and family are the labor force, then heat becomes the biggest expense. Other costs will include soil and growing media, fertilizer, pesticides, pots, seeds, and bulbs. The part-time greenhouse operator must develop a market for his products and skill with attention to details that result in quality plants.

Crops to be grown will be influenced by where and to whom they will be sold. Marketing includes selling wholesale to flower shops and garden centers or selling retail directly to the consumer.

A greenhouse should be on a site that takes advantage of full sun, provides good water drainage and utilizes windbreaks. Electricity and a good water supply are needed. A separate building should be used to store equipment and supplies, to provide work space, and perhaps to house the heating system.

Size of the greenhouse should be well planned. If the hobby or business endeavor proves successful, the greenhouse will soon be too small. Plan the size, location, and layout to permit future expansion.

Larger greenhouses are more efficient and more economical to operate as they cost less per square foot and the environ-

Conrad Link is Professor of Horticulture, University of Maryland, College Park. David Ross is Extension Agricultural Engineer at the university.

ment can be maintained more uniformly. Heating and ventilation systems are the most expensive items needed for a greenhouse. Their costs per unit area are less in larger greenhouses.

Structural Options

Many styles of greenhouse frames exist; select one that is pleasing and practical for you. The frame may be wood, steel, or aluminum. The cover can be glass, plastic sheet, or fiberglass, each available in different sizes and qualities.

A popular low cost greenhouse is the pipe frame or curved roof style. The foundation is a series of pipes driven into the ground to support the curved roof members. Roof members may be made of steel or aluminum pipe or may be a curved truss.

The cover is a single or double layer of greenhouse-quality, ultraviolet-inhibited 6-mil (0.006 inch) thick or heavier plastic sheet. Plastic sheet is good for one or two years, depending on the material quality and weather. An air-inflated double layer of plastic film can reduce heating costs by 30 percent.

For a more permanent cover, use a clear greenhouse grade fiberglass. Fiberglass is available in several grades having service lives of a few years to perhaps 20.

Plans are available through your county or state Cooperative Extension Service for wood frame greenhouses which can be covered by plastic film or fiberglass. Wood in contact with the soil should be pressure-treated or painted with a wood preservative, Copper naphthenate is a safe preservative near plants; creosote and pentachlorophenol are harmful to plants.

A good quality greenhouse can be built with a good foundation and rigid frame, or an inexpensive greenhouse can be built with a temporary frame to give seasonal plant protection.

A glass greenhouse is the third possibility. Glass and aluminum or steel combine to make a long lasting, beautiful greenhouse. The glass, rigid aluminum or steel frame, and a sturdy foundation make the initial investment high. However, annual maintenance is much less.

While the glass greenhouse is a showplace, the beginner will find the less expensive, temporary plastic or fiberglass greenhouse well suited as a first structure.

Two additional structural options are the hotbed and coldframe. These are low-walled frames with cover to give plants protection during cool, windy spring weather. A hotbed has a heat source in the soil. A 3 x 6 foot coldframe or hotbed can be used to advantage for starting vegetable or flowering plants.

Heating Systems

The greenhouse can be heated with steam, hot water, or hot air. The system can be fired by any of the conventional fuels. The heating system should be fully automatic and as free from maintenance labor as possible.

A thermostat is used to control heater operation. The fan on a hot air heater should be wired to run continuously to maintain uniform air temperature throughout the greenhouse.

Two smaller heating units instead of one large one provides some insurance in case a heating unit fails. A small stand-by electrical generator is good to have for a power failure. Heating units must be vented to the outside if there are combustion gases. Provide an air inlet near the heating unit so oxygen is available for combustion.

Heater size is determined by the following equation: heater size (BTU/hr) = (total surface area in square feet) times (night temperature difference between inside and outside, °F) times (a heat loss factor).

The heat loss factor is 0.7 for air-separated double plastic sheet and 1.2 for single layer glass, fiberglass, or plastic sheet. These figures should be increased by adding 0.3 for hobby (small) greenhouses or for windy locations.

For further details read the bulletin, *Hobby Greenhouses and Other Gardening Structures,* available from NRAES, Riley-Robb, Cornell University, Ithaca, NY 14853 for $2.

Ventilation, Shading

Ventilation is essential for producing good quality plants. The temperature must not get too high, and a supply of carbon dioxide must be maintained. Ventilation can be provided by natural convection, using side and roof vents, or by mechanical means using exhaust fans and inlet louvers. Thermostats and electrical motors are used to automate ventilation.

The ventilation system must be able to change the air once each minute in a large greenhouse, and to change it one and a half times each minute in the hobby (small) greenhouse.

Winter ventilation requirements are about one-quarter air change per minute. Two fans, with one having two speeds, are often used; the low speed of one fan is enough for winter. Motorized intake louvers are placed on the opposite wall.

The volume of a greenhouse is length times width times average height and is given in cubic feet. The fan rating will be in cubic feet per minute (cfm).

Shading materials such as saran cloth, movable lath strip covering, lime and water, and dilute white latex paint are used to reduce light intensities and to cut the solar heat load in summer. Light reduction is necessary for those plants which grow best in low light.

Many plant functions are controlled by the length of day. Some plants such as petunia, China aster, or tuberous begonia naturally flower in the long days of summer (long-day plants) and others such as chrysanthemum or poinsettia flower in the short days of fall or winter (short-day plants).

Other plants such as carnation, rose, lilies, and everblooming begonia flower regardless of the day length (day-neutral plants).

A greenhouse operator must protect short-day plants with a lighttight cover to induce flowering when days are long. Artificial light is used on long-day plants to induce blooming in winter months.

Temperature Control

A well designed heating and ventilating system allows the greenhouse operator to maintain the most efficient and economical temperature for plant growth. Greenhouse night temperatures are generally maintained at 50° to 70° F, depending on the kind of plant. The temperature is permitted to rise 10 to 15 degrees during the day before ventilation is started.

The effect of temperature on growth varies with plants. Seedlings of many crops are started at a warm temperature and then grown at lower temperatures. This is true of annual vegetable and flower plants germinated and started at 70° to 75° F, grown at 65°, and finished at 55°.

Water that is safe for drinking is appropriate to use in a greenhouse. Water from ponds and wells is fine, providing it doesn't contain excess amounts of salt.

When plants are watered, apply a sufficient amount to moisten the entire volume of soil plus some that will drain through. This drainage helps prevent buildup of salts from the water or fertilizer used. Frequency of watering is determined by size of the plant, temperature, and the growing medium's ability to hold water.

Water is applied manually with a sprinkling can or hose. Spray nozzles or porous plastic tubing are used for watering cut flower crops. Trickle tubes may be placed into individual pots or plants. Such watering systems may be made automatic using a time clock switch that is set to water at designated times, or by using devices that operate on dryness of the soil.

Capillary watering of pot plants is possible by placing them on a bed of sand kept continually moist. Recently a carpet-like mat of natural or synthetic fibers has been used in place of the sand.

Soils, Growing Media

Plants may be grown in many types of soils, soil mixtures, or mixtures of organic matter and inert materials without soil. The growing media mixture must be uniform in texture, hold sufficient water and drain well, be porous and well aerated, and pest-free. It need not have any available nutrients as these are supplied in fertilizing.

Growing media ranges from fertile top soil with no additions, to a variety of mixtures that may include sharp sand, peat, perlite, bark and wood chips, sludge, or composted leaves. When using soil, select a sandy loam or loam, preferably one containing organic matter. Be careful of soils containing herbicides as they may damage your plants.

Sterilize soils and growing media before use to reduce the problem of soil insects, diseases, and weed seed. Steaming is most effective but certain chemicals may do the job. Growing mixtures are prepared by the greenhouse operator or bought already prepared. Commercial mixtures are often more economical because they are sterilized, ready to use, and may even contain some fertilizer.

Proper application of fertilizers is another part of growing under greenhouse conditions. Have soil tests made of mixtures with a high proportion of soil. Mixtures without soil generally contain little available fertilizer nutrients.

Modern greenhouse procedures call for using soluble fertilizers. These are applied at the time of watering. Fertilizer in this form is available to the plant at once.

Soluble fertilizers are generally of the so-called complete types, supplying nitrogen, phosphorus and potassium. Some may contain other fertilizer nutrients. Mixtures without any soil often need the application of "minor fertilizer nutrients" in very minute quantities to supply plant needs.

Equipment for liquid fertilizer application ranges from simple devices which meter concentrated fertilizer solution into the water hose, to elaborate proportioning devices which are adjustable to specific concentrations.

Chrysanthemums

Commercially, chrysanthemums are produced year round by regulating the day length to encourage vegetative growth or flowering as needed. Many cultivars (cultivated varieties) are available for greenhouse culture.

Flower types vary from singles to full doubles, incurved, thread or spider types, or anemone forms. Flower size varies from less than an inch in diameter to exhibition types 8 to 10 inches or more. Plant habit varies from dwarf compact forms to tall ones suitable for cut flowers.

Both cut flowers and potted mums have year-round sale.

The natural dates of flowering of chrysanthemums range from late August and early September to late December. Cultivars are grouped into response groups based on the number of weeks of short days needed to produce flowering. Response groups range from 7 to 14 weeks long.

The best mums for cut flowers and pot plants are in the 9-, 10-, and 11-week response groups. A crop takes about 16 to 20 weeks from planting to harvest, varying due to the season and response group.

Chrysanthemums grow best in porous, well-drained soil with a moderate level of fertility. Commercial growers use a 60° F night temperature.

Single stem cut flowers are planted on 4 x 6 to 5 x 6 inch spacings. Multistem cut flowers are planted on 7 x 7 to 8 x 8 inch spacings. Top the plant when it is 4 to 6 inches tall, and allow two or three stems to develop.

Potted plants usually have three or four cuttings per 6-inch pot and are topped once shortly after they become established. Commercial producers of cuttings supply their customers with schedules of culture suitable for their area.

Poinsettias

Poinsettias are grown primarily as Christmas plants. They are propagated by stem cuttings from June until mid-September. Rooting under a mist system, in sand or perlite, requires 21 to 24 days. Some growers root directly in pots in a soil mixture.

In summer, poinsettias may be grown under glass that is slightly shaded to reduce temperatures. They should have full sun starting in mid-September. The poinsettias will flower by mid-December if they receive only normal day light and a temperature of 60° to 62° F. If growth is satisfactory and the bracts have developed good color by early December, the temperature may be lowered to 55° to 58°. However, light of any kind at night, even a nearby streetlight, can delay flowering.

Growing media for poinsettias should be porous, well-drained, and slightly acid with a pH of 6.0 to 6.5. Soluble liquid or slow release fertilizers are used with the different cultivars.

Poinsettia cultivars are standard or self-branching types.

Colors are the familiar red and also white, pink, or variegated. Growth-regulating chemicals may be used on early propagated plants to prevent tall growth.

Tomatoes

Greenhouse tomatoes are a possible vegetable crop, especially for local specialty markets. Tomatoes are grown in ground beds and trained upright 6 feet or more. Soils must have a slightly acid pH 6.5, be sandy to silt loam, have good organic matter content, and be well-drained. Tile should be installed in the beds for drainage and for steam pasteurization of the soil before planting.

A regular scheme of fertilizing is established using a complete fertilizer. Avoid excessive fertilizing in November through January when light is the poorest and temperatures more difficult to control.

Special cultivars of tomato have been developed for greenhouse use, as Floradel, Michigan-Ohio hybrid, and Tuckcross 520. Garden-type cherry tomatoes are also heavy producers.

Greenhouse tomatoes are most economically grown as a fall crop to fruit from October to January, or as a spring crop fruiting from March to June. Fall yields of 6 to 10 pounds per plant can be obtained. A spring crop will yield 10 to 20 pounds.

Seed is sown and seedlings are ready for transplanting in 3 to 4 weeks. They are transplanted to pots and, when 3 to 4 inches tall, planted in the ground. Planting distance is 18 to 30 inches apart, giving each plant 4 to 5 square feet. Once planted and established they are mulched to reduce soil compaction, using ground corn cobs, peanut hulls, straw or hay.

Side branches are removed as they develop to produce a single stemmed plant. Flowers are tapped daily to insure they are pollinated so that each flower will produce a fruit.

Tomatoes require 60° to 62° F night temperature. Ventilate freely to keep the foliage dry. Humidity control is important because mature plants produce a lot of moisture.

Give special attention to controlling insect pests such as aphids, spider mites, white fly, and several leaf-eating insects. Foliage diseases, verticillium wilt, and expecially virus diseases may be serious problems.

Annuals—Bedding Plants

Production of annual flower and vegetable plants for spring is an important greenhouse use in the months of January to May. These plants are propagated by seed sown in flats and then transplanted. Greenhouse growers use a variety of soil or growing media for this purpose. The growing medium should be slightly acid, well aerated, and hold water. A medium with this texture allows the seedlings to be removed at transplanting time with little root damage.

Germination time varies with each plant but most plants do well at temperatures of 70° to 75° F. Examples of germination times are 6 to 8 days for marigold and zinnia, 10 to 14 days for petunia and snapdragon, 14 to 21 days for begonia, browallia, and salvia, and 20 to 25 days for impatiens and lobelia. Some seedlings grow slower than others and that must be considered in determining the date of seeding. The greenhouse temperature is another factor in the speed of growth.

Seedlings are transplanted as soon as they can be handled easily, which may be 10 to 14 days after germination. They are watered after transplanting and placed in a 60° to 65° F temperature. If growth proceeds too rapidly, lower the greenhouse temperature or move plants to a cold frame. Apply fertilizers soon after transplanting.

Some pests are aphids, thrip, white fly, spider mite, botrytis, mildew and, in the soil, damping off, rhizoctonia and fusarium.

Other Crops

Forcing of spring flowering bulbs is easily done with a greenhouse. Plant the bulbs in pots in October or November, using a well-drained soil. Tulip, narcissus, hyacinth, crocus and others may be handled this way.

After potting, bury the pots under several inches of sand. An additional cover of straw or other material may be used later to retard freezing. If cold storage facilities capable of maintaining 35° to 38° F are available, potted bulbs may be stored there. Such storage makes it easier to get the pots into the greenhouse for forcing as compared to pots buried outdoors.

After about a six-week rooting period, the pots may be brought into the greenhouse for forcing at 55° to 60° F. The temperature may be raised if necessary.

Tulips handled this way will flower in four to five weeks. In order to have tulips or iris in flower in December or early January, the bulbs must have had some pre-planting temperature treatments to hasten flower bud development. The time required for flowering varies between the different varieties and gets shorter as the normal flowering season approaches.

Narcissus, hyacinths, crocus and other bulbs would be handled similar to tulips.

Beekeeping as a Hobby or Economic Sideline
By Norman E. Gary

There are some 300,000 beekeepers in the United States. No other activity can produce as much food and fun for the same investment or for the few square feet required for a hive.

Honey bees thrive in most areas and climates. In this country there are about six million hives, nearly a hive per square mile. One per cent of beekeepers are commercial, the remainder hobby or sideline beekeepers.

Recreational beekeeping is growing rapidly in popularity. Bee clubs are active in many communities. In addition there are regional organizations, such as the Eastern Apicultural Society and the Western Apicultural Society. Honey bees are the most abundant and common beneficial insect. Few pets are as easy to care for as bees. They collect their own food— and enough extra to harvest.

Learn all you can about honey bees before starting a hive (also called a colony). Many excellent bee books, periodicals and pamphlets are available. County extension offices and state land grant universities often distribute publications on bee-keeping. Beekeeping classes are offered in many areas. Visit beekeepers in your area and ask for their advice and a demonstration of hive manipulation.

Prime Pollinators

Major benefits of beekeeping are enjoyed by everyone who has fruit trees, ornamentals, and gardens or other plants that need insect pollination. Approximately a third of all the food we eat is dependent on honey bee pollination. Honey bees provide about 80% of insect pollination in the United States.

Honey bees are a vital link in the ecology of man-disturbed areas where most native insect pollinators cannot survive in

Norman E. Gary is Professor of Entomology, University of California, Davis.

adequate numbers. For example, bee-pollinated berries and fruits attract birds and provide feed for other wildlife.

In many areas beekeepers rent their colonies for pollination of orchard and field crops. Income from this source can be significant if the hives are moved to several crops. Rental per crop is $10 to $30 per hive. However, unless you are located near such areas the expense of moving colonies can offset the income.

Generally, colonies simply are moved to the crop at the beginning of bloom and moved out when bloom is over. Two hives per acre is a rule of thumb.

CAUTION—be sure pesticides injurious to bees are not used during bloom, and have a simple contract to protect all parties.

Honey Sales

In most seasons a skilled beekeeper can produce up to 100 pounds of honey per hive. Honey may be sold at roadside

Hives among almond trees, which could not produce economically without honey bee pollination.

Norman E. Gary

245

stands and natural food stores, marketed by mail order and local advertising, or sold to commercial honey processors for domestic and foreign consumption.

Baked goods containing honey, such as cookies and cakes, stay fresh and moist for a long time. Many recipe books for honey cooking are available. When honey is substituted for sugar it is usually necessary to reduce the amount of liquid in the recipe by 1/4 cup for each cup of honey. Many fine beverages, such as mead (honey wine), can be made.

Honey is a quick energy food and can replace candy in children's diets. It may be used as a preservative for canned and frozen fruit.

One great advantage in hobby beekeeping is that honey can be harvested fresh from the hive and consumed immediately without further processing. Honey is at its best when eaten in the comb along with the natural beeswax. Yet it can be extracted easily from the comb in liquid form when desired.

Periodic hive inspections are necessary to insure good health for the colony, abundant food, and a prolific queen.

Honey bees are a living lesson in biology for children, and often chosen as projects by boy and girl scouts or 4-H Clubs.

Aside from standard outdoor hives, bees may be kept indoors in a glass-walled observation hive to provide viewing of activities such as egg laying by the queen and communication dances by foragers. And there is no fear of stings by beginners because the hive is never opened indoors.

Properly constructed observation hives are easily detachable and portable for use as living visual aids when giving talks at schools or clubs.

Beekeeping has its social aspects too. Hobby beekeepers by the thousands are forming and joining local bee clubs to share beekeeping experiences. Many vacationing beekeepers attend national or regional meetings where bee scientists and other professionals present the latest developments in bee culture and equipment.

Getting Started

Before starting, determine if bees are permitted in your area. Most localities have reasonable restrictions on the number of hives permitted in residential areas. Unduly restrictive ordinances usually can be changed, especially when the benefits of bee pollination are documented properly.

Beginning beekeepers usually enjoy starting with new hive equipment, sold in kit form by bee supply companies that advertise in bee publications or the telephone directory. Experienced beekeepers may be willing to help you assemble and stock your first hive.

Frames of beeswax comb foundation are the "backbone" of the colony. Assemble them with great care. Paint new equipment and allow it to dry thoroughly before use.

Stock the hive in spring with commercially available packaged bees, containing three pounds of worker bees (approximately 12,000) and a queen. Or if you wish to stock your hive with a swarm, call your local fire and police departments and offer your swarm-catching services during the spring.

Starting a new colony costs less than $100 at this writing.

Established colonies can be purchased from a beekeeper. Be sure the colonies are healthy, well fed, and in standard dimension equipment with movable frames. Keeping bees in miscellaneous containers is impractical.

New colonies should be fed sugar syrup (equal volumes of sugar and water) to supplement natural nectar supplies until

Equipment list

Item	Approximate cost
Protective clothing	
Bee veil	$ 7
White coveralls	20
Work boots (optional)	30
Bee gloves (optional)	5
Tools for hive work	
Hive tool	3
Bee smoker	8
Bee brush (optional)	2
Equipment for one hive	
Hive stand (optional)	4
Standard bottom board	6
Brood chambers (standard hive bodies with frames)	20 (each)
Honey supers (shallow with frames)	12 (each)
Hive cover (need inner cover with telescope style)	12
Queen excluder	4
Wax comb foundation	
Brood chambers	6 (each)
Honey supers	4 (each)
Sugar syrup feeder	2
Equipment for extracting honey	
Electric uncapping knife	26
Honey extractor	100
Honey storage tank (optional)	65

all combs are constructed. Feeding also helps bees store enough food to ensure survival during the first winter. Don't expect to harvest honey the first year if the bees are required to construct new combs from foundation.

The beginning beekeeper ideally should start with only one or two hives the first year. Backyards usually accommodate up to five hives if they are placed strategically near hedges, fences, or buildings to direct bee flight upward. Pets and children should be excluded from the immediate hive area.

Nectar and pollen from flowers of certain plant species supply the total food of bees. They forage within at least three miles in every direction, an area of around 20,000 acres. Nectar and pollen availability is seasonal.

Under most circumstances colonies produce far more honey than the 150 pounds needed annually for survival. The surplus is harvested as liquid honey or can be eaten in the natural honeycomb.

Biology of Bees

Honey bees are highly organized social insects. Each colony contains up to 50,000 worker bees (non-reproductive females), several thousand drones (males) during the spring and summer, and one queen.

In nature, bees nest in cavities, such as hollow trees, crevices in rock cliffs, or the space between walls of buildings. They cluster inside the nest on a series of combs arranged side by side with just enough space between to permit free movement. Combs are made of pure beeswax that is secreted by worker bees and fashioned into precise hexagon-shaped cells.

During brood rearing, worker bees maintain a remarkably constant nest temperature (94° F) even when the hive is covered with snow or exposed to scorching desert temperatures up to 120°. A combination of fanning and distributing tiny water droplets provide efficient evaporative cooling during hot weather. In cold weather warming is achieved by muscular activity and clustering together.

In this controlled, dark environment, the queen lays around 1,500 eggs daily, one per cell. Eggs hatch after three days of incubation, producing tiny larvae that are fed by nurse worker bees for 5 to 6 days. Cells are then capped, and adult worker bees emerge from the cells 21 days after the eggs are laid.

Immediately the cells are cleaned by housecleaner bees and the queen soon lays another egg, starting another brood cycle. Many cycles of brood in all ages and stages are developing from late winter to early fall.

During the first 3 weeks of adult life, some worker bees "guard" the hive entrance. They are alert to any disturbance that signals invasion by any animal that may be attracted to the rich food supply inside the hive.

Stinging behavior is the most misunderstood activity of bees. Guard bees normally are defensive and not aggressive as commonly believed. Stinging is most likely within a few feet from the hive entrance—after provocation.

Bees foraging on flowers away from the hive are non-defensive. Stings received away from the hive are very rare and usually provoked by such acts as grabbing, swatting, stepping or sitting on, or colliding with a flying bee. These are simple accidents, an element of all human activities.

A single honey bee can sting only once and it always dies within a few hours. Stinging away from the hive area contributes nothing to colony defense and makes no sense from a survival standpoint. Honey bee colonies, when properly managed and handled by the beekeeper, are quite safe.

Honey bees often are mistakenly blamed for stings inflicted by many other kinds of insects, especially yellow jackets, wasps, and hornets.

Worker bees surround and provide special care for the larger queen bee (center), the most vital bee in the colony.

Norman E. Gary

If You Are Stung

Less than 1 percent of people are hypersensitive to bee stings and should not keep bees as a hobby. Fortunately, there is a recent medical advance for treating the problem—desensitization by an allergist using pure honey bee venom antigen.

If you are stung by a honey bee, immediately scrape the easily visible sting away with a fingernail before it has time to deliver the venom—thus minimizing the effects greatly. Then wash with water to remove the alarm odor, a chemical signal which directs guard bees to the "enemy."

A typical reaction to a bee sting is minor discomfort near the sting site. An ice cube pressed over the area eases the pain.

Beekeepers enjoy their hobby without special concern for stings. The stinging tendency of guard bees is controlled easily by puffing smoke into the hive before and during examination. Protective clothing is also worn. Slow movements near the hive are essential. Fast movements alert guard bees.

Experienced beekeepers examine colonies during maximum flight when many bees are out foraging and the incoming nectar places the bees in a "good mood." Never examine colonies at night, early mornings, late afternoons, during rainy weather, or other times when they are more defensive.

Managing Hives

In some areas hives can be kept in one location throughout the year. However, in many areas colonies are relocated seasonally near nectar-producing plants to maximize honey production and to "choose" special nectar and pollen sources. For example, to produce citrus honey, hives must be placed near citrus groves.

Consult local experienced beekeepers to determine the kinds and locations of plants most beneficial to bees. In most areas the small-scale beekeeper doesn't have to move hives elsewhere because enough forage is usually provided by a great variety of fruit trees, ornamentals, gardens, lawns, cultivated crops, and wild flowers within foraging range.

Hives are moved best at dawn before bee flight starts. Moving hives to foraging areas is the only practical way for beekeepers to influence the selection of nectar and pollen plants. Hives must be moved at least 5 miles to prevent foraging bees from returning to the original hive location.

Efficient management is based on a thorough knowledge of bee biology and seasonal changes that affect growth of colonies. A summary of management is shown in the illustration.

By fall (I) each colony must have enough stored pollen

and honey to overwinter and to support population growth during spring months, or until nectar and pollen plants bloom again. Other overwintering requirements are an adequate bee population, a young queen (less than a year old), and a protected location.

During winter in cold climates, bees cluster quietly in the broodless nest (II). Don't make inspections during this time.

In late winter, brood rearing begins (III) and the pace quickens in spring, in response to early spring nectar and pollen, especially from fruit trees (IV). At this time the brood cluster tends to concentrate in the upper brood chamber, owing to the natural tendency for upward expansion of the brood nest.

When the upper chamber is crowded with brood (IV & VI) reverse position of the brood chambers (V & VII). This permits continued upward brood nest expansion, increases brood rearing, and diminishes excessive crowding of bees, a condition that favors unwanted swarming.

When both brood chambers become occupied with brood, add a honey chamber (super) on top, above a queen excluder, to accommodate the enlarging population and to receive any nectar and honey that may be crowding the brood chamber space (IX). The queen excluder restricts brood rearing to the lower two brood chambers.

Swarm prevention is a major problem for most hobbyists. Swarming is migration of approximately half the bees to a new nest as a means of colony reproduction. Along the way bees

Typical honey bee colony mangement during seasonal changes in temperate zone areas where brood rearing ceases during late fall and midwinter. One and two are brood chambers. Three and four are honey chambers (supers) for harvested honey. H = stored honey; P = stored pollen; C = clustered bees in broodless nest; B = active brood rearing area surrounded by clustered bees; HS = empty honey combs to receive honey; QX = queen excluder. All views are cross-sectional, as seen from front of hive.

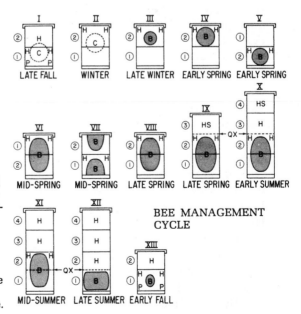

BEE MANAGEMENT CYCLE

may cluster for a few hours or days, sometimes in very inconvenient places.

Swarms cause two problems. First, they are a nuisance to people who may become frightened at the sight of thousands of flying bees, even though swarms definitely are not a sting hazard. Secondly, the loss of colony population greatly reduces honey production.

Swarm prevention is possible by several methods. A simple procedure is to destroy all queen cells at weekly intervals during the few critical weeks of "swarm season," usually in spring or early summer.

Throughout spring and summer, add additional honey supers as needed, usually when about half the cells in the recently added super contain nectar and/or capped honey (X). During peak production, bees may store several pounds of honey daily per colony.

Around midsummer, after the swarming season has passed and when 1 to 2 months of honey production remain, move the queen to the lower brood chamber and confine her there by a queen excluder (XI & XII). As brood emerges in the upper brood chamber (completed in 3 weeks), honey (50 to 60 lbs.) will be stored in this chamber as food for overwintering (XIII).

This simple management program restricts broodrearing to brood combs, and honey storage to honey combs. Honey harvested from brood combs is lower in quality. Also, if operations are timed carefully, the colony is organized for overwintering without sacrificing significant brood or honey production. Harvestable honey is easily accessible on top and can be removed by the super, rather than by sorting individual frames.

Harvesting Honey

Timing is critical in harvesting honey. Harvest honey (a) when necessary to obtain more empty combs for storing incoming nectar, (b) after a major honey flow (production peak) is completed, if you wish to separate that floral type from subsequent honeys, (c) when bees are foraging actively for nectar (this avoids bee robbing or "stealing" honey from other hives because nectar is not available), and (d) before the onset of cool fall weather.

You can remove bees from the honey supers by either smoking and brushing them off individual combs with a bee brush, or inserting a partition (inner cover) with a downward, one-way bee exit (Porter Bee Escape) which transfers bees from the honey super into the brood chamber in 24 to 48 hours.

These unprotected honey supers must have no openings or robber bees will "steal" all the honey. Harvest only surplus honey—leave a generous supply as bee food.

Newly harvested honey supers should be placed in a warm room (85° to 100° F) and extracted within several days. Delays are risky. Honey may granulate in the comb, absorb excessive moisture in humid climates, or the combs may become infested with destructive wax moths.

Basic equipment for extracting honey is an uncapping knife to cut cappings from the comb cells, a small honey extractor to remove honey centrifugally from the combs, and a small honey storage tank. Honey does not require further processing. Empty combs are stored or recycled to hives to be filled again.

Honey varies greatly in color, flavor, thickness, and tendency to granulate, according to the predominant floral nectar sources and conditions of storage. Most kinds of honey eventually will granulate, some within a very short time after extrac-

Beekeeper, wearing protective clothing and a bee veil over the head, examines brood comb from hive. Hive entrance is at hive front (lower left at bottom) where a small "porch" is provided for alighting and departing bees. Hive tool in right hand is used for prying combs apart (bees "glue" them together). Bee smoker, lower left, generates smoke to calm bees.

Kenneth Lorenzen

tion, especially if stored at the ideal temperature for granulation (50° to 60° F).

Some granulated honeys get rock hard but when warmed to room temperature become a smooth spread of butter consistency, sometimes called "creamed" honey. Others form a coarse, "sandy" consistency that is not as pleasing to eat. The granulation process can be controlled easily.

Granulated honey can be liquefied by immersion in hot water. Minimize the time of heating to preserve quality. All honeys become darker in color and change flavor during prolonged heating and/or storage at room temperature. Long term storage (years) is best in the freezer.

Place freshly extracted honey for home use in small containers before it granulates. (However, not all kinds of honey will granulate.) Storage of honey in the refrigerator is unnecessary and usually quite inconvenient because it becomes too thick for easy pouring.

Pesticides are a major risk to bees, especially in farming areas. Avoid locating hives near frequently treated areas such as orchards, except during fruit bloom. Never store or use pesticides of any kind in the same building occupied by stored comb or wax comb foundation. Be sure to obtain current information on the protection of bees from pesticides by contacting your county agricultural agent or agricultural college.

Learn symptoms of the most serious bee diseases, especially the bacterial disease, American Foulbrood, which kills brood. Stored combs will be destroyed by wax moth infestation if not protected. Other dangers are floods, vandalism, theft, animals (especially bears and skunks), and fire. Choose hive locations carefully.

Africanized Bees

A hybrid honey bee, produced by the interbreeding of African bee stock imported into Brazil and the European honey bee, common to the U.S., became established in Brazil around 1957. This hybrid bee is more sensitive to disturbance near the hive and defends a greater area around the hive. This caused minor problems in South America until beekeepers learned to alter beekeeping practices, primarily by locating apiaries at greater distances from people and animals.

Stinging incidents have been publicized in the press and dramatized in books and films, creating an impression that Africanized bees are not beneficial and are out of control. Contrary to popular opinion, the bees are being managed productively in Brazil at this writing.

Africanized bees are migrating toward the U. S. at about 200 miles per year, and currently have reached Venezuela. However, the Africanized hybrid is not a temperate zone bee. Instead, it is adapted to tropical conditions.

At most, only the southern third of the U.S. ultimately may receive this bee during the next 15 to 30 years. By that time selective breeding combined with new information from research should make it feasible to use this new hybrid for beekeeping in the U.S., especially in the warmer climates.

In any event, there is no cause for alarm concerning the so-called "killer bees." They are not present in the U.S. at this time and indeed may never reach this country.

The greatest effect of the "killer bees" (a name coined by the press) is to alarm some farmers and homeowners who may refuse permission for beekeeping in rural and suburban environments. This is unfortunate because all of us have to pay when pollination is not efficient. In time these fears may subside as more information reaches those people whose fear of bees usually is based on a misunderstanding of normal bee activities and behavior.

Further Reading:

American Bee Journal, Dadant & Sons, Hamilton, Ill. 62341. Monthly. $6.50 per year.

Gleanings in Bee Culture, A. I. Root Co., Medina, Ohio 44256. Monthly journal. $7 per year.

Hive and the Honey Bee, The, Dadant & Sons, Hamilton, Ill. 62341. 1978. $9.95.

Speedy Bee, The, Rt. 1, Box G-27, Jesup, Ga. 31545. Monthly newspaper. $4 per year.

Woodlots Offer Wide Range of Benefits to the Owner

By Burl S. Ashley

Property owners who have a portion of their holdings in trees are indeed richly endowed. Possessing a few scattered trees can be beneficial, but owning a bona fide woodlot can result in numerous advantages.

A woodlot, or small forest, can be defined as a dense growth of trees on an acre or more, usually under 100 acres. A few scattered trees cannot be construed as a forest. There must be many trees, each depending upon the others and all forming a community. This close association is important to maintenance of productivity through planned management.

Benefits and pleasures of owning such a tree community are many. However, they will be greatly magnified if you practice good forestry. Not only will planned care of a woodlot insure monetary rewards, but it can be an enjoyable experience.

Some owners with large holdings derive their entire livelihood from the sale of forest products. Others sell trees occasionally to supplement their regular source of income.

It takes an average of 30 to 50 years for a tree to yield a merchantable log. Value of these trees varies greatly because of the difference in demand for them and the quality of the wood. Black walnut is presently the most valuable with the possibility of a 40-year-tree selling for as much as $500. In contrast, a pine tree of the same age might bring $25.

Many owners cherish their forests for the multitude of intangible riches they yield, and accept the wood harvests as an extra benefit. Others consider the wood yield as the primary objective.

Some landowners are blessed with an already established forest when they purchase their property. Some begin with bare land and "build" a forest by planting seedlings.

Burl S. Ashley is a Resource Management Forester
with State and Private Forestry, U.S. Forest Service,
Morgantown, W. Va.

256

Establishing a plantation usually requires the planting of 400 to 1,000 seedlings on each acre. In 10 years, these small trees should reach a height of 10 to 20 feet, depending on the species. Their crowns should touch and the entire plantation begin to appear as an established forest community.

Forest Planning

No matter what stage of development a forest is in, whether a new plantation or an old, mature community of trees, you need a plan for its future management.

This plan of action should consist of activities which will increase productivity of the area and produce a forest that fulfills your desires.

You need not be a forester to prepare this plan, but advice from one would be extremely helpful. Foresters are available to prepare a complete plan—some who are government-employed are available without charge.

Many woodlots are producing less than half their potential capabilities because of the lack of plans and appropriate resulting activities.

A management plan can be very simple. It should list your objectives and a timetable for carrying out activities needed to achieve these objectives. Your plan also should include a detailed map of the area and an inventory of the trees by species and size.

Because of the constant changes occurring within a forest, improvement in forestry techniques, flexibility of markets, and other varying factors, the plan cannot be absolutely rigid. It should be periodically amended and updated as necessary.

Forest Renewal

Just as the human population must be perpetuated, so must the tree population in a forest. Trees become established, grow to maturity and are harvested or eventually die. You should provide for continuous replacement of these trees.

The theory of planting a tree for every one that is removed from a forest doesn't always apply; in fact, other means normally are used.

Where the forest is to be started by planting, or harvested trees are to be replaced by this means, the process is accomplished with special hand tools or machines. Seedlings for planting are available from private nurseries and State Forestry organizations at nominal costs. One person can plant about 500 trees in an 8-hour day, but with a planting machine the daily output can be upped to nearly 10,000 trees.

Most forests can be restocked naturally. This system requires careful planning. The species of trees involved dictates the methods that can be used since each has different characteristics and requirements.

Many species need an abundance of light to develop, and won't survive beneath larger trees. Others prosper in the understory. Because of such critical differences, you must create conditions that will stimulate the well-being of each species.

Forest Improvement

Once a woodlot is established, it needs attention and care. If it turns into a neglected area, its productivity and usefulness will be greatly reduced, and sometimes it becomes a liability.

All woodlots don't require the same care. The needs depend upon the species of trees, markets, past treatment, age of the forest, and your objectives.

Improvement activities often needed are:

Thinning. As trees increase in size, their requirements for space, moisture, nutrients, and light increase. If these requirements are not met, individual tree growth slows and vigor is lost. Under these conditions, it is impossible to grow high-quality trees and realize the maximum profit.

For example, a pine forest may begin with 1,000 trees per acre and after several thinnings at 10-year intervals, the final harvest may yield only 100 large, high-quality trees. If no thinnings are made in such a forest, many trees will die and the final harvest will yield several hundred small, low-quality trees.

Trees removed in thinnings can usually be sold at a profit. However, when a thinning is made in a stand of very small trees, they may not be salable. Nevertheless, you will profit by having higher-quality trees to market in subsequent sales.

Cull Tree Removal. All trees don't develop into beautiful specimens. Some grow crooked, others decay and become hollow, still others suffer root or top damage. With such defects, trees lose their ability to yield merchantable products. They are referred to as culls and most should be removed from the forest community to permit better trees to develop. These culls can be compared to weeds in a garden.

If a hollow cull tree is being used by wildlife, such as squirrels, and you have wildlife production as one of your objectives, do not remove this otherwise undesirable tree.

Sometimes a flowering species of trees, such as dogwood, is considered valuable for esthetic reasons, but otherwise it is

258

a cull. It should, of course, be left to grow, even though it will never yield a commercial product.

Pruning. Trees that are knotty are regarded as low in quality and don't sell for high prices. You can improve the quality of your trees by pruning off limbs from the main stem at an early age. Never remove the limbs from more than half the total height of the tree and only from chosen "crop trees"— trees which will form the final harvest.

This practice is usually confined to some evergreen species and high quality hardwoods. At today's labor rates, it costs about 30¢ to prune a 5-inch diameter pine tree to a height of 10 feet.

When trees are grown close together, the mutual shading prevents development of many side branches, thus reducing the need for pruning. Such dense stands, however, prevent fast diameter growth of the main stem.

When growing a very valuable species such as black walnut, where you want maximum diameter growth, don't allow the stand of trees to become dense enough to cause natural pruning. Do the pruning manually.

Although many types of tools are used for removing the side branches, a pruning saw is best and can be purchased for $15 to $30.

Selecting for Harvest. One of the most crucial periods in the life of a woodlot is when trees are selected for harvest. The improper choice of trees can result in a substantial reduction in future productivity. Forests have been completely destroyed by harvesting the wrong trees.

A pruned pine plantation.

One of the most-used reasons for harvesting individual trees is maturity; however, this is only one of the reasons that may be used. Consideration for the establishment of new trees is always important. It is best to operate under the guidance of a professional forester when selecting trees for harvest.

Protection

Forests can be decimated by a number of enemies, and as the owner you must be on guard at all times. One of these enemies can quickly wipe out results from many years of good care. Protective procedures should be part of your management plan so quick action can be initiated.

The greatest and most feared enemy of the forest is fire. A single catastrophe can kill trees, scar others, damage the soil, and even bring about the invasion of insects and diseases. Many years, sometimes as many as 100, are often required to recuperate from one fire.

Prevention is the key to protection from this nemesis. Many forest fires start from carelessness with trash burning, campfires, and smoking.

Become familiar with the local forest fire control organization. Knowing how it operates, the services it offers, and how to obtain help in case of fire can be important.

Small woodlots will receive some protection from cleared firebreaks that are at least eight feet wide. Also, having fire-fighting tools available for easy access is a good idea. With knowledge of how to use these tools you can control small fires before they cause damage.

Livestock trampling has damaged a valuable walnut tree.

Although in some portions of the country controlled grazing by domestic livestock is permissible, intensive grazing is a very harmful practice in most areas. And, even though it doesn't destroy a forest as quick as fire, grazing can accomplish the same disastrous results. It is sometimes referred to as "creeping devastation".

In general, woodlots are a poor place for livestock. The forage is sparse and poor quality. Livestock not only trample and browse young trees and injure the larger ones, but compact the soil. This compaction reduces tree growth and increases water runoff during heavy rains, thus promoting soil erosion. Remember—cows make poor foresters!

Trees are subject to attacks from many kinds of insects. A poorly-managed forest of weak trees is usually more susceptible to insect damage than a well-managed one composed of vigorous trees. Insects attack by defoliating, boring into the twigs and roots, and by girdling the stems. Some insects also attack the fruit and seeds, which reduces reproductive capacity of the trees.

Close observation and the use of early control measures is the best way to prevent excessive insect damage.

Many diseases attack trees. Some cause only minor damage, while others have been so damaging that species of trees have been almost eliminated from American forests. The chestnut blight and the Dutch elm disease are examples of very devastating diseases. Early detection and prompt control are important.

Harvesting, Marketing

After protecting, managing, and caring for the forest for many years, you will reach the point when wood products should be harvested. It's time to reap a profit from the investment of effort and money. This is a period of caution, because without good marketing procedures, full monetary returns may not be realized.

Marketing assistance is available from private or government-employed foresters. They can greatly increase the profit from timber sales in most instances. If you aren't knowledgeable about marketing procedures, secure the services of a forester.

The first step in marketing is to select the trees to be cut and mark them, preferably with paint. Tally each tree by species, size, product, and volume.

Determine current prices for similar quality material. Most states have periodic price reports available to the public free of charge. Properly informed, you are ready to bargain with buyers. A common practice is to obtain bids from several buyers before selling.

After deciding on a buyer, a contract should be prepared which provides penalties for violating certain provisions. Several items of concern are: amount, method and time of payment, logging damages, cutting unmarked trees, and time allowed for removal.

If you have the necessary equipment and the basic knowledge, you can harvest your own trees. So, instead of selling them as they stand, they can be logged and sold at the roadside or mill. This results in higher returns.

Since you will often need lumber, your harvestable trees can be cut, then sawed into usable lumber at a local mill. Drying and planing services may also be available if you desire finished lumber.

Never hurry when marketing your forest products. Many years were required to grow them, so a few extra days will be of little consequence when trying to realize the maximum return.

Side Benefits

Although the principal product from the forest has usually been wood, it yields other products and benefits which are very important. In fact, wood production is considered to be secondary by many owners. Fortunately, our forests are quite adaptable to adjustments and may be manipulated so more than one objective can be fulfilled.

Many species of wildlife use the forest. Among the most common are deer, squirrels, bears, wild turkeys, grouse, and many species of songbirds. Timber and wildlife grow quite well together and a high production of both can be realized from the same area.

Methods of timber management must be adjusted according to the species of wildlife desired. For example, deer and squirrels require different management techniques, but you can create desirable conditions for each in separate portions of the woodlot.

Occasionally, an owner will sell hunting privileges to his woodlot, thereby realizing an annual income. Many, though, prefer to manage for game so they will have their own private hunting preserves. Others prefer to establish and perpetuate an abundance of wildlife just to observe and not hunt.

Wildlife may become too abundant and cause damage to the forest. An overpopulation of deer, for example, can destroy many young trees. Familiarize yourself with good wildlife management principles so you can assist game agencies in preventing such occurrences. ˙

Some trees produce edible nuts as well as wood. Black walnut, pecan and hickory trees are examples. Although many nuts are produced in orchards, as a forest owner you can take advantage of the versatility of your nut trees. Some forest-grown walnut trees produce more than 100 pounds of nuts in a single year.

In the Northeast the sugar maple tree abounds. The landowner who has a forest containing this species has the option of operating a maple syrup production business.

Trees are tapped in early spring and the sap is gathered. This sap, with a low concentration of sugars, is then "boiled down" until the remaining product has the high sugar content of syrup.

About 40 gallons of sap are required to produce one gallon of syrup. The content of sugar in the sap varies from tree to tree; therefore, sap from "sweet" trees requires less boiling and fewer gallons of sap are needed per gallon of syrup.

Each size of tree should have a definite number of tap holes. For example, a 15-inch diameter tree should be tapped with two holes. Each tap hole will usually yield enough sap to produce one quart of syrup. Recent price reports indicate that the producer is receiving $5 per quart for good syrup.

An often overlooked benefit from well-managed forestland is watershed protection. Woodlots not only yield clean water for personal and wildlife use, but also prevent soil loss and flooding by reducing water runoff. In certain critical areas, the forest's principal purpose is watershed protection. In all instances, it is of primary importance.

A forest provides a place for numerous recreational activities such as camping, hiking, picnicking, and hunting. The building of hiking trails, camping sites, and picnicking areas within even a small woodlot can result in many hours of pleasure for you the owner and others. When utilizing a forest for recreation, take care to protect the trees from damage.

Many owners feel that the benefits from observing the beauty of a forest exceed all others. Although a well-managed woodlot has eye-appeal throughout the year, the person who can view the colorful eastern hardwoods during October or the dogwood and redbud flowers in spring is blessed with an example of the beauties of forestland.

Although many esthetic benefits are available without extra management efforts, they can be increased by special practices. Conifers may be planted in hardwood stands to give green color in winter, flowering shrubs and trees may be released from competition of large trees, and vistas may be opened by cutting groups of trees.

A woodlot can produce wood for fireplaces and wood-burning stoves. With energy conservation a necessity, this source of heat can be of great value to the owner. If your supply exceeds your needs, ready market exists for the surplus. Consider culls and otherwise undesirable trees for use as firewood, with the better trees left for other purposes.

Many miscellaneous benefits, pleasures, and products are available to woodlot owners. Some common ones are mushroom hunting, wildflower observing and picking, birdwatching, herb gathering, Christmas trees, and many others. Of course, the same things aren't available in every woodlot, but vary by species composition, section of the country, and differences in local conditions.

Technical Help

Professional technical management assistance is available to anyone who owns a woodlot or desires to establish one. The Federal Cooperative Forest Management Act of 1950 provides services to private landowners. A governmental organization, usually the State Division of Forestry or Cooperative Extension Service, administers these services in each state.

The nearest office can be located by contacting either the State Forester, who is usually located at the State Capitol, or any U. S. Department of Agriculture office. Services provided are free and include guidance from establishing a woodlot to harvesting forest products.

Private consulting foresters are also available to give a full spectrum of forest management help. They often offer more services than the government-employed forester, and, of course, charge a fee.

A wealth of publications are available to assist you with your desire for more knowledge. Since there is quite a variance among states relating to forest types and growing conditions, you should secure reference material for a specific area. The Cooperative Extension Forester or the State Forester can supply applicable publications.

Christmas Trees Pay Off, If You Can Wait 5 Years

By Maxwell L. McCormack, Jr.

Christmas tree production can provide supplementary income from land which otherwise might be idle. A relatively high financial return can be achieved where the emphasis is on quality trees in a quantity which can be handled by the producer.

Careful management and a commitment of time throughout the production period is necessary. During initiation of a production program, a delay of 5 to 10 years is likely before there is a return on investment.

Though Christmas tree production can provide work during off-season periods when a landowner has time available, there are also periods during growth of the trees when it is essential that certain cultural practices be carried out. Precision of timing can be critical in doing this work, and the trees must be observed on a regular basis. A grower should gradually develop production to a level capable of supporting a consistent marketing program.

Besides monetary gain, Christmas tree production provides other benefits. It is an excellent way to maintain, or improve, abandoned farm fields and to inhibit invasion by undesirable brush species.

The relatively low growth is a valuable way of maintaining open areas as part of a desirable land use pattern. Open areas contribute to scenic beauty and provide vantage points for scenic vistas. Such patterns contribute to recreational uses and also provide desirable habitat for many wildlife species.

In some cases Christmas tree production can be combined with growing other products, but it is usually advisable to designate areas for tree production only.

Generally, an acre of suitable land can produce 700 to 900

Maxwell L. McCormack, Jr. is Research Professor in the Cooperative Forestry Research Unit at the University of Maine, Orono.

marketable trees over an 8 to 10 year period. There is a tendency to try to produce too many trees per unit of land area. A spacing of 5 feet by 5 feet (1.5m x 1.5m) is the minimum, with wider space between the rows recommended.

Though production rates will vary according to the type of tree grown and the quality of the land, over 1,000 trees per acre within a growing period should be considered excessive and it is unlikely that a major harvest would be possible in less than 6 years.

Most efficient production and best returns on investments are achieved through developing uniform plantations of trees. Always emphasize quality. Rarely have there been sufficient high quality trees to satisfy market demands throughout the history of the Christmas tree industry.

Suitable Land

A readily accessible site is essential. Physical characteristics of the site are easiest to evaluate. More detailed analysis of soils and other environmental factors should then follow.

A gentle slope, free of frost pockets, is best. Drainage of air and excessive moisture is important for good growing conditions. Direct exposure to prevailing winds, especially during winter, must be avoided.

North-facing slopes are definitely preferred over those exposed to the south. Moisture conditions are usually better on northerly slopes. Undesirable droughty conditions are more prevalent on south slopes where trees tend to initiate growth earlier in spring, making them more susceptible to frost damage. Though frost injury rarely kills established trees, it deforms the shapes of Christmas trees.

Evaluate soil conditions by submitting samples for testing to an appropriate laboratory in the area. Then use the relative levels of important nutrients as guides for fertilizer applications. Much can be learned from a history of the land and by observing plant cover presently growing on the site.

Though you must remove competing vegetation in establishing a Christmas tree stand, its healthy condition usually indicates that trees will grow well.

Avoid areas with soil conditions which do not support good natural plant growth. Identifying the plants present on a prospective site will indicate characteristics of drainage, nutrient levels, and other aspects of soil quality.

There should be sufficient soil depth to allow good development of tree roots. Moisture relationships are the most important to consider in evaluating an area. Excessively drained, dry sites are undesirable and those with too much moisture should be avoided as well. Very light-textured, sandy soils with low levels of organic matter, as well as heavy clay soils, do not support good tree growth.

Also consider general weather conditions for the area. Temperature extremes, especially late spring and early fall frosts, are undesirable. Precipitation should occur at times which support good tree growth. Moisture input well-distributed through the growing season is important.

Good winter snow cover serves to protect trees and reduce winter movement of large animals. However, winter conditions must also be considered with respect to harvesting operations because of the product's seasonal nature.

Other considerations center around accessibility and security. Christmas tree production requires ready accessibility for efficient administration of cultural practices and harvesting. Yet the final product is susceptible to theft. Have adequate security and supervision to minimize such losses.

Trees to Grow

A wide variety of trees is available. Restrict your choice to species readily marketable in your region. The potential species, determined from market evaluation, should then be matched against the characteristics of each site available for production. Do this through a study of site requirements of the species being considered, and observe trees already growing in the area or nearby. If necessary, get advice to assure suitability.

Once you settle on realistic candidates, evaluate them with regard to production problems. Problems include common insect and disease pests.

Additional considerations are special requirements for producing quality trees of a given species, and any cultural practices which involve scheduling during a restricted season. Practices such as shearing pines only when the new growth is soft can often present problems if your personal schedule does not allow the time needed.

Species selections can usually be placed in three groups. One group includes cedars and cypresses, which have scaly and awl-shaped leaves and are produced in areas in the South. The remaining two groups make up most of the Christmas tree species marketed and can be classified according to needle length.

Long-needled trees are the pines and include such common Christmas tree species as Scotch pine and white pine. Pines only produce branches in annual whorls, which result in sections of bare stem along each length of annual growth. Consequently, they require a more rigid shearing schedule so as to develop desirable shape and foliage density.

Short-needled species include firs, spruces, and Douglas-fir. Several are species available across our northern and central regions.

The short-needled species not only have branches in annual whorls but produce shorter branches along the length of each annual stem growth between the whorls. Because of this branching habit, shearing requirements are not as demanding as with high quality pine trees.

Genetics is important in choosing the kinds of trees to grow. The most popular species such as Scotch pine, Douglas-fir, and balsam fir exhibit a wide range of genetic variation.

For efficient production of high quality trees it is essential to have a suitable genetic strain, variety, or recognized seed source of the species to be grown. Use a genetic source specifically suited for the production site, when available.

Establishing the Stand

In some cases, improving the trees existing in natural stands on the property can provide early yields of marketable trees. Resulting income can support the establishment of plantations specifically for Christmas trees. It also allows for development of markets while the plantation trees are growing.

Intensive cultural practices in natural stands can help sustain production of quality trees over long periods of time. Provide natural seed supplies to maintain regeneration of desired species. Sometimes you need to interplant to keep the area fully stocked.

Some species are suited to a practice known as stump culturing. This involves keeping a stump alive by retaining some live branches after a tree is harvested. A bud, or turned-up branch, is then cultured into a second tree on the same stump. This is usually a slower, inefficient procedure; use it only where there is no better choice.

Successful stand establishment involves achieving the highest possible percentage of trees living and developing well in their first growing seasons. This means assuring good establishment of proper stock in quantities which you can handle within the time available.

Develop an annual planting schedule aimed at sustaining an appropriate, consistent level of production.

Prepare the site before doing any planting. Though areas

should not be exposed to erosion and other forms of deterioration, suppression or elimination of vegetation which will compete with the trees is most important.

You can mow or cultivate, but some form of herbicide treatment is the most efficient and effective way. Preparation treatments will depend on the nature of the site as well as the vegetation present. Woody brush species require different treatment than vegetative cover composed of grasses and herbaceous weeds.

Safe, effective herbicides which favor Christmas tree species are usually applied as sprays. In some cases a cover crop such as rye will help hold the site in manageable condition.

Evaluate nutritional levels in the soil prior to planting as a guide to determine the need for nutrient supplements. Apply fertilizers to correct any nutrient deficiencies before planting trees.

Once the site is prepared, plan the desired planting carefully with an adequate access road system before any trees are put in. Actually mark out the planting areas and the roads beforehand. The road system is necessary for cultural operations and harvesting.

Planting Stock

Obtaining suitable planting stock can be a problem because of short supplies of good quality material of the most desirable species. Low-priced stock usually is not the bargain it appears to be. Good stock helps assure survival and early growth; it costs less in the long run. Evaluate stock quality carefully and inspect it before purchase when you can.

Seedlings are stock which has been grown only in the location where the seed originally germinated. Transplant stock is material which has been moved from the seedbed to an area of wider spacing to provide for a secondary period of development. Seedling stock of the pines is usually satisfactory for outplanting. Short-needled species perform much better as transplant stock.

Root systems and tops should be balanced so there is adequate root mass to support above-ground portions of the trees. A compact root system with many fine rootlets, rather than coarse heavy roots, is desirable. The tops should have good caliper (stem diameter) and bud development, since the buds will form the basis for the first growth in the field.

Planting stock should be packaged, shipped, and handled in a way to assure maintenance of good fresh condition. Long shipping distances and exposure must be avoided. Obtain only quantities of stock which you can handle readily in the planting operation.

You can get suitable planting stock from tree nurseries. Where only seedlings of short-needled species are available, it is advisable to prepare your own transplant beds. In this way you can minimize exposure of the trees and lift them at the exact time they are needed. Two growing seasons in a properly managed transplant bed can result in excellent root and top development.

Natural Seedlings

Another possibility is to lift natural seedlings of desired species from nearby woodland and transplant them in beds adjacent to future planting sites. Direct outplanting in the field of such seedlings is not recommended.

Use of natural seedlings allows you to develop planting stock, which is available when needed, at relatively low costs. In the absence of recognized desirable genetic sources, this approach assures having trees suitable for the local area.

Christmas tree growers can develop a home planting stock nursery. Seed collection programs, using superior trees in the area, are also possible. But management of a home nursery, though convenient and productive, is complicated and requires special knowledge and skill.

Stock should be outplanted while dormant and when a period of root growth can occur before the first winter. There are many acceptable methods of planting. Numerous suitable hand tools are available. Spade-type tools are common and post-hole augers effective. One person can plant 600 to 800 trees in a reasonable workday.

Where the terrain allows, and equipment is available, you can use tractor-drawn planting machines. In some mechanized procedures, herbicide and fertilizer applications can be combined with planting.

The planting method used should insure good distribution of the roots since they tend to remain in their initial position during subsequent growth periods.

Depth of planting should approximate the position occupied by the trees in the nursery. Pack soil firmly around the roots to eliminate all air spaces. Avoid exposure throughout the planting of young trees.

Tending the Crop

Frequent inspections should be made during the early years of development. Maintain a continuous program of weed control and necessary fertilizer applications. Where healthy trees develop double tops, do early corrective trimming.

Be ever watchful for early stages of development of insect and disease pests. Examine foliage, twigs, and buds for symptoms. When abnormal conditions exist, collect specimens for examination by a qualified forester. Recommended pest control treatments change frequently; obtain the most up-to-date recommendations before carrying out treatments.

As the trees begin to develop their basic Christmas tree frame, you can improve shape and density by some type of shearing treatment. Shearing practices depend on personal experience and preferences. Tools such as hedge shears and special knives are commonly used. Best results are attained when the trees are a size that is easily reached and they are exhibiting good health and vigor.

Combinations of cultural practices produce best results. For example, best responses to shearing will occur on trees which are well-nourished, free of competing vegetation, and with ample room to develop. Old practices, such as scarring the stems, produce negative responses and suppress growth, and are undesirable.

When trees of inferior quality exist in the early stages of plantation development, remove them to avoid efforts wasted on individuals that will never be marketable. Do not hesitate to cull out poor quality, defective specimens.

Produce trees in the shortest possible time to get maximum return on your investment and to minimize risk and exposure to harmful agents.

Preparing to Harvest

Locate buyers before cutting any trees. Assistance often can be obtained from other tree growers in the area or from a marketing coordinator in the local Christmas tree growers association. To sell trees most effectively, have an exact inventory of trees available for sale.

A pre-harvest marketing inventory should include species, size classes, an indication of relative quality, and exact location of the trees. Sale is usually made on the basis of trees cut, packaged, and collected at a truck-loading location.

It is best for you to cut your own marked trees and to package them as soon as possible after cutting. Packaged trees are easier to handle, suffer less breakage, and remain in better condition.

Time of harvesting depends on weather, available labor, processing methods, number of trees to be cut, storage conditions, and marketing requirements. Keep trees cool, with some air circulation, protected from direct wind and sunlight.

Foliage of late-cut trees and those stored completely under cover often has lower moisture content than the foliage of trees cut earlier and stored in cool outside locations. Very low temperatures at harvesting time result in brittle branches and excessive breakage.

Assemble harvested trees according to size and grade at a location readily accessible to the vehicle they will be loaded on. Growers should be familiar with the type of transportation which will be used and plan for easy loading.

A marketing alternative is choose-and-cut sales, direct sale of single trees to individual customers who come to the farm. This can often be combined with a recreational experience for the buyer's family and will yield a much higher rate of return.

Such a system involves marketing a smaller quantity of trees from locations within a convenient travel distance from population centers. It also gives an opportunity to sell wreaths and other products such as maple syrup and honey.

Sizes of trees sold depend on market demand and can range from small table-top trees to household specimens as tall as 12 feet (3.6m) or more. In regions where winter conditions are not severe, there are some opportunities to sell trees with roots balled in burlap for future outplanting by consumers. Digging and root preparation require extra work, however.

As trees are harvested, strive to maintain uniformity in the tree production areas. Uniformity in size classes, species, and growth rates provides for more efficient production and eliminates injuring small trees during the harvest of larger ones.

Mark trees sold by any system before harvest. Keep records of all production practices, harvesting operations, and marketing. These records serve to guide future operations and for accounting purposes.

Keep in mind the personal preferences of Christmas tree buyers throughout the production process. The objective is to sustain a consistent supply of quality trees which satisfies consumer desires, while the land benefits from proper cultural practices.

Sources of additional information include: Your State Extension Forester. The Service Forester, State Forestry Agency, at your State Capital. State or Regional Christmas Tree Growers Association, which can be contacted through the National Christmas Tree Association, 611 E. Wells Street, Milwaukee, Wis. 53202.

Bringing Home the Bacon, by Raising Your Own Pigs

By Vernon Mayrose, James Foster, and Betty Drenkhahn

Raising a few pigs can be interesting, fun, and a learning experience. It may also provide some income on a small scale for families who live on a few acres.

Pigs are very intelligent and can even become pets. However, they grow fast. Most pigs grow from about 3 pounds at birth to market weight at 225 pounds in about 6 months. It takes some 10 months from the time the sow conceives until her pigs reach market weight.

Pigs can be sold alive at a livestock market or perhaps processed into pork for home use at a local livestock slaughtering facility. The most important products from hogs are hams, roasts, chops, bacon, and sausage.

Before acquiring pigs, get additional information from your county agricultural agent, and check on local regulations about keeping animals.

The best ways of getting started raising pigs are: (a) buy a bred sow or gilt and produce a litter of pigs, then sell the litter as weaned pigs or grow them to market weight; (b) buy weaned pigs (feeder pigs) and feed them to market weight.

Although there are several breeds to choose from, it's best for the small operator to select crossbred animals. Crossbred sows are usually better mothers than purebreds. They farrow more pigs and faster growing pigs. They are more vigorous, and there is less death loss. They may also be lower in initial cost than purebreds.

The quickest way to produce a litter is to buy a bred gilt, or an older sow that has produced one or more litters. Select sows or gilts that have 12 to 14 well-spaced teats without deformity. Try to obtain breeding females that are themselves

Vernon Mayrose and James Foster are Extension Swine Specialists, Department of Animal Sciences, Purdue University, West Lafayette, Ind. Betty Drenkhahn is an Information Specialist, Department of Agricultural Information, at Purdue.

from litters of eight or more pigs. They should have structurally sound feet and legs. Select pigs that walk free and easy.

The pregnancy or gestation period is about 114 days. Usually 8 to 12 pigs weighing about 2-1/2 to 3-1/2 pounds each are farrowed. A gilt, a young sow in her first pregnancy, usually has fewer baby pigs than older sows that have produced one or more litters.

On the average, producers lose about 25 per cent of live pigs farrowed before they are weaned. With certain diseases, losses may reach almost 100 percent. Mortality from weaning to market is usually less than 3 percent.

If you're raising only a few litters, it will probably not pay to buy a boar. Buy bred females or make arrangements with another swine producer to have females bred.

Another method of mating is by artificial insemination (A. I.). This is desirable for disease control but should be used only if good technical help is available, such as an A.I. technician or a producer who has had experience with A.I.

When buying feeder pigs, select pigs from a reliable source where pigs are raised under sanitary conditions. Pigs should be healthy, weaned, and started on feed. Buy pigs of uniform age and size that weigh between 35 and 60 pounds. Choose females or castrated males (barrows).

Usually, 8 to 12 pigs weighing 2-1/2 to 3-1/2 pounds each are farrowed per litter.

Shelter, Equipment

Pigs require shelter that is dry and free of drafts. The place where they will be kept should be completely ready before you bring them home. You will need an appropriate building, a shady place in summer, a good hog-tight fence, a self feeder or feed trough, and a waterer.

A simple house can be used for swine if it keeps out drafts, snow, and rain, provides shade in hot weather, and has a dry floor.

The hog shelter may be all or part of an existing older building or a small individual house. The simplest would be an A-frame that has a watertight roof which forms two sides of the building, and a rear wall. The front of such a house is usually open but can be fitted with a door. If your house is movable, face it away from the wind and don't place it where water puddles.

Keep the inside of the house dry, clean, and well-bedded with straw, peanut hulls, or wood shavings. Remove the bedding when it gets wet and dirty, and spread it on a field or pasture. To avoid complaints from neighbors about unpleasant odors, do not locate hog houses or haul manure within 500 feet of your neighbors.

In hot weather, hogs need protection from sun and heat. Hog houses should be made so they can be opened for good ventilation. Keep hogs out of airtight structures in hot weather. Trees give good shade; however, livestock should be fenced away from valuable trees.

Another method of providing shade is to place four posts in the ground, connect them at the top with a framework of poles, lumber, or wire fence, and cover with material such as straw that provides shade. The shading materials should be about 4 feet above ground.

Hog lots must be fenced hog-tight. For larger lots of several acres, woven-wire fencing (32 inches high), with a strand of barbed wire at the bottom of the fencing or just above the ground, works well. For smaller lots, temporary or permanent board fence (1 x 6 inch boards) or wire panels (about 35 inches high) will be easier to construct. Attach the boards or panels to steel or wood posts. Electrical fencing is satisfactory once pigs are trained to it.

Hogs can be fed in a trough, pan, or self feeder. Make the trough long enough so all hogs can eat at one time. If a self feeder is used, provide a feeder hole for each four to five hogs.

Hogs should have plenty of clean water at all times. A 35-pound pig drinks about a half gallon per day; a 225-pound hog, about 1 to 1-1/2 gallons; and a brood sow suckling a litter, about 5 gallons. You can use a heavy trough or pan that pigs cannot upset, a homemade waterer made from a steel drum, or an automatic or nipple waterer connected to a water line.

Feed Needs

Feed is the biggest expense in raising hogs, about 70 to 75 percent of the total cost of production. Swine need a balanced ration or diet each day. The complete ration, which can be purchased from a feed supplier, should contain energy, protein, vitamins, and minerals.

Corn is the standard grain (the energy source) for hogs, but barley, wheat, grain sorghum (milo), and oats also can be fed. The protein, vitamins, and minerals are provided with a complete protein supplement available from a feed supplier.

The grain and protein supplement can be ground and mixed together as a complete feed, or the corn and supplement can be fed separately after the pigs weigh about 75 pounds.

Another way is to buy the protein such as soybean meal, a mineral premix, and a vitamin premix separately, and then mix these with the grain source.

Whichever method you use, be sure the ration has the correct amounts of nutrients for the age of the pig being fed. Follow the mixing directions and any regulations on the feed tag. Get more information on feed sources from your county agricultural agent.

You can lower feed costs by providing a good environment and selecting animals that gain fast and efficiently. It will require 3-1/2 to 3-3/4 pounds of feed to produce a pound of live pork. Therefore, a hog fed a complete ration will need about 650 to 700 pounds of feed to grow from weaning weight (40 pounds) to market weight at 225 pounds.

To estimate potential profits, compare your total feed cost to the expected market price for live hogs. Besides feed costs, take into account any other production costs such as buildings and equipment, utilities, veterinary expenses, and bedding.

Feed can be supplied in a self-feeder where hogs will have access to it at all times. Or pigs can be hand fed all they will consume in about 30 minutes twice a day.

Hogs do well on pasture. About a fifth of an acre of good pasture is recommended for a sow and litter or for three to five growing pigs. Alfalfa and ladino clover are considered the best pasture. However, red clover, alsike, white clover, and lespedeza also make good pastures for hogs.

Rye, oats, wheat, cattail millet, rape, soybeans, crimson clover, and cowpeas can be used for temporary pasture.

Feed the same ration on pasture except for pregnant sows, which will require up to 30 percent less feed depending on the pasture quality. Hog rings may be used in the noses of sows and older pigs (over 40 pounds) to prevent them from rooting and destroying pasture.

Sow and Litter

During gestation, feed about 4 (summer) to 6 (winter) pounds of complete ration each day. Do not let sows and gilts get too fat. During gestation, gilts should be fed so they will gain about 75 pounds, and sows about 30 pounds.

A sow will farrow around 112 to 115 days after she is bred. On the 109th day of gestation or about five days before she farrows, move her into a cleaned and disinfected farrowing house or pen.

If a farrowing crate is not used, install guard rails, if possible, to prevent the sow from lying on her pigs.

Place a layer of bedding in the house or pen. Use straw, peanut hulls, or wood shavings. Remove wet bedding and manure daily to keep the pen dry.

If weather permits, wash the sow with soap and warm water before moving her into the farrowing facility. Be sure to

Pigs grow fast. This pig weighs about 50 pounds at 10 weeks of age.

Thomas DeFeo

wash her teats. Washing removes worm eggs and other organisms that infest baby pigs.

To prevent constipation in sows, add wheat bran or other bulky ingredients to the ration at a level of 15 percent, or feed the regular diet containing a tablespoon of Epsom salt or Glauber salt.

For 24 hours after farrowing, give the sow water but little or no feed. On the second day, start feeding about 3 pounds of feed and increase the ration each day. She should be on full feed, about 10 to 12 pounds, when the pigs are a week old. As soon as she is on full feed, the sow may be self-fed.

Normal, healthy sows and gilts usually farrow without trouble. Farrowing normally takes 2 to 5 hours. If possible, be on hand to help. Remove immediately any membranes that cover the head of newborn pigs to prevent suffocation. If a newborn pig appears lifeless, breathing can sometimes be started by rubbing or slapping its sides.

Newborn Pigs

If pigs are piling or shivering and have rough hair coats they are probably cold. When possible, use supplemental heat lamps or hovers to prevent chilling. Make sure sows or pigs cannot reach the lamp or cord. Do not hang the heat source by its cord; hang it securely. Make sure it is impossible for the lamp to touch bedding or other flammable material.

After delivery, paint the navel cords, if still wet, with a tincture of iodine (U.S.P. 2 percent solution). Clip off the tips of the eight tusklike needle teeth of the pigs.

If pigs do not have access to clean soil, they will need an iron injection or iron orally during the first three days to prevent baby pig anemia. Once pigs start eating, the ration will provide enough iron.

Pigs can be weaned between 4 and 8 weeks of age. Male pigs that are not to be sold for breeding purposes should be castrated any time between birth and 4 weeks of age.

At weaning, reduce the sow ration to about 4 or 5 pounds per day. The sow will come into heat, and then she may be rebred 3 to 6 days after pigs are weaned.

Internal parasite control begins with deworming the sow before farrowing. Deworm young pigs before 7 to 8 weeks of age. Also control external parasites, lice and mange. Follow all directions and heed all precautions on labels of products used.

For information on swine health and vaccinations, check with a veterinarian or your county agricultural agent.

Plant Nursery Business Opportunities on the Rise

By Francis Gouin and Ray Brush

Opportunities are on the rise for rural residents to use their land and available family labor to operate a plant nursery. The demand for fruit plants, trees, shrubs and vines is expected to grow as our society continues placing increasing emphasis on environmental improvement.

Many kinds of plants can be produced. There are approximately 2,000 species common in the nursery trade and thousands of cultivated varieties. For example, in roses, azaleas, and camellias there are over 100 cultivated varieties. However, most of the commercial trade is concentrated in a limited number of varieties.

Producing some of the more unusual cultivated varieties is one possibility for the small nursery. Many of these varieties are especially adapted to the soils and climate of a local area.

Of course, there is a difference between growing plants as a hobby and as a business venture. Growing plants that you find pleasure in growing, on land and in an area where the climate is best adapted for those plants, can result in a profitable small business that is enjoyable for the individual or family. But if you are not willing to grow plants which are most adapted to your soil or climate, then you cannot effectively compete in the marketplace. Your investment of capital and labor will produce a low return, showing an unsatisfactory net profit or even a loss. In such cases you are better off devoting your time to hobby gardening.

Producing nursery plants requires a lot of work. And they need special attention at various stages of growth. Yet the knowledge and skill to produce quality plants is well within the realm of possibility for the average individual.

Francis Gouin is Associate Professor and Extension Specialist in Ornamental Horticulture, University of Maryland, College Park. Ray Brush is Secretary, American Association of Nurserymen, Washington, D.C.

You need not necessarily have been born and raised on a farm to succeed in this business, but it helps. At any rate see how the successful producers do it, talk with them seeking advice, and read current journals and literature for recommendations based on new research.

A small nursery operated on a part-time basis does not require extensive acreage. Depending on the kinds of plants produced and the size to which they are grown, two to ten acres could be adequate. Because of the great number of kinds of plants produced in nurseries, it should be easy to select plants well adapted to the land and climate of the area as well as plants you will enjoy growing.

Depending on the kind of plants and the region of the country, there is a choice between producing the plants in field rows or in containers. Containers are pots of plastic, metal or other material which will hold the soil or growing medium in which the roots develop.

Container production provides the opportunity for producing a higher number of plants in a limited area. A nursery of this type will probably produce smaller plants which may be sold retail at the site to home gardeners, or sold in quantity to the larger wholesale producing nurseries, or to garden centers and other retail outlets.

Licensed by States

Producing nurseries are licensed by the state department of agriculture in every state, with all nursery stock inspected for freedom from insects and diseases at least once a year. Many kinds of nursery stock must be inspected two or more times during the year before they can be marketed.

Nurseries with plants found infested with insects or diseases may be restricted from selling any plants or have portions of the nursery "tagged" for non-sale until that kind of plant or that portion of the nursery has been treated and found free of the hazardous insect or disease.

As a rule the nursery inspector is considered a welcome counselor by nurseries. In most states the inspectors hold bachelors degrees in a biological science, preferably entomology or plant pathology. Their specialty is identifying plant pests. For specific remedial action they will refer the nurseryman to the proper agricultural Extension specialist for recommendations.

If the nursery specializes in propagating plants, those plants may be sold as "liners" to other producing nurseries for growing on to sizes salable to the general public. "Liners" are young plants either grown from seed or by rooting cuttings. Cuttings are portions of the stems of growing plants. Depending on the kinds of plants and the facilities for protecting and sheltering them while the roots are developing, cuttings may be the tips of young shoots that are just beginning to harden or any stage between that and woody stems.

Nursery production is unique in agriculture because plants grown in the nursery are sold complete, including the roots. Often larger plants are sold with a ball of earth around the roots in which they were growing. When that ball is wrapped in burlap or a similar material to hold the soil and root mass intact, it is called a "balled and burlapped plant" (B&B). If the soil and root mass is in a container, usually a fiber pot, it is referred to as a "balled and potted plant" (B&P).

Over the years the nursery industry has developed a system for standardizing plants to facilitate sales to nurserymen, landscape contractors and others. In the case of B & B or B & P plants, the standards specify minimum depth and diameter measurements of the root ball according to the height or caliper (average trunk diameter) of the plant.

This system of standards is contained in the publication *American Standard For Nursery Stock ANZI Z60.1-(year)*. The Standards are sponsored by the American Association of Nurserymen and approved by the American National Standards Institute. The current edition, approved in 1973, is being revised. The revision will be submitted to the standards institute for approval in 1978.

Both landscape and fruit plants are high value crops and must be grown with minimal losses.

Soil Needs

Nursery crops require deep, rich, well-drained soils for developing healthy and vigorous root systems that will overcome the shock of digging and transplanting into the new site.

It is estimated that 100 to 150 tons of topsoil per acre are removed with each crop dug B & B or B & P. Smaller plants which are dug bare root remove only limited quantities of soil. However, they suffer more severe transplant shock and may have low survival rates if not carefully handled.

Nurserymen carry on soil building programs to maintain both the organic and nutrient content of the soil at high levels. That is done through production of green manure crops (corn, grasses, or legumes not subject to soybean cyst nematode). These crops are carefully fertilized and limed so as to produce the maximum vegetative plant growth. They are then plowed under to enrich the soil.

281

For both field production and container production, the nurseryman desires level land, and often grades the fields to level them when establishing a nursery. In both production practices, service roadways are maintained through the nursery. Wet areas are tiled so as to remove excess water, permitting air to penetrate to the roots. Often in refitting nursery fields, a subsoiler is used to penetrate and loosen the subsoil to encourage greater depth of root growth.

In field production nurseries that are not level, nurserymen frequently use strip or contour farming methods to prevent water erosion. The nursery rows are made to follow the contour of the hill, and wide strips of sod are left undisturbed between cultivated blocks of plants.

It is now becoming more common for nurserymen producing shade trees, for landscape purposes, to grow them in sod and control the weeds in a small area about each tree. The grass is mowed periodically both to control undesirable plants and to avoid providing rodents nesting areas near the plants. During winter when other food is not abundant, rodents feed on the bark of plants, girdling and killing them.

The length of time required to produce nursery plants for landscape use may vary from 3 to 15 years. Some of the bulb and tuber plants and flowering biennial and perennial plants require only one year to produce a salable crop.

Root-Pruning

To assure that B & B and bare root plants transplant well, many trees and shrubs are root-pruned two and more times. Root-pruning is accomplished by pulling a cutting blade, often "U" shaped, beneath the plants—cutting the tips of the roots. This practice stimulates development of a branched root system. The number of times depends on the kinds of plants, how long they are grown in the nursery, and the type soil they are growing in.

Shallow, fibrous-rooted plants, such as azaleas, seldom need to be root-pruned. Deeper rooting plants such as yews, many conifers, and the shade trees produce the most desirable root system for transplanting when they are root-pruned as often as every three years.

Although root-pruning retards plant growth, plants that have been root-pruned are easier to dig. Such plants have a well branched root system and survive transplanting shock better.

Training environmental plants to make them conform to specific shapes and sizes, or to their natural habit of growth, is a continuous process. Prune most nursery plants frequently either to stimulate branching, or to eliminate undesirable branches.

Because each kind of plant has a unique habit of growth, you must be familiar with these characteristics. By knowing plant peculiarities and following proper pruning practices, you can enhance the plant's individual characteristics. Yet it is no simple task to grow a block of shade trees with straight trunks and well-formed heads, or to grow uniformly branched shrubs.

Gardeners soon learn that not all plants grow well under the same conditions. Weeping willows, maples, and some hollies grow satisfactorily even when planted in poorly drained soils. However, juniper, yews and oak trees require well drained soils. Azaleas, rhododendrons and andromedas grow best in acid soils. Yews, maples, roses and forsythia prefer soils that are only slightly acid. White birch, mountain ash and spruce prefer colder regions while crape myrtle, camellias and magnolias grow better in warm climates.

These differences enable a nurseryman to specialize, based on personal preference to produce a given kind of plant, the local market, and local soils and climate. Many nurserymen specialize in growing only certain kinds of plants, such as azaleas, rhododendrons, shade trees, ground covers, dwarf conifers, or herbaceous perennials. For the nursery to be profitable, you must become proficient in growing your chosen plants economically. With specialized equipment, you can develop efficient methods and facilities for propagating and growing these plants.

Container Growing

In recent years there has been an increase in the number of nurserymen producing in containers. One advantage of container growing is the number of plants grown per acre may be several times greater than when plants are grown in nursery rows or in beds. The blend of soil and soil additives in the container media also enables the nurseryman to specialize in a plant which might not be efficiently produced in nursery fields in the area.

Often container-grown plants grow faster because the blend of growing media is more suitable for that kind of plant. Also, container-grown plants can be sold any time of year regardless of weather.

Although container culture offers many advantages, there are disadvantages. You need a dependable source of quality soil and soil additives to have a continuous supply of satisfactory potting media. Because container-grown plants must be irrigated frequently, an irrigation system with an ample supply of high quality water is essential.

In the North, East and Northeast, most container-grown plants must be protected from low killing temperatures during the fall and winter months. Furthermore, plants only up to a certain size can be economically grown in containers.

Weeds, diseases and insects are as much a problem in growing plants in the nursery as they are in growing vegetables, fruits, and turf. Weeds are frequently more difficult to control in nurseries than in other crops. This is due to the wide variety of plants being grown, the slow rate of growth of most environmental plants, and the limited number of herbicides available. Some nursery plants are more sensitive to herbicides than others, making it impossible for the nurseryman to use one herbicide over the entire nursery.

Because most environmental plants tend to grow slowly, especially during the first two to three years, there is little natural weed control from shading. Perennial weeds frequently become a serious problem, especially in fields where plants must remain five years or more. With only a limited number of herbicides to choose from, nurserymen who grow a wide variety of plants must be familiar with the tolerance of each type and the herbicides that can safely be used around them.

Since most nurseries grow many different kinds of plants, insects and diseases are controlled whenever potential problems are observed. Therefore, you must continually monitor your nursery and select the proper pesticides after the insect or disease has been properly identified. The material or materials selected must not cause damage to plants treated or to the environment.

Nurserymen have been using bark and wood chips, waste products from sawmills and paper manufacturers, and compost made from leaves collected by neighboring cities. They were the first to use slow-release or time-release fertilizers. Although these fertilizers are 2 to 3 times more expensive than common agricultural grade fertilizers, they are more efficient and less wasteful.

Information Trips

In planning your nursery business, several sources of information will be helpful. The Small Business Administration has a very useful publication in its "Counseling Notes" series on the nursery industry. Another of their publications which is a must is "Small Marketer's Aid" No. 71, *Checklist for Going Into Business*.

The American Association of Nurserymen also has an information sheet listing its publications which will be helpful. An important feature of the sheet is the name and address of the Extension Horticultural Specialist at the land grant university in your state who will be able to counsel you on cultural recommendations for your area. Also listed are the name and address of the State Department of Agriculture official responsible for inspecting and licensing nurseries in your state, and the state nursery association.

After you review this material and before starting your nursery, consult with your County Cooperative Extension Service. Extension specialists in your county office will be able to assist you in interpreting the technical information as well as supply you with additional bulletins, leaflets and other helpful information. Your County Agent, perhaps in cooperation with the State Extension Specialist, can help you select the most suitable land and the species of plants you should consider growing.

Since the nursery industry is varied, take every opportunity to learn from others. The Cooperative Extension Service in most states conducts workshops, short courses and tours, often in cooperation with state or local nurserymen's associations. These provide an excellent opportunity to learn about the industry, talk with experienced nurserymen, and to visit facilities.

Many land grant universities and community colleges now offer one- or two-year programs in nursery operations. These generally are open to the public and designed to teach the practical side of managing a nursery, with opportunities for on-job training. There is no better way of learning the business than working in a nursery.

If you start without any formal training or nursery experience, start small. Use no more than one acre of land if the plants are to be grown in nursery rows, one-half acre if the plants are to be grown in beds, and no more than a quarter-acre are to be grown in containers.

Testing Water, Soil. Have your soil or potting media and water supply tested. It is simpler and less expensive to correct soils and potting mixes before the crops are planted than after they are established. An early test on the quality and availability of water may also save you considerable grief in the future. Many nursery problems have been traced back to irrigation water that was too acid, too alkaline, or contained excess salts.

Limit the number of species you grow to a dozen for two

to three years. Learn to identify these plants immediately upon receiving them, label them properly when planting, and study their growth habits and cultural requirements. Give each plant ample room in the nursery to grow. Beginning nurserymen have a tendency to crowd plants close together because they are small.

Group species of plants with similar soil requirements and with similar growth habits. Remember, some plants prefer growing only in mildly acid soils. Growing these plants together with other varieties in one common soil will generally result in some species growing poorly while others prosper.

Growing upright plants with low spreading plants will often reduce the quality of the low spreaders. The uprights will most often shade out the others, resulting in poorly formed, low quality plants, more susceptible to insects and diseases.

Growing quality environmental plants requires a thorough understanding of the growing needs of many kinds of plants, and their different growth habits. To keep soils productive, follow good soil conservation practices and invest some of your earnings into building new topsoil.

To protect plants from insects and disease, provide ideal growing conditions for each kind of plant, and use the proper pesticides only when necessary. To survive in the nursery industry, you must become a taxonomist (able to identify plants), a pathologist (able to identify diseases), an entomologist (able to identify insects), and a good farmer.

The market for thrifty, healthy plants free of insects and disease is expected to remain good. The small nurseryman who is able to meet these demands may be richly rewarded in pleasure and profit.

Growing English Walnuts, Pecans, and Chestnuts

By R. A. Jaynes, G. C. Martin, L. Shreve, and G. S. Sibbett

What could be simpler than to buy a nut tree seedling, plant it in a sunny location, and wait for a bountiful harvest? Unfortunately, that is not the way it goes.

Local climate and site have to be right for the species grown. Grafted trees of proven selections, as opposed to seedlings, are usually essential. The common crop-growing problems of weeds, fertilizer, and water must be dealt with. Plus pruning, pollination, insect, disease, rodent and bird problems. And finally, you need knowhow in harvesting, storing, and marketing.

Attention to establishment and care of nut trees is even more important than for shorter-lived crops, because of the 5 to 10 year wait from planting to bearing. A mistake in the vegetable garden can be recouped the same or the next year, but with nut trees eight years may be needed just to learn of a mistake, such as planting the wrong variety. Yet, this challenge in growing nut trees captivates many people.

Nut trees are a highly nutritious food source and traditionally grown for their nut meats. They are also valuable as shade trees and as a source of food for wildlife.

The small landowner should consider growing seedlings or grafted trees for sale. Most nut-tree selections, especially away from geographic areas of commercial production, are difficult to obtain. Grafting and budding trees is a skilled technique, but can be learned by the interested layman.

Marketing nuts is another area where imagination can greatly enhance return. Wholesaling at the local farm market or co-op may be practical but will give a relatively low return.

R. A. Jaynes is Geneticist, Connecticut Agricultural Experiment Station, New Haven. G. C. Martin is Pomologist, University of California, Davis.
L. Shreve is Area Horticulturist, Texas Agricultural Extension Service, Uvalde. G. S. Sibbett is a Farm Advisor, Tulare County, Visalia, Calif.

Consider direct retailing, pick your own, and mail order sales. Attractive packaging boosts sales and increases the dollar return. Even unfilled nuts have value when made into novelty items for sales at craft shops and church fairs.

Only three nut-tree species are discussed in this chapter; pecan, English walnut, and chestnut. There are numerous other possibilities including filbert, black walnut, butternut, hickory, pistachio, and macadamia.

Pecan Orchards

The pecan tree is native to North America and commonly found in the valleys of the Mississippi River and its tributaries. Pecans also are grown in the southeastern states north to Virginia and in western Texas, New Mexico, Arizona, and California.

Even though outside the natural range, more pecans are commercially produced in planted groves in Georgia than in any other state. Still, over 50% of all marketed pecan nuts are produced from native groves managed intensively for nut production. The trend, however, is away from native stands and toward planted grafted trees of proven performance.

The first commercial varieties were selections from native stands, whereas most new selections are from breeding programs. The U.S. Department of Agriculture's Pecan Research Center at Brownwood, Texas, has been a leader in releasing new varieties, all distinguished by their Indian names.

Optimum conditions for growing pecans include deep, well-drained soils and a warm growing season. Different varieties are adapted for frost-free growing seasons of 150 to 210 days. Although the season is long enough in parts of the Northeast and Northwest to grow "northern" varieties, it is not hot enough for the kernels to develop.

The following examples illustrate how pecan growing can be profitable on small acreage with good management.

Annual production and gross return from a 10-acre (30 trees per acre) pecan orchard planted in 1970

Year	Lbs. per tree	Total yield lbs	Price per lb	Total received
1974	15.5	4,650	$1.25	$5,812.50
1975	17.9	5,386	1.25	6,732.46
1976	21.6	6,487	1.50	9,730.90
1977	15*	4,129	1.50	6,193.00

* Low yield attributed to tree damage to Wichita variety in a shaker demonstration.

288

Annual production and gross return from a 10-acre (35 trees per acre) pecan orchard planted in 1972

Year	Lbs produced [1]	Price per lb	Gross return
1975	300	$1.25	$ 375.00
1976	400	1.50	600.00
1977	3,643	1.50	5,464.50

[1] Severe hail damage to the young trees depressed yields in 1975 and 1976.

Bill Perry, a former barber, and his wife, Virginia, planted a pecan orchard in 1970 on ten acres at Quemado, Texas. Through the State Extension Service they obtained assistance on spacing, irrigation, fertilization, pruning, and insect, disease, and weed control. In 1972, another 10 acres was planted with hybrid pecan varieties.

The pecans are sold directly to consumers. The only advertisement is a sign near the mailbox, "Perry's Pecan Orchard, Pecans for Sale." Customers who appreciate their high quality pecans have been their best advertisement.

Annual production and gross returns from the two plantings are given in the tables.

Costs that prospective pecan growers need to consider include:

* Land purchase
* Tractor, disk, and sprayer
* Irrigation, including water, leveling, and water lines
* Planting stock
* Operating costs including labor (tillage, chemicals, fertilizer, and harvesting)

Bill and Virginia Perry sell their pecans directly to consumers, depending on word-of-mouth advertisement.

Barry W. Jones

The Perrys' advice to prospective pecan growers is to find a good banker, have an understanding family, and don't be afraid to work.

Another couple, Horace and Lorine Brown, Ricksprings, Texas, purchased land containing 14 acres of native pecan trees. The undergrowth was removed by shredding in 1973. In addition, they planted 1.75 acres to selected varieties of pecan.

The native stand required thinning, spraying, and fertilization. With advice from a local grower and the Extension Service, they have begun to graft small native trees to paper-shell varieties.

The Browns work in their pecan orchard on weekends and holidays—spraying, fertilizing, weeding, grafting, and thinning. Production from 1972 through 1977 has gone from less than 500 lbs a year to 7,000 lbs. Loss of nuts to squirrels is a problem, especially in years when the hundreds of acres of surrounding, unmanaged native groves have no crop.

Native pecan nuts usually sell for about half the price of paper shells. However, because of the high quality of nuts from the managed grove, the Browns should receive 67 cents per pound instead of the usual 40 to 45.

The Perrys and Browns have demonstrated that pecan growing does not require vast acreages to be profitable as a vocation or hobby.

English Walnuts

English walnuts received their name because they were brought to this country on English ships. They probably originated in Persia, hence their alternate name, Persian walnuts. California accounts for more than 97% of the domestic commercial acreage, and for more than 55% of all English walnuts grown in the world.

English walnut trees may grow very large under ideal conditions, reaching heights of 90 feet with a 60-foot spread. But in orchards they commonly achieve heights of 40 feet and spreads of 30. In home plantings, walnuts provide shade and, under proper climatic conditions, an edible crop.

Because of land value and cost of production and management, walnuts are grown intensively, under sophisticated cultural management. However, productive trees and small plantings occur throughout much of the Midwest and eastern U.S., in addition to the west coast states.

Walnut trees require good-textured, deep, well-drained soils for optimal growth and production. Extensive root development can occur to 15 feet in deep soils.

The best sites have good air drainage and moderate temperatures. During winter, about 1,000 hours of temperature below 45° F (7.2° C) are needed to complete the winter chilling requirement.

At the same time, bear in mind that walnut trees can be damaged by winter temperatures as low as 14°F (–10° C), especially when followed by a warm period. However, some of the hardier Carpathian strains of English walnut regularly withstand temperatures of –20° F (–29° C).

In the present commercial orchard areas, the growing season should have 200 or more frost-free days, but the trees will thrive and produce where the frost-free season is as short as 150 days.

Grafted varieties of proven performance are recommended instead of seedlings. Rootstocks used in California are *J. hindsii* and Paradox (a cross of *J. hindsii* and *J. regia*) and in the East, eastern black walnut, *J. nigra*. Nurserymen or local nut growers should be consulted for the best variety for a given location.

Franquette is the standard old variety. It is a late leafing and harvesting variety, with small nuts of high quality. Yields are moderate and requirements for cultural care minimal. Hartley is the preferred variety for the inshell trade, and important for export.

Market Preferences. The domestic American market prefers packaged walnut kernels to inshell walnuts. For this market, new varieties have been developed which have high kernel yields of good quality, such as Payne, Ashley, Serr, Vina, and Chico. These varieties bear early in the orchard's life and yield heavy crops, but in general need more cultural care than Franquette.

Hanson is a productive, well filled, small nut selection favored in the East.

Preparations before planting might include leveling the ground for irrigation, ripping the soil to eliminate restrictions in the soil profile, and fumigating to eliminate weeds, pathogens, and nematodes. Poor site preparation will result in less than optimal production at increased expense throughout the orchard's life.

Specific information on suggested methods of planting, irrigation, pruning, and nutrition can be obtained from your state Extension Service.

Walnuts are mature and ready for harvest as soon as the tissue between the kernel and the inner lining of the shell turns brown. Shortly after that time, nuts can easily be shaken from the tree. Any harvest delay will result in a darker colored kernel, and also allow time for mold and entrance of the navel orange worm.

The homeowner can easily harvest walnuts as they fall from the tree. However, economic farming enterprises must, in our present labor market, harvest mechanically. Special equipment required for harvest includes a tree shaker, a wind-rower or sweeping device to pile the nuts into a neat, long row, and a pickup machine.

Once harvested, nuts are taken to the huller, where the remaining hulls are mechanically removed and the shells brushed clean. From there, the nuts go into a drying facility and are treated with forced air at 109° F (43° C) for 24 to 36 hours.

Even with mechanization, harvest costs comprise nearly a third of total cash costs for producing a walnut crop. The dried nuts will store for at least a year under fairly variable conditions, up to two years if the storage temperature is kept below 40° F (4.8° C).

Insects infesting fruit and wood of walnuts reduce quality and yield of nuts. Those which directly affect the current season's crop include codling moth, walnut aphid, dusky-veined aphid, spider mite, walnut husk fly, and navel orange worm. They either cause infested nuts to drop prematurely (primarily codling moth), or affect the nuts' quality by reducing kernel color or directly infesting the kernels, making them worthless.

Spider mites, aphids, and scale pests cause production cuts the next year. By infesting wood and leaves, they stress the tree, resulting in less fruit wood or fruit buds being produced for the next crop. These insects can be controlled by complex programs integrating either chemicals (codling moth and scale), biological control (walnut aphid), or adequate cultural practices (navel orange worm). Consult a trained pest control advisor on these matters.

Insect and disease control, perhaps more than anything else, limits the success of small plantings of walnut and other nut trees. The knowledge and equipment required for pest control, as well as other operations, often exceed the small landowner's capabilities. These limitations should be recognized before the crop is planted.

The small landowner often needs as much knowhow to grow 3 acres as the grower with 3,000 acres, and expensive equipment used by the big commercial grower often is not justified with the potential return from small plantings. However, in at least some instances, ingenuity, hard work, and elimination of middlemen can give the grower with small acreage a competitive edge.

Chinese Chestnut

Of the several species of chestnut, the Chinese chestnut is the best one for nut production in the U. S. It does well in areas where peaches can be grown. The American chestnut was an important timber species in the eastern U. S. but the chestnut blight fungus destroyed it.

Although numerous selections of chestnut have been named—including Crane, Eaton, Nanking, and Orrin—they are not normally available from nurseries. Thus the purchaser of trees usually has to settle for seedlings offered by mail-order nurseries. These are generally satisfactory for the homeowner, but lack the uniformity and performance that could be expected from proven varieties.

Chestnuts prefer a well-drained, acid (pH 5.5-6.0) soil. Fertilization, cultivation, use of herbicides, and mulching are essentially the same as for other temperate climate crops. Pruning, in the early years, should only be enough to develop a single trunk and basic scaffold. Excessive pruning of young trees delays the onset of bearing.

Chinese chestnut bears at a younger age than most other nut trees, usually 3 to 4 years after transplanting. Mature trees may have to be spaced 40 feet apart — but to increase unit area yields, a 20 x 20 foot spacing initially is more practical. Just be prepared to remove trees when they begin to shade each other.

Chestnuts are fairly regular bearers compared to many of the other nut trees, which are prone to a biennial cycle. Common yields are 1 to 2 tons per acre.

The most common and ubiquitous pests are the chestnut weevils. Adult weevils lay eggs in the nuts as the kernel fills. The larvae are well developed about harvest time. Some insecticides are capable of controlling the weevil, but none are presently registered for use. Another pest is the chestnut blight fungus; it is not generally a problem on well grown Chinese chestnut trees.

The gall wasp was recently introduced into the U. S. and is a threat to chestnut. It destroys new shoot growth. So far the wasp has been found only in Georgia.

Chestnut trees have a place in the home planting, as seen by the thousands of trees people have in their yards. Small plantings may be profitable where there is a local market for the nuts, and if problems such as squirrels and weevils can be handled.

Poultry Puts Eggs on Table and Provides Meat Supply

By Hugh S. Johnson

A small poultry flock can provide an excellent source of fresh eggs and meat for the flockowner and family. In addition, the surplus eggs and meat can be sold to friends and neighbors.

In many locations, however, any kind of livestock or poultry is either restricted or inappropriate. Check your zoning regulations. If you have close neighbors, avoid offending them with the noise, odors, and flies resulting from a poultry flock.

Flock size will vary with each individual situation. To determine your correct size consider the purpose for which the flock will be used, the space and time available for poultry keeping, and the market potential for surplus production.

A family of four will be well supplied with eggs from a flock of 15 to 20 layers. It takes about four hens to furnish one person with two eggs a day.

For meat, decide how much poultry your family will consume. You can have several different ages and kinds (broilers, turkeys, geese, etc.) so that fresh dressed poultry is available at all times. Or you may want to raise all of them at once and put the extras in the freezer.

Some flockowners like to produce both eggs and meat for home use. The so-called "dual-purpose" breeds are good for this type flock. Breeds such as Rhode Island Reds, New Hampshires, and Plymouth Rocks are good layers, and the cockerels (males) of these breeds are satisfactory for home-meat production.

Getting started. An egg-producing flock can be started in one of three ways: with hatching eggs, day-old chicks, or started pullets (young hens).

Hatching eggs, besides being limited in availability, involve certain skills not required by the others. Incubation facilities

Hugh S. Johnson is Professor and Poultry Extension Specialist in the Department of Animal Science, University of Illinois, Urbana.

and brooding equipment are needed, and there are risks and problems in obtaining and successfully hatching fertile eggs. But many people receive a certain amount of pleasure and satisfaction from hatching and rearing their own chicks.

Day-old chicks share some of the same disadvantages as hatching eggs, but are more practical. Incubation facilities are not needed, but brooding equipment is. Again, risk is involved due to the possibility of losing chicks by disease outbreaks and management problems. Also, if straight-run chicks are purchased, about half will be cockerels

Started pullets are the most expensive ($2 to $3 each) but the easiest to handle. Normally they are purchased at 20 weeks of age. A distinct advantage is that they have been vaccinated for most troublesome diseases and in general have less risks of loss from disease or mortality.

Most layers are White Leghorn or Leghorn-type crosses. Availability will be a factor in choosing. A few local hatcheries, feed stores, and mail-order farm and suburban catalogs may be sources.

It is difficult to estimate costs for a small laying flock because people have widely varying attitudes about investing in their flock.

Some will have housing for little or no cost; others will want to use their handyman skills to build an attractive unit that fits well into the landscape. A shelter cost of $25 to $30 per bird could result. The range of costs for equipment can be similar. The commercial poultryman has a housing and equipment investment of $5 to $7 a bird.

Baby chicks. Those who decide to start with baby chicks must first locate a source and then order 40 to 50 straight-run chicks or 25 pullet chicks. In that way, you are assured of housing 15 to 20 pullets at 20 weeks of age. Order the chicks in advance and be fully prepared when they arrive. Spring or early summer brooding is safest.

Electric hover-type brooders are frequently used for small flocks, although gas brooders may be used, too. They are available from several companies in various sizes. Battery-type brooders and infrared brooding lamps are satisfactory. Also brooders may be constructed that use ordinary light bulbs as the heat source.

Start brooders at least 24 hours before the chicks arrive. Adjustments can be made during this time and the house will be warm.

Temperature under the brooder at chick level should be 95° F during the first week. Lower the temperature by 5° each week until 70° is reached. With infrared lamp brooding, place lamps at least 18 inches above the floor to minimize the chances of fire. If the brooder temperature is incorrect, the chicks will either huddle close to the heat source (too cold) or stay away from the heat source (too hot).

Sanitation is important in any phase of poultry production. The brooder house and all equipment should be cleaned, disinfected, and allowed to dry several days before housing chicks.

Brooder houses can vary from the standard type poultry house to the outdoor electric brooder used in warmer climates.

Corrugated cardboard can be used to confine chicks to brooding area. Arranging feeders like spokes, with sufficient water fountains between them, makes it easier for chicks to find their way back to heat. Place chicks in brooder house as soon as you get them.

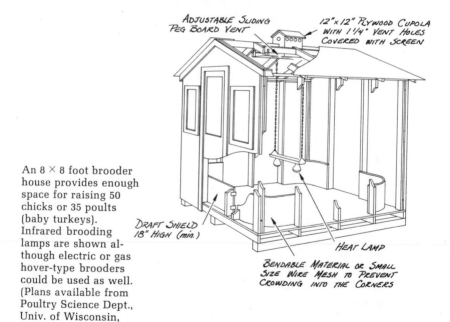

ADJUSTABLE SLIDING PEG BOARD VENT

12"x12" PLYWOOD CUPOLA WITH 1¼" VENT HOLES COVERED WITH SCREEN

DRAFT SHIELD 18" HIGH (min.)

HEAT LAMP

BENDABLE MATERIAL OR SMALL SIZE WIRE MESH TO PREVENT CROWDING INTO THE CORNERS

An 8 × 8 foot brooder house provides enough space for raising 50 chicks or 35 poults (baby turkeys). Infrared brooding lamps are shown although electric or gas hover-type brooders could be used as well. (Plans available from Poultry Science Dept., Univ. of Wisconsin, Madison, Wis. 53706).

The brooder house has several basic requirements regardless of the type construction. Sufficient floor space is essential; one square foot per chick is recommended. Ventilation should provide plenty of fresh air but prevent the chicks from being exposed to drafts. Tight construction is necessary to make the house economical to heat.

Cover the brooding area with 4 inches of absorbent-type litter. Wood shavings is the most common type poultry litter. Other materials can be used—such as sawdust, peanut hulls, ground corncobs, and wood chips.

If straight-run chicks are raised, the cockerels will be ready for processing between the seventh and tenth week, depending on the size wanted and kind being reared.

Lighting Programs

Both natural and artificial light are powerful stimulants to birds. Light influences such things as sexual maturity, behavior, and egg size. Using light effectively can help the flockowner achieve higher production and greater income.

A number of lighting programs give satisfactory results. But two cardinal rules must be followed: Never increase light on growing pullets, and never decrease light on laying hens.

Provide all-night lights in the brooder house for the first week. They do not need to be bright, but should provide sufficient light for the chicks to find their way around.

After that, pullets being brooded between July 1 and Jan. 1 require no additional artificial light, provided they are being raised in a house with windows.

Pullets brooded between Jan. 1 and July 1 will be maturing when the day length is getting longer. These flocks should be on a declining light schedule.

This is done by determining the day length when the flock will be 22 weeks old and adding 5 hours to it. Provide this amount of light for the second week. Then, reduce the total light period 15 minutes per week and at 22 weeks the pullets will be on natural day length.

Laying birds (22 weeks and older) need a constant or increasing amount of light for maximum production. Normally 16 hours is optimum.

One 40-watt incandescent bulb for 200 square feet of floor space is sufficient.

Laying House

The laying house does not need to be elaborate or expensive. In fact, the first time chickens are kept, the brooder house

—if chicks are raised—can be converted to a laying house. Basic requirements are the same for the laying house as for the brooder house.

With floor-type operations, the laying house should provide at least 2 square feet per bird for light breeds such as Leghorns, and 3 square feet for the heavier breeds.

Laying cages are excellent for a small home flock. Cages can be installed in any small building, or a house can be constructed especially for cages.

Equipment in the laying house will include feeders, waterers, nests, and lights. Besides these items, the furnishings may include containers for granite grit and oystershell, and roosts.

The small flock owner will probably get best results with commercial feed. No attempt should be made to get by with just table scraps and grain.

Buy prepared feeds in quantities that will be used in 2 or 3 weeks, particularly in warm weather. It takes about 19 pounds to grow a pullet to 20 weeks. Hens eat about a quarter pound a day, or 80 to 90 pounds a year.

Follow the manufacturers recommendations for the particular type feed being used. If instructions are not given, use those in the table as a guide.

Feed is the major expense in producing eggs. It represents at least 60 percent of the total cost. So, avoid feed wastage.

LAYING HOUSE
This house is adequate for 15 to 20 layers. Individual nests, designed to accommodate one hen at a time, are usually about a foot square. Porvide a nest for each four layers. Clean, fresh water should be available at all times. With trough-type waterers, provide one linear inch per layer. With fountain-type waterers, provide enough fountains to have 7 to 10 gallons for each 100 layers. (Plans available from Poultry Science Dept., Univ. of Wisconsin, Madison, Wis. 53706).

Feed and weight guides for growing and laying Leghorn-type chickens

Age of birds	Type ration	% protein in feed	Feed Consumption (lbs.) per bird for period listed	Avg. bird weight (lbs.) at end of period
0-8 wks.	Starter	20	4	1½
9-20 wks.	Grower	16	15	3¼
Laying (1 yr.)	Layer	16	80-90	4

Don't expect every hen to lay an egg every day. Even the best flocks don't do that. But you can expect layers in a well-managed flock to lay 230 eggs per bird in a year's time. That's over 60 percent production.

Normally egg production is not at the same level year-round. Pullets will start laying when they are between 20 and 24 weeks of age. Production starts slowly and includes many small eggs. Layers "hit their peak" when they are about 30 weeks old. After that, production drops off.

Egg Handling. Eggs should be gathered at least twice daily. More frequent gathering is recommended, especially during extremely hot or cold weather. Leaving eggs in the nest increases the chances of shell damage and quality loss.

Dirty eggs should be dry cleaned or washed with an egg washing compound as soon after gathering as possible. Use wash water that is between 100° and 120° F. It is better to leave the eggs dirty than to wash them in cold water.

Allow washed eggs to dry. Then, place them in clean cartons and refrigerate.

Welded-wire mesh roosts 16 to 24 inches above floor keep hens out of droppings and prevent manure buildup.

Producers may sell eggs from their own flocks directly to consumers. However, producers with over 3,000 hens who sell eggs to retail stores, bakeries, and restaurants are subject to Federal regulations. Check with your State Department of Agriculture about the current egg marketing laws.

The owner of a laying flock should be prepared to process his own chickens. Neither custom dressing nor the live sale of birds is normally available.

Meat Production

A well-planned and well-managed flock can be a good source of fresh poultry meat. But large scale commercial broiler and turkey production techniques often result in market prices difficult to match.

In addition, producers are limited in the number of birds that may be sold without Federal inspection. So, check with your State Department of Agriculture on the latest marketing regulations and make your plans accordingly.

The most economical meat production is obtained with commercial meat strains developed from breeds such as the Cornish, Plymouth Rock, and New Hampshire. These crosses have been bred for top feed conversion; they feather rapidly and mature early.

The same commercial strains of chicken meat birds can be raised for several purposes. Modern-day broilers or fryers are grown to 4 pounds in 7 to 9 weeks. The same bird processed at 5 or 6 weeks is marketed as a Cornish Game Hen; grown to 12 to 14 weeks it makes a delicious roaster. Males are often caponized (castrated) and sold as capons.

Broilers are brooded and reared the same as chicks raised for layers. A broiler starter ration is used. It can be fed throughout, or switched to a broiler finisher when the birds are 5 to 6 weeks old. No additional feeds or supplements are desirable or necessary.

Light is not restricted on broilers. They may be raised on 24 hours of light for the entire growing period or placed under 14 hours after the third week. Use 24 hours for the first 3 weeks in either case.

Broiler chicks are usually available from a local hatchery, although commercial strains may not be readily available to the owner of a small flock.

Capons, Turkeys

Male chickens are caponized at 3 to 5 weeks to make them fatten more readily and produce higher quality meat. If straight-run chicks are being raised, the males can be recognized from the females at this age by their larger combs, slightly larger size, and greater aggressiveness.

Capons are raised to 5 or 6 months for best weight and finish. Their live weight at that time may be over 10 pounds.

The feeding and management program for capons is the same as for broilers. Use a broiler starting ration until the birds are 12 weeks old. After that, use a broiler finisher.

Cracked corn can be supplied in the afternoon after changing from the starting mash. Gradually increase the grain until the birds are getting equal amounts of corn and finishing mash at about 15 weeks. Feed a small amount of grit once a week whenever cracked or whole corn is fed.

Turkeys can be raised successfully on small farms, but they require special care and equipment. Raise turkeys away from chickens and other birds to prevent disease.

The most common turkey variety is the Large White. Similar in size, but less available, is the Broad Breasted Bronze. Hens of both varieties will grow to live weights of about 15 pounds at 22 weeks of age; toms, 25 pounds at 24 weeks.

Small Whites are commonly used to produce roaster-size turkeys. At 22 weeks, hens will average 9 pounds and toms 14 pounds.

Flocks are usually started with day-old poults (baby turkeys). The poults must be kept warm and dry. Feed and water the poults as soon as possible after bringing them home, dipping their beaks in water to help them learn to drink. Brood young turkeys the same as broilers.

Poults on litter need at least 1-1/2 square feet of floor space per bird to 6 weeks of age. After that, place them in a yard or larger pen, allowing at least 30 square feet per turkey. Provide a shelter outside to protect them from sun and rain.

Since turkeys are fast growing, be sure to buy correctly-formulated turkey feeds. It takes about 80 pounds of feed to raise the average large turkey to market weight.

For the first 8 weeks, the poults need a turkey starting ration containing 26 to 28 percent protein.

At 8 weeks, a growing mash with 22 percent protein is recommended. Whole or cracked grain, such as corn, can be fed after 11 or 12 weeks. Equal parts of growing mash (22 percent protein) and corn (10 percent protein) provide an overall ration containing about 16 percent protein.

Ducks and Geese

Small duck flocks are raised primarily for meat. The commercial duck industry is built around the Pekin breed. Pekins reach market weight in 7 to 9 weeks and are fairly good egg producers, but are poor setters and seldom raise a brood.

The Rouen is a popular farm flock breed. It is slower growing than the Pekin, but reaches the same weight in 5 to 6 months. Its growth rate and colored plumage make it undesirable for mass commercial production.

The Muscovy, a breed unrelated to other domestic ducks, is also used to some extent in farm flocks. It is a good forager and makes a good setter. Muscovy males are much larger than the females at market age.

Select potential breeding stock from the flock when the birds are 6 to 7 weeks old. At that time, the sounds that ducks make can be a clue to their sex. Females have a definite sharp quack, while males have more of a muffled sound. One male to six females is recommended for breeding. Bring birds into egg production at 7 months of age.

Natural methods of incubation are frequently used on small farms. Muscovy eggs require 35 days to hatch; all other domestic duck eggs require 28 days.

Start ducklings on crumbled or pelleted chick starter, if duck feeds aren't available. Feed starter for two weeks. After that, ducklings can be fed a pelleted grower ration plus cracked corn or other grain.

Small flocks of ducklings raised in the late spring with access to green feed generally have few nutritional problems. While ducks are not as good foragers as geese, they do eat some green feed when allowed to run outdoors. Water for swimming isn't necessary for successful duck production.

Geese are excellent foragers and can live almost entirely on good pasture. Good succulent pasture or lawn clippings can be provided as early as the first week. By the time the birds are 5 to 6 weeks old, a good share of their feed can be from forage.

The Emden and Toulouse are the two most popular breeds. Emden are white in color; Toulouse vary from dark gray on their back to light gray on their breast. Normally the Emden is a much better sitter than the Toulouse. Other common breeds are the African and White Chinese.

Geese do not do well if enclosed in a house. They should be confined to a yard with a house for shelter during winter storms.

Larger breeds of geese usually mate best in pairs and trios (one male and two females). Ganders will mate with the same females year after year Do not change these matings, unless they prove unsatisfactory. Geese 2 to 5 years of age normally give the best breeding results.

Artificial incubation can be used to hatch goslings. The incubation period varies from 29 to 31 days. Follow manufacturer's directions for the machine being used. Goose and duck eggs require more moisture during incubation than chicken eggs.

Provide feed and water for goslings within 36 hours of hatching. Feeds formulated for geese are normally not available. So, start the goslings on a crumbled or pelleted chick starter. Wheat bread or corn bread moistened with milk is a reasonable substitute for a complete ration.

After the first 2 or 3 weeks, a pelleted chick grower ration can be fed, supplemented with cracked grain.

Considerable attention has been given to the use of geese for controlling weeds in cotton, strawberries, and some truck crops. For best results, start with 6-week-old goslings and provide shade and waterers throughout the field. Normally 2 to 4 geese per acre are sufficient.

Most geese are processed for home use or sale when they are 5 to 6 months old and weigh 11 to 15 pounds.

Goose feathers are valuable, if properly cared for. As much as a pound of feathers can be plucked from three geese.

Guinea Fowl

Prime young guineas are served in homes and restaurants the same as game birds. Their flesh is tender and has a flavor resembling wild game.

There are three principal varieties of guinea fowl—Pearl, White, and Lavender. The Pearl and the White are the most highly prized.

The sale of guinea hatching eggs, guinea chicks (keets), and guinea fowl for breeding is very limited. Only a few hatcheries have taken up baby-keet production, although several million guineas are raised annually.

At maturity, both sexes range from 3 to 3-1/2 pounds in weight. Males and females differ little in appearance. Sex can be distinguished by the cry of the birds after they are about 2 months old. The cry of the female sounds like "buckwheat, buckwheat" and is quite different from the one-syllable shriek of the male.

Guineas raised for breeding should have a growing diet in fall and winter prior to egg production, a breeder diet during the laying season, and a maintenance ration after laying is over.

Start feeding the breeder ration, containing 22 to 24 percent protein, about a month before eggs are expected. Commercial chicken or turkey rations will give satisfactory results.

Guineas will mate in pairs, if equal numbers of males and females are present in the flock. On most general farms, however, one male is usually kept for every 4 or 5 females.

The incubation period is from 26 to 28 days. Natural methods of incubation are generally used in small flocks. For larger flocks, incubators are more satisfactory.

Chicken hens are commonly used for hatching eggs, because guinea hens will usually set on eggs only in an outdoor nest. Also, chicken hens make the best mothers for the young keets. The guinea's attachment to the chicken hen tends to control its natural wild instinct and simplifies production and management.

Keets may be raised in confinement with the same kind of equipment and brooding methods used for turkeys. They also are fed much the same as turkeys.

Guinea fowl are ready for market in 15 to 18 weeks. At this age, their live weight is about 3 pounds with a dressed weight of 2-1/2 pounds. Most buyers prefer a dressed weight of at least 2 pounds.

Bantams. The word bantam means a small, miniature fowl. Today it may include any one of nearly 350 varieties. Most, but not all, of these miniatures are the likeness of a larger variety of domestic chickens.

Generally bantams are bred for beauty of color and form rather than ability to produce eggs or meat. But don't confuse them with unhealthy, midget, or unproductive dwarf chickens.

Bantams should be small. Generally adults will weigh between 16 and 30 ounces.

To get started with bantams, purchase at least a trio (one male and two females) of a breed and variety recognized by the Standard of Perfection. More than one breed or variety may be kept but a minimum of a trio is suggested. All adult birds should be separated by varieties.

Bantams often are exhibited at shows or fairs. The awards offered, the sport of competition, and the thrill of having produced a winner all add to the excitement and benefits of the bantam show. Bantams are interesting and many are relatively unusual. They are always a popular attraction.

How to Market Vegetables, Fruit, Some Other Items

By Martin Blum, O. Ray Stanton, and Raymond Williams

In preceding chapters production of fruits and vegetables was discussed. This chapter deals with ways to sell surplus products grown beyond immediate family needs.

Producers of farm products in less-than-wholesale quantities may have a number of sales options. Despite the mass distribution system of our economy, there can be opportunities for small producers to move their products directly to consumers at a profit.

By-passing wholesale marketing channels does not ensure that the seller will get a better price. Rather, in many instances it may be the only way that backyard gardeners or weekend farmers can market their produce.

Most marketing methods outside the mainstream of our conventional distribution network entail additional costs. In such cases the grower assumes the "middle man" function of marketing. Thus, a great deal more labor and management capability may be required than for wholesale marketing.

Consumers may not be willing to pay enough to cover the extra costs associated with some marketing alternatives. So any proposed operation must be evaluated in terms of its potential to increase income.

We have attempted to rank the various methods of marketing discussed on the basis of popularity with home gardeners and others who have a volume of production in excess of personal needs.

Roadside Marketing. Selling produce and related products

Martin Blum is Senior Agricultural Economist, Fruits & Vegetables, Cooperative Marketing & Purchasing Division, Economics, Statistics and Cooperatives Service (ESCS). O. Ray Stanton is Agricultural Economist, Natural Resource Economics Division, ESCS. Raymond Williams is Senior Cooperative Development Officer Regional Representative, Cooperative Development Division, ESCS.

at roadside represents the most typical method used by growers to market their production direct to consumers.

Development of roadside marketing has soared in recent years due to urban growth, rising incomes, expanding highways, increased customer mobility, and the desire to obtain fresh and high quality produce at reasonable cost.

Roadside marketing offers the grower an opportunity to sell at near retail prices and thus get a larger share of the consumer's dollar than by selling to processors, wholesalers, chains or supermarkets. The venture is relatively easy to start with limited capital. It appeals to individuals wishing to operate their own businesses, and provides family employment.

As with any type of business, roadside marketing has drawbacks. Some potential entrants may not have the personality to deal successfully with people or possess the managerial skill to operate a business, others may not be prepared to put in the long hours often required for six or seven days a week. The individual operator must also be prepared to bear the cost of promotion, merchandising, quality control and customer service.

"Roadside marketing" may range from placing a few tomatoes on a card table under a shade tree to permanent facilities along the lines of a supermarket. Experienced roadside marketers would recommend that the beginner start off in a small way and "grow" into the business.

Most roadside markets are part-time seasonal operations. They specialize in a limited number of fruits and vegetables, locally produced, and rely mainly on repeat sales to customers in the area. Transient trade usually accounts for only a small percentage of their volume. Some operators may purchase produce and additional items from neighbors or others to supplement their home-grown offerings.

Roadside produce stand.

John Messina

Location appears to be a key factor. Most operators tend to locate on roads fronting their property. But before you start up a market, determine the site's accessibility to potential consumers.

Ideally, roadside markets should be near towns or cities, on heavily traveled highways. Many operators recommend locating on the inbound side of a road to a city, or on a straight stretch offering good visibility.

Attention to ease of entry, parking, and egress is important. Neatness and attractiveness of the site and structures is also a plus.

Regulations

State and local zoning regulations, licensing requirements, sales taxes, labeling, sanitary, weight and Sunday operating policies vary. Check these requirements with the proper authorities. For example, construction of buildings near rights of way may be restricted, and access permits required.

Because customers prefer to make all their purchases at one stop, it is important to carry several items.

Top sellers among vegetables are tomatoes, sweet corn, green beans, and melons. Any fruits—particularly apples, peaches and strawberries—sell well and should be handled in season if possible. More commercial operations may include eggs, flowers, honey, Christmas trees and other specialties among their offerings.

In any event, customers expect to find locally produced items available in season. This calls for scheduling planting and harvesting to ensure adequate supplies for sale.

High quality, fresh home-grown fruits and vegetables are the backbone of a roadside market. This implies careful grading and sorting procedures, coupled with measures to maintain condition of the products on display.

However, customers often differ in their requirements for quality and their ability to pay. Market operators often find it good practice to offer two or more grades of the same commodity. Thus both quality shoppers and price shoppers can be satisfied.

In any case, represent products honestly to foster repeat sales. This is especially important with prepackaged items.

Preservation of quality while on display may entail refrigerating certain commodities. Highly perishable items like sweet corn or strawberries should be cooled. Others, particularly leafy greens, should be sprinkled periodically to compensate for moisture loss.

The means for maintaining condition—icing or mechanical refrigeration—will depend on sales volume of the operation. A small volume, of course, will not justify a large investment in mechanical equipment and facilities.

Pricing policy among roadside stand operators varies, but prices charged tend to fall between those in nearby city retail stores and wholesale markets. Selling above the retail price may be justified where quality is exceptionally superior. However, customers expect a somewhat lower price to compensate for the transportation and delivery services they assume.

Advertising

Most roadside marketers stress the importance of advertising to attract customers. The extent depends on the type of market, location, commodities and volume handled and other factors.

The best advertising is "word of mouth" by satisfied customers. But, being based on performance, it takes time to establish.

Signs placed on roads leading to the market are often effective on well traveled routes. Newspapers and radio are frequently used by operators to good effect. Direct mailings and handouts are other methods.

There are many other facets to the organization and operation of these markets. For example, many operators have found a pick-your-own operation a good adjunct to a roadside stand business. Others, especially those with small volume and few items to offer, have experienced satisfactory results using non-attended self-service stands. Honor selling, while not without hazards, can under some circumstances be profitable and provide a start to a full-fledged marketing operation.

Fortunately, much help is available to the prospective roadside marketer. Many State Extension Services and universities provide information through meetings and publications. Some State departments of agriculture also are active in this area. Experienced roadside operators can provide guidelines.

Roadside marketers in a number of States have organized into associations to provide themselves with information and services they could not provide on their own. Major services of the cooperative groups are advertising, promotion, centralized purchasing of marketing supplies, and setting industry standards.

State departments of agriculture often assist by providing inspection and certification services to the group. Members who qualify can display trademarks or certificates attesting to their compliance with the association's code of ethics.

Roadside marketing associations will likely play increasingly important roles in the future as recognition of the advantages of membership become more widespread.

Pick-Your-Own

The pick-your-own method of selling, as the name implies, is where the customer does his own harvesting in the field or orchard. This concept of consumers bypassing the usual marketing system has been growing in popularity.

From the customers' viewpoint, pick-your-own offers a chance to select the desired size and quality of fruits and vegetables. Assured freshness of product is, of course, a major motive for consumers to do their own harvesting. Expectation of reduced prices is another.

From the grower's standpoint, the pick-your-own method greatly reduces the need for harvest labor. In addition, grading, packing and storing functions are eliminated. Thus, net returns may be increased.

Consumer harvesting is especially adapted to crops which tend to mature and be harvested all at once, or where ripeness may readily be determined by color or size. Fruits and vegetables that are commonly canned or frozen in homes are good candidates.

Pick-your-own crops include apples, peaches, grapes, cherries, cranberries, strawberries, peas, green beans, potatoes, and even Christmas trees. Items whose stage of maturity are more difficult to judge, such as melons and sweet corn, appear less suitable for consumer harvesting.

Location and distance from population centers may be less critical to a pick-your-own operation than for other direct marketing approaches.

Location must necessarily be tied to the land where the crop is grown. Experience indicates that customers are more likely to travel longer distances to harvest their favorite crops, and often look on the activity as a recreational outing or a means for savoring their rural heritage.

A chief concern of pick-your-own growers is the possibility of damage to plants or trees by inexperienced handlers. However, experience has shown that with proper planning, excessive damage to the crop can be minimized.

Most customers, given a little instruction, will treat plants or trees with care, and reductions in yields—if any—should be minimal. Many experienced pick-your-own operators report that customers have a tendency to pick all the fruit or vegetables before them even though the items of better quality may have been spot picked beforehand.

Apparently, the satisfaction gained from picking one's own produce often overrides grade or quality considerations.

Another concern is the risk of financial loss or costly litigation from personal injury to customers while on the property. Although the possibility of accidents may seem small, the consequences of such occurrences can be devastating.

It is essential to make your property as safe as possible. Preventive steps might include prepicking fruit from tall trees prior to public entry, keeping ladders and other equipment in good repair, fencing off ponds, ditches and other possible hazards, supervising children—perhaps in a designated play area, and posting ground rules promoting safety.

Accident and liability insurance coverage is a "must" in any pick-your-own operation. Consult legal and insurance advisors before starting operations.

Attracting Customers

Potential customers need to be informed when crops are available for harvest and the times when picking will be permitted. Providing directions to the property is also vital.

A number of methods for presenting and publicizing the operation, including ads in daily newspapers and local radio commercials, might be considered. However, tailor the coverage of activity to the volume of production available and the capacity to handle the anticipated clientele. Otherwise, chaos can result.

For established small operations, you might develop a mailing list of prior customers. Mailed announcements of crops and picking dates may be sufficient to move the available volume.

Road signs are also useful to announce picking in progress as well as to direct customers to the site. On well traveled highways, signs alone may be sufficient to attract the desired number of customers. Placement of signs must conform to State and local regulations.

Parking for customers' cars deserves special attention. Provide safe and convenient space for the peak volume of cars anticipated. Weekends are usually the busiest periods. Keep in mind that pick-your-own operations require a longer stay by customers compared with other retail selling methods.

Plan the relationship of the parking area to picking sites with a view to reducing the carrying distance of often bulky and heavy produce. Locating the check-out point close to the parking area makes security and theft prevention easier.

Prices, of course, need to more than meet production costs yet be competitive for the area. Customers should sense that the operation is sharing the picking cost savings with them.

One basis for establishing prices might be to use retail store prices less harvesting and delivery costs, plus the additional marketing costs incurred under the pick-your-own operations. You might offer discounts for volume purchases.

Basis of sale can be weight, volume, or count. The types of produce sold and the check-out system used may influence your selection.

Many operators find selling by weight a fair basis for pricing most items. Often customers wish to use their own containers to maximize savings. Selling by weight facilitates this preference and eliminates possible disagreements.

The pick-your-own operator may have a number of sources to draw on for further insights into the business. In many States, the land grant college, Extension Service, or State department of agriculture have publications.

Some States sponsor workshops or conferences on direct marketing methods. Others may issue directories listing current operators and their offerings.

The serious operator would do well to explore availability of these services. Your county agricultural agent can be a first contact as to what assistance is available.

Farmer Markets

Roadside and pick-your-own selling may not work for growers because of their location. Farmers' retail markets situated in more distant, heavily populated areas may be more practical. Backyard gardeners and weekend farmers who are not located in areas where an adequate number of customers will come to them may choose to go to the customers instead.

Farmers' markets are one of the oldest forms of direct marketing. However, with improvements in transportation and refrigeration, their popularity had declined.

Currently there is renewed interest in this form of marketing, primarily by downtown business associations, community development groups, urban developers, historical societies, and real estate interests. Many view farmer markets as one way to rejuvenate downtown areas.

Preference by an expanding group of consumers for less processed and packaged foods and more fresh and "natural" foods has spurred the farmer market movement.

Both consumers and producers are being encouraged by State departments of agriculture and Extension Services to increase their participation in this form of market.

In a farmers' market, growers are provided a place to congregate and retail their products to consumers. Markets vary from simply a place where growers can park their trucks, to enclosed buildings with booths or stalls for produce and other types of retail establishments that attract more customers. Sponsors exercise differing degrees of supervision of trading dates, times, and sales practices.

To cover costs of facilities and operations, most farmers' markets charge a rental or service fee, usually quite moderate. Variations may exist because of the number of stalls or vehicles used, location within the market, seniority of the vendor, volume of products sold, and whether the market is used on a seasonal or daily basis.

Most markets operate seasonally. Some may open year-round, including those which combine wholesaling with retailing or provide space for other kinds of retailing businesses.

The typical farmers' market operates 2 to 3 days each week. Fridays and Saturdays are usually big market days.

Good management is a key to overall performance of a farmers market.

Aggressive advertising and promotion of the market, coupled with a neat and attractive appearance, are essential to draw and hold both customers and sellers.

Many markets are owned and managed by States, municipalities, or quasi-governmental agencies. Some are owned privately. Others are run cooperatively by the producers who use the facilities.

Many markets try to limit sellers to those who grow or make their own items. Regulation of peddlers or dealers who buy low quality merchandise from wholesalers and sell at a mark-up is often difficult. Unresolved, this problem can ruin the image of the market and lead to its eventual closing.

Most markets are located in the central city or close suburbs. Congested parking conditions usually prevail.

Popularity of farmers' markets is due to the wide selection of fruits and vegetables as well as the kinds of other items offered. Successful markets often include vendors who sell craft items, baked goods, ethnic specialties, honey, maple syrup, eggs, cut or potted flowers and plants.

Judgment and experience play big roles in setting prices. Besides checking wholesale and retail prices, you need to consider prices charged by competing vendors.

Farmer market customers tend to be price sensitive, and bargaining for produce may be more prevalent than under other selling methods. The competitiveness of these markets suggests somewhat lower price levels than for other types of retail operations.

Before venturing into a farmers' market operation, visit the market ahead of time to observe operations and gain some idea as to its potential. Since many markets open only one or two days a week, affiliation with several markets may be necessary to gain more sales days if this is desired.

Also study the costs associated with this type of marketing. Usually, considerable distances to market are involved, requiring outlays of time and money. In addition, figure out the time spent at the market to handle sales and the cost of providing packaging or containers in small-sized retail units.

Renting Garden Plots

A growing number of urban residents are interested in home gardening to offset higher vegetable prices and to gain the satisfaction of growing part of their own food. Owners of land near population centers may realize additional income by renting them plots.

Nearness to prospective renters is especially important because of the frequent working visits required.

A prime function of the owner, besides providing suitable land, is usually preparation of the site for planting. This could require a substantial investment in equipment. Therefore, carefully assess beforehand the number of potential renters reasonably expected.

Separate areas might be set up for renters engaging in standard gardening practices and those following an organic approach. The two types of gardening require different cultural practices and must be planned for.

Organic gardening involves applying of natural waste materials for fertilizer and natural methods of pest controls. It is not compatible with the usual techniques involving commercial fertilizers and chemical herbicides and insecticides.

Plots for the organic gardeners should be separated from the others by a buffer area.

The size of a standard plot needs to be determined. For example, a 1,000-square-foot parcel may be decided upon as representing a basic unit. This would allow for about 40 plots per acre and some space for access and parking.

Gardeners may provide their own tools and equipment, or the owner may have them on hand for rent. Likewise, supplies of seed or plants may be made available for sale by the owner if the renter prefers.

The owner who has a knowledge of cultural practices and is willing to spend a little time to provide assistance to those who require it may enhance his business.

Vandalism or theft can be a problem. Surveillance by the owner and fencing may help.

Determine the price to charge per lot on the basis of costs incurred in setting up and operating the enterprise. Cost factors include land preparation, equipment, fertilizers, pesticides, advertising, liability insurance, taxes and imputed interest on the land itself.

A written lease for the rental of garden plots can make clear what is expected from each of the parties involved and provide for their mutual protection.

Terms should cover location and size of plot, duration of the lease, times at which the gardener may gain access to the plot, rental fee and when due, services and facilities to be provided by the owner, protection against theft or damage to be provided, and cultural practices permitted.

Renting plots has profit potential for operators who have a good grasp of the costs involved and the ability to attract a clientele.

Rent-a-Tree. An interesting variation of the rent-a-plot concept is rent-a-tree, although the idea may have limited use.

Under the plan the owner designates a tree as belonging to the renter for a given period. The renter has a right to harvest the tree's entire output, and assumes all production risks. However, the owner provides maintenance services.

Its appeal to renters, usually families, is the opportunity to spend a day or two in the country and at the same time gather a quantity of fruit meeting their needs. Good orchard location is important to minimize losses due to frost. Exceptional promotional abilities on the part of the owner are vital.

Food Buying Groups

In the late Sixties a number of people who wished to avoid the traditional marketing channels, and who desired less expensive produce, joined to form food buying clubs. These clubs often purchased directly from farmers. Thus they were not faced with grading regulations, and were able to buy more fully field-ripened produce for their members.

Many clubs were loose neighborhood organizations. Farmers were contacted and asked for the names of other farmers who might be interested in direct selling. In some cases, most of the purchasing took place at a local farmers' market.

Many groups were formed on a preorder buying basis. One member collected the orders of others along with payment. The food was purchased from farmers and taken to a distribution point where members picked up their orders.

The groups began to band together as cooperative buying units. Through minor capital accumulation, food could be sold to nonmembers or on a walk-in basis. Some groups went on to form food federations and cooperative stores. Presently, these federations have a number of warehouses in various parts of the country.

One concept underlying many of these buying cooperatives is support of the small farmer, and another is questioning the consumer value of marketing orders. The influence of these cooperatives has helped place consumer members on some marketing boards.

Sales to such buying groups eliminate the need for a broker and partially solve the problem of crops not grown under contract which might otherwise be left without a market. In many cases, the farm prices paid are greater than those the farm family would realize under the formal market structure.

As the size of buying groups increased, the volume of purchases has allowed growers in some cases to enter into more common forms of sales contracts. The major difference is that several steps have been removed between farmer and consumer.

The extra-contract sales and individual grading may allow more complete sale of the production from small farms. This coupled with the increase in selling price caused by removal of some marketing steps can combine to increase the income of cooperating small growers.

Retail Routes

Retail home delivery routes in cities or towns at one time were commonly used by growers to sell produce direct to consumers. This method is no longer as popular, although a few delivery routes still operate.

Prospects are not bright for a general resurgence. Several factors work against it, including the time required to deliver small amounts of produce, the increase in working wives—making customer contact more difficult, and rising costs of labor and transportation. Also, municipal ordinances may be restrictive.

However, it might be possible to tailor a home delivery

program to a select clientele. Development of door-to-door selling to consumers on established routes on a recurring basis may have potential, particularly in large apartment complexes or resort areas.

Due to the high costs of this method, it would appear that only the handling of top quality products justifies the attempt.

Sale to Retail Stores

Most direct sales to retail stores are of seasonal crops, many of which have high handling and shipping costs. Perishability or weight may be major factors; examples are flowers, strawberries, cantaloupes, peaches, and pumpkins. The grower will incur a greater portion of the handling and shipping costs.

This form of sale may be accepted by even major retailers because it cuts handling and warehouse costs, supplies better quality produce, and supports local production.

An advantage to the grower is that such sales may provide a stable season-long outlet. Containers can be reused after delivery. An individual grading system may allow premium prices to be paid. If the grower's name can be placed on the retail shelf, this advertising may lead to increased farm or outlet sales.

There are problems. The initial contract for store delivery may be hard to get if the store does not have experience with the grower's reliability. And labor needed for transportation may require hiring an additional worker if a member of the grower's family is not available.

Under some arrangements the retailer may give the grower only a short lead time before stating the desired volume and time of delivery. The arrangement may demand precise timing of the picking or extend the harvest scheduling.

Adverse weather conditions may cause production to fall outside the contracted period. Problems also arise because the commodity is usually being handled at a riper stage than in other marketing channels.

Growers may tend to underestimate costs and set the contract price too low. The grower should estimate the additional capital necessary for packing and delivery. If small unit packing is necessary, allowance must be made for the packing unit specified and the increased labor involved.

Take into consideration the return on the investment of your time during formulation of the contract, handling and delivery, mileage driven in relation to volume sold, and the price difference over other market outlets.

Don't disregard such seemingly minor points as the time

of day of delivery, as well as the possibility of marketing numerous products at the same time.

Most supermarkets do not make cash payment. The delivery ticket is sent to headquarters where it is matched against an invoice from the grower before a weekly check is sent.

A preseason conference is needed to make sure the contractual agreement is understood by both parties. A postseason conference is usually desirable to maintain smoother working relationships. Some supermarket chains have field buying offices.

Organic Products

Many local retail stores strongly desire or demand certification of the commodity sold as organic. This may be done by personal inspection by the retailer or through certification of the grower by a regional organic farmers' association.

Organic gardening is a way of life which extends far beyond the marketplace. To some growers the market system is a way to distribute produce, not a place for personal gain. Their goal is a personal reward of enriched soil and good food rather than dollars.

The sale price of fresh organic products varies greatly. The retail price may be substantially increased because of the small scale of many retail outlets. Commodities which cost no more to grow organically than with petrochemical inputs may not, in some cases, bring a higher price in specialty outlets.

What is being sold often is taste and a belief. In some cases, this may conflict with trained consumer attitudes toward exterior appearance. In other cases, belief may waiver if the difference in selling price is great.

It is estimated that there are at least 172 organic gardening clubs with about 13,000 members, and 22 regional organic farming groups active in 27 States. About 40 major distributors service some 1,500 retail organic food outlets.

Over a million people subscribe to the magazine *Organic Gardening and Farming.*

Marketing of organic products is competitive. Individuals interested in such marketing should investigate the ramifications. A starting point could be to contact their regional organic farmer's association.

Local Restaurants

Most supply arrangements with local restaurants are handled on a personal basis and may differ greatly. High quality produce is a must. Due to variation in the menu, not all commodities may be sold at the same location every day.

Prompt delivery is essential. If the commodity is on the menu, the restaurant owner must purchase it somewhere. If the grower neglects to deliver, this has to be done on the restaurateur's own time. Volume of delivery also may vary greatly between days of the week.

As the seller, remember you are competing with the frozen and canned food industry. Not all restaurant patrons or restaurant owners care to discriminate between the produce from different marketing channels.

In relation to the volume sold, consider mileage driven, number of stops, length of time at each stop, and the need for dealing on a personal basis with the restaurant employees or owner at each stop. At times, arrangements may be made to trade the produce for restaurant meals.

Constant contact with the restaurant owner or purchaser for a small institution is necessary to ensure future sales.

Due to the loose nature of the arrangement, you may face the problem of competition from other growers, some of whom may be relatives or close friends of the restaurant owner.

Supply to the larger institutions and restaurants may be on a much more formal basis and call for a written contract. Many State-run institutions have their own source of supply.

Canning Foods

Community canning centers enable a number of neighbors to operate a facility to can surplus vegetables or fruit.

While most produce canned is subsequently consumed by the individual family, some is sold to others in the immediate community or to local stores or roadside markets (such sales may be closely regulated, however).

Normally the canning center is equipped with more sophisticated equipment than the individual family can afford. Each family may be charged a fee per quart or gallon, etc., for use of the facility and equipment.

Community canning centers may be operated by a private company, an individual, or people of the community. The community, town, county or State may sponsor the facility. At one time there were approximately 3,600 canning centers in the United States. A good number are still operating and a few new ones have begun operations.

Besides the usual fruits and vegetables, it is also possible to can fish, poultry and red meats. Due to contamination problems, however, professional supervision and assistance may be necessary.

The Center for Community Economic Development, Cambridge, Mass., has published a book, *Community Canning Centers* ($2.75), with very complete information on cannery costs and uses. The Ball Corporation of Muncie, Ind., can provide information on cannery equipment it sells.

Home Canning. Some families take advantage of marketing surplus home produce by canning vegetables and fruits as specialty items. Generally, the specialties are sold to local novelty stores, restaurants that have gift shops, or roadside stands.

In a few instances, a number of producers have formed marketing associations or cooperatives to market their products under one label.

A community canning center can also be used.

Home specialty items are usually canned in small volume glass jars or cans. Some customers buy the glass jars of fruits or vegetables for decorative purposes.

When canning, exercise caution to prevent contamination. Comply with local, State or Federal regulatory laws.

Mail Order shipping for retail is not regulated by the U.S. Department of Agriculture unless the volume of sales is $100,000 or more per year. Some States have regulations concerning commodities allowed into the State and the times of year when shipping may take place. An example is the restriction upon shipping avocados into California.

Salability is improved if the grower is from an area nationally known for quality production of the commodity sold. Examples are Michigan apples and Vermont maple syrup. Seasonal oddities are an exception to this rule and may receive a very high price in restricted markets.

Many purchasers are reluctant to buy from smaller concerns. This is especially true of highly perishable commodities.

Shipping and packaging costs may add substantially to the grower's initial production cost. In some cases, these costs may put the price of the final product way above the supermarket price.

Remember that regardless of the condition of the produce when shipped or mailed, the marketer is judged by its condition on arrival.

Cost of advertising also must be taken into account. Becoming a member of a direct mailer's association should be considered.

Auction Markets

Of all commercial marketing outlets, the produce or fruit

auction probably offers the best opportunity to the small grower. Most auctions are open to the public. Quantities may be small or large.

The typical auction market has facilities that permit one or more lines of vehicles such as pickup trucks or cars to pass through the "auction block" at the same time. For large producers, the fruit or produce may be a sample of that in the field or on a tractor trailer nearby.

Buyers can observe the quality of the fruit or produce as the handler opens the baskets, boxes, etc., for their observation. The buyers are looking for quality items, so the small growers can compete for price and is not discriminated against.

Chain stores, operators of small locally owned stores, and private individuals are often among the buyers. Revenue to operate the auction comes from a "charge per unit" or a percent of the sales price.

There are numerous other ways to market fruits and vegetables. Although most packing sheds, grower-shipper arrangements, processing plant contracts, or sales through a broker are geared toward commercial operators, sometimes small producers can use them.

The grower-shipper arrangement generally represents a producer with sufficient fruit or vegetable volumes to ship directly to terminal markets or to other large wholesale buyers such as chain stores. A small grower would need to establish a personal working relationship with a local grower-shipper to participate.

Packing sheds can be operated by individual entrepreneurs, cooperatives, corporations or other organizations. They basic-

Left, members of marketing co-op examine tomatoes being grown for early fresh market sales. Right, tomatoes being graded and packaged for market.

C. R. Roberts

ally are central assembly points where fruit and vegetables are washed, graded, packed and shipped to market.

The sheds may purchase the produce or ship it on consignment, in which case the producer is paid after the sale takes place.

Production and sales to processing plants are on a contractual arrangement that provides little opportunity for the small grower to market fruits and vegetables. There are exceptions in certain concentrated vegetable areas, particularly in the South and East.

Fruit and vegetable brokers assemble huge volumes of produce to satisfy the needs of large buyers. Brokerage sales generally are on a consignment basis. Normally the grower's identity is maintained even though the produce may be commingled for large orders.

Marketing Co-ops. About 25 percent of all fruits and vegetables are marketed through some type of marketing cooperative. These co-ops are organized for the benefit of members. Members are limited to individual producers, whether they have a small garden in the backyard or a large farming operation of many acres. By forming co-ops, these members are attempting to attain maximum prices for their produce.

Services by the co-ops range from simple pooling to operations that include many of the following: harvesting, grading, packing, cooling, storage, transporting, and providing marketing and production supplies. Some cooperatives also are engaged in processing.

Members bring their produce to a central location, where

C. R. Roberts

the co-op usually commingles products, thus reducing each member's risk of loss due to spoilage, market decline, etc.

Some of these co-ops are linked to larger regional co-ops in hopes of establishing a reputation for high quality products and demanding a price for them. To this end, many have their own brand names that are marketed regionally or nationwide.

As a member, an individual has voting privileges in establishing operating procedures and also via capital stock ownership of the co-op.

Any excess profits (net savings) made during the operating year, after allowances for operating reserves, are returned to members in the form of patronage refunds based upon the individual's participation in the co-op.

Further Reading:

A Review of Roadside Marketing Literature, A. E. Departmental Series 362, Ohio Agricultural Experiment Station, Columbus, Ohio 44691. 1964. Free.

An Economic Analysis of Roadside Marketing in Georgia, Research Report 254, Department of Agricultural Economics, University of Georgia, College Station, Athens, Ga. 30602. 1977. Free.

Community Canning Centers, Center for Community Economic Development, Cambridge, Mass. 02139. 1977. $2.75.

Direct Marketing of Produce, The Shelby County Farmers' Market Case, Bulletin 569, The University of Tennessee, Agricultural Experiment Station, Knoxville, Tenn. 37901. 1977. Free.

Economics of "Pick-It-Yourself" Vegetable Production and Marketing, Research Report 274, New Mexico State University, Las Cruces, N. M. 88001. 1974. Free.

Farm Roadside Marketing in the United States, Food Business Institute Cooperative Extension Service, University of Delaware, Newark, Del. 19711. 1965. $2.50.

Farmer-to-Consumer Direct Marketing Act of 1976, Committee on Agriculture and Forestry, Report No. 94-1022, Calendar No. 965, Bill S.2610, June 30, 1976, House Documents Room, U. S. Capitol, Washington, D. C. 20515. Free.

Management of Pick-Your-Own Marketing Operations, Cooperative Extension Service, University of Delaware, Newark, Del. 19711. 1975. $2.50

Managing the Roadside Farm Stand for Fun and Profit, Extension Publication 25, Cooperative Extension Service, University of New Hampshire, Durham, N. H. 03824. Free.

Rabbits Suited to a Few Acres, and Capital Outlay Is Small

By H. Travis, R. Aulerich, L. Ryland and J. Gorham

Rabbit raising is well adapted to a few acres. Capital investment and land required are small compared to other livestock enterprises.

About 200,000 producers raise 6 to 8 million rabbits in the United States each year. About 8 to 10 million pounds of rabbit meat are eaten annually, some of it imported. Laboratories use 600,000 rabbits yearly.

Before starting to raise rabbits, there are two points to consider—how large will your operation be and what is your market?

If you intend to raise just a few rabbits to supplement the family meat supply, you can consider a unit of 3 or 4 does and a buck. One doe will produce 25 to 50 rabbits a year, or about 50 to 100 pounds of meat if they are raised to fryer size, more when raised to roasters. If you are interested in a part-time business, you might expect to establish a herd of 50 to 150 does.

The amount of space and cost of equipment are not great to start a small rabbitry. Breeding animals cost $15 to $25 apiece. Capital invested per pen for all-wire cages, feeders and automatic waterers is about $20 to $25. Cost of housing varies with the climate.

If you have no previous experience, start small with a buck and a few does.

Before making any commitment to raise rabbits, investigate your potential markets. There are three general markets—meat, laboratory supply, and breeding stock. There also is a small market for hides and Angora rabbit wool.

Hugh Travis is Research Animal Husbandman, U.S. Department of Agriculture (USDA). Richard Aulerich is Associate Professor, Department of Poultry Science, Michigan State University, specializing in fur animals and rabbits. Lennox Ryland is a Research Assistant, Department of Veterinary Science, Washington State University. John Gorham is Research Veterinarian, USDA.

The opportunity to sell animals for various purposes varies with different areas of the country. Areas of concentrated rabbit production are California, the Ozarks, Florida and the East Coast. Processors are found in 14 or more states and Canada.

Animals sold for meat may be sold alive to processors or to customers who kill and dress the rabbits themselves.

Potential meat producers can obtain information on voluntary grading and inspection of rabbits by writing to the Food Safety and Quality Service, U.S. Department of Agriculture (USDA), Washington, D.C. 20250. State laws governing the sale of dressed rabbits vary, and you should be aware of the laws of your state.

If you plan to sell rabbits to laboratories, you need a license from USDA. You can only expect to sell breeding stock after you have become known as a producer of exceptional animals.

Market Income

In descending order, the price received for your product will be for breeders, laboratories, dressed, live retail, and live to processors.

Selling to laboratories requires an established reputation and the ability to supply numbers of certain types of animals at specific times. You may be able to start out by selling to a middleman or breeder who supplies laboratories. Prices for laboratory animals to the middleman are about a dollar a pound.

Retail selling of dressed or live animals calls for establishing a local market. Selling to a processor is the easiest way but will also yield the lowest return (about 50¢ a pound live weight). In selling to a processor, it is desirable to have a year-round contract.

Every two months the American Rabbit Breeders Association (1925 S. Main Street, Bloomington, Ill. 61701) publishes a list of commercial processors and current market prices for fryers (1-1/2 to 3-1/2 pounds, under 12 weeks of age) and roasters (over 4 pounds, over 8 months of age), along with the caution that any meat producer living more than 150 miles from a processor should investigate local markets to avoid expensive shipping costs.

Beware of buy-back schemes. These are deals in which people offer you breeding stock, usually at exceedingly high prices, and in return promise to buy back all the animals you produce.

Fryers weighing 4 to 6 pounds live weight bring 40¢ to 60¢ a pound, or about $2 for a 4-pound rabbit. Dressout percentage is about 50% to 55% so that you have 2 pounds of dressed meat. Feed costs about 10¢ a pound, and it takes 12 or more pounds—or about $1.20 worth of feed—to produce a 4-pound fryer and maintain the doe.

This leaves you 80¢ to pay for labor, breeding animals, equipment, etc., for the three months from conception to marketing that it takes to raise the animal.

Selecting a Breed

There are about 38 breeds and many more varieties of rabbits raised in the United States. Breeds can be categorized by size. Mature animals of the smaller breeds weigh 3 to 4 pounds each, those of medium breeds 9 to 12 pounds, while adults of larger breeds weigh 14 to 16 pounds. Select a breed based on the purpose your rabbits will be used for.

Animals best suited in size and conformation for producing meat are the medium-sized breeds. These will produce meaty, fine-boned fryers weighing 4 pounds at about 8 weeks of age. New Zealand Whites are the most popular breed raised for meat, followed by Californians.

You can obtain information on where to buy rabbits from local breeders, rabbit clubs, ads in rabbit magazines, and the directory of the American Rabbit Breeders Association.

Before attempting to sell to a laboratory, determine its needs. Check with nearby hospitals, laboratories and health department offices to find out the type, age and size of animals desired.

Two popular breeds of meat rabbits are New Zealand White (left) and Californian.

Angora rabbits are raised for their wool, which is spun into yarn used for making garments. Usually hand spinners raise their own rabbits rather than purchasing wool from rabbit breeders.

Housing Equipment

Locate the rabbitry on a site with good drainage. Check local zoning regulations first.

Housing varies with the climate. In mild areas, hutches can be placed out-of-doors in shade, or provided with shade by open shed-type buildings. During very hot weather, you may need to cool the rabbits by overhead sprayers or foggers placed within the building.

In more severe climates, put hutches in buildings that give protection from the prevailing winds. During stormy weather, use drop curtains or panels. Where weather is extremely cold, extra protection is needed.

Rabbitries in the northern tiers of states usually supply supplementary heat with space heaters to about 40° F in areas where the young are kindled (born). Supplemental heat may also be supplied to the young by suspending light bulbs over nest boxes. Proper ventilation is important when the animals are raised in enclosed buildings.

Worm raising is often successfully combined with rabbit raising in moderate climates or indoor operations because worms will consume the feces and any spilled feed, thereby eliminating odor, waste and some labor.

Earth floors permit absorption of urine into the soil. Alternatively, concrete floors with drains will facilitate waste disposal and lend themselves to easy cleaning with a hose.

Where wire-bottom cages are arranged in double or triple tiers, metal collecting pans must be placed underneath all but the bottom layer of cages to collect both urine and feces. Empty and clean the pans at least twice a week.

Hutches for mature rabbits are about 2 feet high and no more than 2-1/2 feet deep. A length of 3 feet is recommended for small breeds, 4 feet for medium, and 6 feet for large breeds. All animals should be raised on wire floors of 1/2" x 1" mesh to reduce potential disease problems.

All-wire hutches are the most satisfactory, the most expensive, and are recommended for commercial rabbit raising. For smaller rabbitries, a combination wood and metal hutch can be used.

Cages constructed in units of two or three have better resale value. They may be arranged in single, double or even triple tiers if space is limited, although a three-layered arrangement makes cleaning difficult and interferes with observation of the animals. A few extra cages should be available to be used as hospital or isolation units for new or sick rabbits.

A nest box placed in the hutch prior to kindling will supply seclusion for the doe and protection for the litter. During cold weather, these can be insulated with bedding such as straw, wood shavings, or sugar cane waste, with two or three layers of corrugated cardboard on the sides.

Feed and Water

Rabbits may be fed from feed crocks, hoppers or hay mangers. Feed hoppers of the proper design and size save considerable time and labor.

Watering equipment includes crocks and automatic watering systems. Use crocks with the bottom smaller than the top so that ice will not crack them. Automatic watering systems can be set up with electric heating cables to provide water during freezing weather.

The rabbit is herbivorous and lives on grains, greens and hay. A majority of commercial rabbit raisers feed pellets produced by commercial feed manufacturers, which are formulated to contain all the nutrients required for a balanced ration.

Requirements vary with the life stage of the animal. Dry does, herd bucks and junior does need rations that will keep them in good breeding condition. Pregnant does and does with litters need more nutrients to fulfill their added requirements.

Pregnant females should be allowed all they can eat unless they become overly fat, in which case 6 to 8 ounces of commercial pellets a day will supply adequate nutrition while limiting weight gain.

High-quality hay and grains such as oats, wheat, barley, rye and ground corn may be substituted for pellets. In this case the hay should be cut into 4-inch lengths and supplemented with 2 ounces of grain per day. You must decide whether the increased labor of obtaining and feeding these materials is balanced by the decrease in food cost.

Green feeds may cause digestive upsets and should be limited. Unless self-feeding hoppers are used to save labor, only as much feed as can be consumed in a day should be provided in order to avoid wastage and contamination.

Rabbits re-ingest part of their food. They excrete two kinds of feces, one hard and one soft. They eat only the soft. The habit (coprophagy) starts at about 3 to 4 weeks of age. This process is normal and enhances value of the food consumed.

Rabbits require large quantities of fresh clean water. Salt should be provided either in the pellets or from salt blocks placed in the hutch.

Breeding, Care of Young

Proper age for the first mating depends on breed and individual development. Smaller breeds develop more rapidly and are sexually mature at 4 to 5 months, medium breeds are bred at 6 to 7 months, and giant breeds at 8 to 10 months. Males may mature later than females.

The average gestation period lasts 31 or 32 days. Length of the nursing period varies with systems of management. If the young nurse the mother for 8 weeks, a doe can produce 4 litters a year.

Some commercial breeders use weaning intervals of 21, 28 or 35 days after kindling to obtain 5 or 6 litters a year. Backyard operators in northern areas may find it more practical to have 3 litters and skip kindling during the winter months.

When in heat the female will show a red coloration of the vulva and may become restless, rubbing her chin on the hutch. It is not necessary to depend on external signs to determine when a doe is to be bred. Set up a definite schedule and follow it whether the doe shows signs of being ready or not.

Move the doe to the buck's hutch for service. Mating should take place almost immediately and the doe can be returned to her own hutch.

Because ovulation occurs only after copulation, some breeders leave the doe with the buck for two successive matings to insure adequate stimulation for egg release. Other breeders prefer to remove the doe immediately after the initial mating to avoid the possibility of fighting and injury.

Does may be reintroduced into the male's cage 2 to 7 days after mating. If she rejects him, she's probably pregnant.

For greater accuracy in determining pregnancy, does may be palpated 10 to 14 days after breeding. While restraining the doe's head and shoulders in your right hand, place your left hand in front of the pelvis between the hind legs and with a gentle pressure, feel to detect any marble-sized embryos.

Nonpregnant does should be rebred immediately. Commercial breeders may keep one buck for up to 10 does. Smaller breeders need more bucks.

Place a nest box in the hutch on the 20th day after breeding. The mother will usually pull fur from her body to line it.

In warm climates, an inch of bedding material such as sugar cane should be sufficient. In colder climates, additional bedding material and an electric light bulb over the nest box will help provide warmth for the young.

Most litters are kindled at night. As soon as the doe has become quiet, inspect the litter and remove any dead or deformed young. Baby rabbits from large litters may be placed with foster mothers having smaller litters at this time. Average litter size is 7 or 8.

Watch litters closely for the first few days to make sure they are being well cared for and fed. Keep records on all production does and litters so that low producers and poor mothers may be culled. Good does will continue to produce maximum-sized litters for 2 to 3 years.

Rabbit Diseases

In all cases of infection or parasite infestation, isolate affected rabbits from the rest of the herd and consult a veterinarian. Prompt and accurate diagnosis, followed by proper treatment, can save you time, labor, and money. Profits rapidly shrink when disease kills or stunts even a few rabbits.

Rabbits are susceptible to a variety of bacterial, viral, parasitic and fungal diseases which may pose real problems.

Species of Pasteurella, for example, are responsible for some of the most common and varied bacterial diseases, including uterine infections, mastitis, orchitis, otitis media, conjunctivitis, subcutaneous abscesses, "snuffles", pneumonia and septicemia. Of these, the upper respiratory diseases, such as "snuffles", may be the most dangerous because of the threat of widespread contagion.

Most common parasitic disease of domestic rabbits is coccidiosis. The causative protozoa may invade the liver or the intestine, and can be identified microscopically in the feces.

While rabbits may tolerate small numbers of some coccidial species, large numbers of the pathogenic intestinal varieties cause diarrhea, loss of appetite, weight loss and even death. Severe infections of the hepatic form cause loss of appetite, pot-belly, and occasionally death.

Either form of the disease may be effectively treated by adding 0.025 percent sulfaquinoxaline to the feed following diagnosis. But since the parasite may eventually develop a tolerance to this drug, disease control depends primarily on management practices which reduce fecal contamination of feed, water and housing.

Young, susceptible rabbits should not be overcrowded. Where coccidiosis has been diagnosed, disinfect cages with a 10 percent ammonia solution.

Mink Require Savvy to Raise; Market Is a Roller Coaster

By Hugh Travis and Richard Aulerich

Fur farming began as an extension of the trapping of wild animals prized for their fur. Species that have been successfully and profitably raised for their pelts in the United States are the silver fox and mink. Imports from South America—the chinchilla and the nutria, or coypu—also have been raised.

Mink is currently the primary fur animal raised for its pelt, making up about 75% of the dollar value of the U.S. fur trade. So we will devote our chapter to the raising of this species.

Mink have been kept in captivity as far back as 100 years, but were not raised in large numbers until the 1930's. Numbers increased until the mid-1960's when more than 8 million pelts were produced each year in the United States. Starting in the late 1960's, the supply of mink exceeded the demand and prices dropped below the cost of production.

From 1963 to 1973, the number of U.S. mink farms decreased from over 5,000 to 1,300. Those that went out of business were primarily the small producers or hobbyists. At the same time, average production of mink farms increased from 1,800 to 2,900 pelts per year. Currently there are about 1,000 mink farms producing 3 million pelts.

During the 1970's, prices received by mink farmers have averaged $15 to $30 per pelt. Cost of production is about $20. Capital investment per animal for equipment is about $30 to $50 for an average size farm, excluding land and operating costs.

Most fur farms are located in the northern states. Wisconsin is the number one state producing about 30% of the mink raised, followed by Minnesota, Utah, Illinois, and Washington.

Hugh Travis is Research Animal Husbandman, U.S. Department of Agriculture. Richard Aulerich is Associate Professor, Department of Poultry Science, Michigan State University, specializing in fur animals and rabbits.

While fur animals can be raised on a small acreage, their husbandry is different from other enterprises described in this book.

Raising fur animals requires specialized knowledge that is more difficult to obtain than on other species of livestock. Few state or federal experiment stations have personnel qualified to advise on their raising.

Capital investment required to conduct a profitable business is large. Odors from mink and fox farms are objectionable and may require a buffer area between you and your neighbors. And unlike most enterprises in this book, producers cannot use the products for home consumption.

Market values are based on the whims of the fashion industry. There may be several years of good prices followed by several unprofitable years. These cycles are impossible to predict. They can be very discouraging if you happen to enter the business on a down cycle.

Starting Out

So—if after reading the previous information, you are still interested in entering the fur business, how do you go about it? The best way to learn to raise fur animals is to get a job on an established farm for at least a year.

You also can start by buying a few animals and gradually

Mink are kept in individual wire pens most of the year so they won't bite each other and damage the pelts.

building up the size of your herd. This is difficult if you don't have previous experience or a qualified person to guide you.

The best place to start a fur farm is in an area where other farms are located. This will allow you to learn from others and be able to cooperate in buying materials needed for fur farming. In several areas of the country there are cooperatives that mix and distribute feed to fur farmers on a daily basis (Pennsylvania, Minnesota, Utah, Washington, Oregon).

Purchase animals from a reputable, established breeder who has obtained high average pelt prices over the years. Field days, live animal shows, and pelt sales at auction houses are excellent places to observe animals and pelts of known quality. Prices for breeders are usually two to three times the pelt price.

There are many colors of mink produced from various mutations that have occurred from the original dark mink. It is advisable to start with the easier-to-raise color types such as darks or pastels.

Raising Mink

Mink are kept outdoors in raised wire cages, one to a pen during most of the year. Usually the cages are covered with a roof to form an open shed.

For raising young, the mother is supplied with a nest box about one foot in each dimension, which contains bedding for a warm nest. Pens for mothers and litters are about 30 to 36 inches long x 18 inches high x 18 to 24 inches wide. Individual pelter pens are about 24 to 30 inches long x 15 inches high x 14 to 18 inches wide.

Cages have a water cup or automatic waterer and a door. Sometimes a feeder is supplied.

Besides pens and sheds, mink raisers need a place to pelt and to store mink pelts. They also need an area for feed preparation and storage. Size depends on the numbers raised, type of feed, and length of storage.

Mink usually are fed a combination of fresh or frozen packing-house and fisheries byproducts along with cereals and vitamin and mineral supplements. This is mixed to a hamburger consistency and fed in a feeder or on top of the cage. Mink are also fed dried or pelleted feed which can be purchased from feed companies.

Raising mink is a vigorous outdoor activity. Mink are savage, difficult to restrain, and have a musky odor unpleasant to many people. They must be fed and watered every day, and the caretaker must expect to perform physical labor in all kinds of weather.

There is a yearly cycle of husbandry duties. The breeding season occurs in March. Young are born in May, usually four to six in a litter. Generally one male is kept for each three to five females. The young start taking solid food at about three weeks and may be weaned after six.

Young mink grow their marketable pelt during the autumn months, and the fur becomes prime around Thanksgiving. They are graded at this time, the better animals being kept for breeders.

Pelting consists of killing, skinning, scraping excess fat from the skin, and drying and stretching the pelts so they can be easily stored. After the pelts are sent to market they are further processed to be suitable for garments.

Pelts generally are sold through auction houses, which match the pelts into uniform bundles suitable for making garments. The bundles are then sold at public auction.

Mink are healthy animals, seldom sick when properly fed and cared for and if reasonable sanitary methods are followed. The pens are raised off the ground, which reduces chances of disease.

Mink can contract food-borne and infectious diseases such as botulism, distemper, and virus enteritis, which can be prevented by vaccination or antitoxins.

They also are subject to a slow-acting virus which produces Aleutian disease. While this disease cannot be treated, it can be diagnosed. Thus it is possible to eliminate infected animals from your herd, and to purchase animals free of Aleutian disease.

Mink kits are born in May and reach mature size by Thanksgiving. Adult males weigh about 4 pounds and females about 2.

Beef Cattle May Be Easy Way to Put Pastures to Work

By Kenneth G. MacDonald

If you have pasture and hay land on your few acres, raising some cattle for beef can put meat on your table and a little money in your pocket perhaps more easily than any other way of farming. You can use land not suited for other purposes, and in some cases equipment and buildings that otherwise would just gather dust.

Beef cattle thrive on a wide variety of land types, forages and pasture. A lot of small acreages are rolling and must be kept largely in grass. A beef cow herd is one means of selling this grass.

You can start small. Three beef cows with calves can be expected to develop into 30 females in ten years if all the good heifers are kept. Besides that, there have been some steers to sell and the family meat supply provided for with a minimum investment.

A reliable supply of water is of prime importance. If you don't have it, don't try to raise beef calves.

Each cow should have at least two acres of good pasture.

If your pasture land is covered with brush and scrub trees or rock, five or more acres per cow will be required. Shelter for beef cattle can be kept to a minimum.

Good hay will be needed. If hay is the only source of winter feed, a cow requires about 20 pounds of mixed hay (grass-legume) per day or about 2,400 pounds of hay during a four-month winter feeding period.

In most cases the management time spent with a cow herd is at best minimal, so it's wise to buy cows of breeds that are unlikely to have much calving difficulty.

In other words buy the traditional British beef breeds, Hereford, Angus, Shorthorn or crosses from them. The num-

Kenneth G. MacDonald is Associate Professor of
Animal Sciences, Purdue University,
West Lafayette, Ind.

ber of cattle of these breeds or their crosses are also more plentiful.

Cattle of the so-called exotic breeds are apt to have more calving difficulties than the British breeds. Selecting cows of these breeds would not be advisable for the small part-time operator who in all likelihood will be away during calving time.

In addition the continental breeds—such as Charolais, Limousin and Simmental—are not as well adapted to marginal management and feeding situations. But if a small herd owner is willing to accept these facts and provide the extra management needed, the continental breeds can be very productive due to their additional growth and muscular ability. To obtain the best results, these larger and more muscular breeds should be crossed with bulls of the British breeds.

Artificial breeding in a small herd makes crossbreeding more practical because it does away with the need of keeping bulls of different breeds. Crossbreeding can improve pounds of calf produced per cow by 10% to 20%.

Whatever system you pick, crossbreeding or straight breeding, use the best bulls available.

Fencing, Housing

Fences should be planned carefully and constructed properly. Woven wire, barbed wire, a combination of these, or boards can be used for permanent fences. Woven wire normally is used in situations of high animal pressure or where cattle and sheep may share the same pasture. Barbed wire will control cattle under most farm conditions.

The secret of good fences is having well set corner posts to which the fence can be fastened and stretched.

A two-strand electric fence makes an ideal temporary fence. Electric fences are best used to subdivide fields for improved pasture management rather than as line fences or perimeter fences. Electric fence wires can be attached to wood or steel posts by using insulators, or fastened directly to fiberglass or plastic posts. For safety, use fence chargers with the Underwriter Laboratory (U.L.) seal. Locate gates in or near fence corners rather than in the middle of line fences, to facilitate the movement of cattle out of a field.

Beef cows need only minimum facilities and do best if kept outside under most conditions. The digestive process that takes place when forages are fed produces large amounts of heat that is used to maintain the cow's body temperature.

The critical temperature of a mature beef cow adapted to

a cold environment is 0° F or below. When the temperature goes much under zero the cow will need extra feed to furnish the energy needed to keep her warm.

Weather conditions during which a beef animal will seek shelter are wind and/or cold rain. Access to a windbreak, a woodlot, or an old barn will usually take care of this need.

Have a calving pen or two under a roof where cows with new calves can be held one or two days until it is apparent the calf is being taken care of. And if an animal needs treatment it can be handled or observed much easier when confined to a pen of this type. A 10 foot x 10 foot pen is adequate.

One of the most serious mistakes a small herd owner can make is to feel sympathy for the cattle and shut them up in a barn. This brings on the problem of scours and contributes to respiratory trouble.

Managing Pastures

Permanent pastures are basic to a cow-calf enterprise. Bluegrass usually makes up the forages, along with wild white clover, if it is encouraged by adding fertilizer and is kept closely grazed. A good stand of grass alone will respond to the application of nitrogen.

Permanent pastures can be improved by good management, liming applying manure and fertilizers, and seeding with pasture legumes such as alfalfa.

Good management includes mowing to control weeds, and no overgrazing. Permanent pastures vary greatly in their carrying capacity. Some produce less than 50 pounds of beef per acre. With improvement, these pastures can be made to produce five to ten times that much.

The problem with permanent pastures is they go dormant in July and August. During these months a few acres of improved grass-legume pasture can be invaluable. In years of short summer rainfall, feeding hay may be the only way to carry cattle over this period of short feed supply until fall rains bring permanent pasture back so they may be grazed until early to mid November.

Rotation pastures that are well-managed, fertilized and composed of productive grass-legume species will have a carrying capacity much higher than a permanent pasture. Alfalfa is the most commonly used legume in conjunction with grasses such as Brome, Orchardgrass or Fescue.

Legume-grass pastures remain productive throughout the summer. Grass pastures without legumes peak in spring and early summer and are not too productive in late summer. When

336

alfalfa is used in mixtures it should be rotationally grazed because it cannot withstand constant grazing pressure.

Generally, straight grass pastures should be fertilized with nitrogen, and grass-legume pasture with phosphate and potash. However, use soil tests to determine exact fertilizer requirements.

Permanent pastures such as bluegrass, Brome, or fescue that have not been pastured during the late summer or fall provide much winter feed for such cattle as replacement heifers and brood cows. Winter pastures should have good natural drainage to lessen the damage from trampling in wet weather.

It may be advisable to harvest the first cutting of hay from fields that are intended to be left for winter pasture. If round bales are used, the second cutting can be baled and left in the field. Strip grazing these round bales will increase carrying capacity of the winter pasture.

Timber or hills provide protection from wind, and cornstalk fields located near winter pasture will furnish additional winter feed.

In the Corn Belt, cornstalk fields may be grazed during late fall and winter providing the snow does not get too deep. Two acres of good cornstalks that yielded 100 bushels of corn per

Good pasture management will produce more beef. A rotation pasture is sometimes preferable to a permanent pasture.   George Robinson

acre will carry a cow for about 80 to 100 days. A loose mineral supplement along with trace mineralized salt must be furnished. In addition, a protein supplement fortified with Vitamin A must be fed at the rate of one pound per head per day using a 40% supplement as the basis of this recommendation.

Cows will graze more palatable portions of the corn plant first. They will go after whatever grain is present, followed by the leaves and husks, and then the cobs and stalks.

You cannot expect to recover 100% of the corn plant residue by grazing. From 15% to 30% of the potential dry matter present will be recovered by the cow in grazing a stalk field. Thus if the yield of residue dry matter is about 2 tons per acre, expect to recover 0.3 to 0.6 tons per acre of feed.

Feeding Brood Cows

During lush pasture growth, brood cows obtain nearly all the nutrients required for carrying an unborn calf or for suckling a calf. During droughts or long severe winters, however, pregnant or lactating beef cows may not get enough nutrients, thus injuring the cow, the calf, or both. At these times a small amount of good alfalfa hay is valuable to beef cows.

The cow, as a ruminant, manufactures many but not all of the specialized nutrients required for good health. Certain basic requirements must come from her feed. A brief discussion of the nutrients required and how they can be supplied follows.

The largest portion of the feed is used for energy to maintain body heat, for muscular activity, and to repair body tissue. A cow needs additional energy to nourish an unborn calf or produce milk for a suckling calf.

About 10 pounds daily of total digestible nutrients (TDN) should be furnished to winter a mature pregnant beef cow. Since average alfalfa hay contains 50% TDN, a bred cow would obtain her 10 pounds of TDN from 20 pounds of hay.

On the other hand, oat straw contains about 45% TDN. Therefore a cow would need to consume 23 pounds to furnish the necessary energy. Weathered range pasture has about the same energy value as oat straw.

Protein is most apt to be deficient in a bred cow's ration during the winter months. A pregnant cow needs about 0.9 pounds of digestible protein per day. Since average quality alfalfa hay contains 10.5% digestible protein, 9 pounds of average alfalfa hay would fulfill her protein requirements. Of course additional feed would be necessary to meet her energy needs.

A cow would have to eat 150 pounds per day of oat straw or mature western prairie hay, which has only 0.6% digestible protein, to meet her protein requirements. But since her dry hay capacity is less than 3% of her live weight, a 1,000-pound cow can consume only 20 to 25 pounds of hay per day. Therefore, an oat straw, mature western prairie hay, or weathered range grass ration would be deficient in protein.

Supplementing the ration with 5 pounds of alfalfa hay or 1 pound of soybean or cottonseed meal will provide a suitable protein ration. This can be fed free choice if one part salt is mixed with two parts soybean or cottonseed meal.

Vitamins, Minerals

The cow meets all her vitamin needs, except vitamins A and D, from what she eats and manufacturers in her rumen. Since vitamin D is obtained through exposure to the sun, beef cattle rarely suffer from a vitamin D shortage. Only vitamin A is apt to be deficient.

Vitamin A deficiency lowers resistance to colds and other related infections. Cattle deficient in this vitamin often will water at the eyes sufficiently to moisten either or both sides of the muzzle.

Feeding green hay or green silage is one way to provide vitamin A. Five to 10 pounds of green leafy alfalfa hay, bright clover hay, green non-legume hay, or legume silage will provide the daily vitamin A requirements of pregnant beef cows.

If the forage fed is weather-damaged or late cut and is lacking in green color, it probably is low in vitamin A. When that's the case vitamin A can be added to the salt-mineral mixture, or the vitamin may be injected intramuscularly at levels of 1 to 3 million International Units per head. This injection will supply the necessary vitamin A supply for about 100 days.

In most parts of the country only 5 of the 13 mineral elements need to be supplied for beef cattle. The rest are consumed in sufficient quantities in natural feedstuffs.

Sufficient quantities of calcium and phosphorus required for bone manufacture and maintenance are usually found in hays and grains. However, a mineral mixture containing calcium and phosphorus should be available free choice at all times. Either steamed bone meal or dicalcium phosphate are excellent sources of these two elements.

Ordinary feedstuffs are always deficient in salt (sodium and chlorine). Therefore, salt should be included in the free-choice mineral mixture.

339

Beef cows whose rations are deficient in iodine will give birth to hairless, dead calves with enlarged thyroids (called "big necks"). Iodine, the one element deficient in the Middle West in particular, can be provided in the form of iodized salt.

Feeding trace mineralized salt is cheap insurance to insure that your cattle get all the minute mineral elements that may be deficient in your area. Cobalt, copper, iron and iodine are known to be deficient in some areas of the country. These should be supplied.

If you are in doubt about mineral deficiencies in your area, contact the county agricultural agent, who can furnish the proper information.

A free-choice mineral mixture of two parts steamed bone meal to one part iodized salt, or one part dicalcium phosphate and one part iodized salt, is usually an adequate mineral mixture for beef cows. Along with this free-choice mineral mixture, a supply of loose trace mineralized salt should be available.

Give the herd access at all times to a free-choice salt-mineral mix, in a covered feeder to protect it against the weather.

Heifers, Bulls

Growing heifers should be fed to gain from 1.0 to 1.5 pounds per head daily. Replacement heifers should not be fat. They require protein, mineral and limited amounts of energy. Too much energy will make them fat.

A 500-pound growing heifer needs about 0.9 pound digesti-

Suggested Wintering Rations for Beef Cows

Ration	Pounds per cow daily
1. Legume hay or mixed legume-grass hay	20
2. Good quality legume hay	6-7
Oat straw or prairie hay	12-15
3. Prairie hay or other grass hay	20
Oilmeal (soybean, linseed or cottonseed)	0.75
4. Grazing weathered grasses	Free Choice
Oilmeal	1
5. Oat straw	20
Oilmeal	1
6. Ground corn cobs	15
32% protein supplement	3.5
7. Corn or sorghum silage	50
Oilmeal	1
8. Corn or sorghum silage	35
Legume hay	6-7

ble protein and about 7.75 pounds of TDN. Ten pounds of alfalfa hay would supply her protein requirements while 15 pounds of alfalfa hay would fill her total digestible nutrient needs. Since a 500-pound heifer can eat only about 14 pounds of feed daily, feeding 4 pounds of grain and 10 pounds of alfalfa hay would be one practical way to meet the heifer's nutrient requirements.

Another typical ration for a 500-pound heifer is 2 pounds of whole oats and 2 pounds shelled corn plus 10 pounds mixed hay and free choice minerals. If the hay contains no legume, a pound of soybean meal should be substituted for a pound of the oat-corn mixture to provide supplemental protein.

In summer, good pasture will provide all the protein and energy needed. A free-choice mineral mixture plus loose salt should be available to the growing heifer at all times.

It is common practice for the herd bull to run with the cows during the summer pasture-breeding season. Good pasture can provide all the nutrient requirements for bulls. But during late fall and early winter, the bull should be conditioned for the next breeding season. Feed young bulls more liberally than mature bulls.

The amount of concentrate fed will depend largely upon the amount of flesh the bull is carrying. For example, a mature bull in good flesh can be conditioned on good legume hay and silage. Mature bulls that are run down and young bulls should be fed high quality roughage: 5 or 6 pounds of equal parts whole oats and shelled corn and 1.5 to 2 pounds soybean oil-meal.

Bulls should have plenty of exercise during the conditioning period.

Equipment, Facilities

Every beef cattle operation needs certain minimum items of equipment to function efficiently.

Four basic items are so obvious it seems foolish to list them—a pencil, pocket notebook, sharp pocket knife, and an animal thermometer.

The notebook is to write down breeding dates, calving dates, health treatment, etc. Writing down pertinent information, even with a small herd, is better than trying to remember every detail. Use of the pocket knife is obvious, and the thermometer, next to your powers of observation, is the best diagnostic tool you can use.

Safe, efficient, economical handling facilities should be high in the list of equipment needs. For it is virtually impossible to vaccinate, pregnancy check, dehorn or do any of the many other management practices that need to be done without minimum handling and restraining facilities.

The working corral system should be carefully planned before starting construction. A good reference for ideas and plans is the Midwest Plan Service book, *Beef Housing and Equipment Handbook,* that can be obtained through your County Agent or purchased for $2.50 from Midwest Plan Service, Extension Agriculture Engineer, Iowa State University, Ames, Iowa 50010.

You also should have a rope halter, nose lead, obstetrical chain with two handles, syringe and needles, and some plastic sleeves.

This is the most basic list of equipment and facilities needed for beef cattle management.

Herd Health must be planned and constantly maintained. Each herd owner is faced with problems peculiar to his own operation. Size of herd, physical environment, water and feed supplies, exposure to neighboring herds, and past history of diseases must be considered in disease control and prevention.

Prevention is the key to reducing disease-caused production losses. There is no better method of prevention than effective vaccines properly administered.

Vaccination to prevent these diseases is recommended: Blackleg and Malignant Edema, Leptospirosis (Lepto), Infectious Bovine Rhinotracheitis (IBR or rednose) and Parainfluenze 3 (PT₃), Bovine Virus Diarrhea (BVD), Brucellosis (Bang's Disease: Contagious Abortion), Vibriosis (Vibrio).

The vaccinations recommended often are modified to adapt to general management, geographic, and climatic conditions. More specific modifications may be in order to adapt to conditions of particular herds. Use the services of a good local veterinarian and follow his directions to the letter. Remember, only healthy livestock make profits.

Further Reading:

U. S. Department of Agriculture, *Beef Slaughtering, Cutting, Preserving, and Cooking on the Farm,* Farmers Bulletin No. 2263, on sale by Superintendent of Documents, U. S. Government Printing Office, Washington, D. C. 20402. 1977. $2.

Part-Time Sheep Producers Have Several Options Open

By Larry Arehart

Types of sheep production on a few acres include purebred production, feeder lambs, and a small flock of ewes.

Many part-time producers of purebred sheep are found in New England, the Midwest, and the West. Most of these flocks consist of 6 to 50 sheep, and generally can be classified as hobbyist in nature. Many are 4-H programs which are involved in showing sheep.

The sheep show has played an essential part in encouraging the purebred phase of the industry. It provides educational value and the best means of advertising for purebred breeders. Showing sheep is good for young people as they learn to compete with others. Furthermore, planning for a sheep show allows parents to cooperate with youth on subjects of mutual interest.

Smaller purebred flocks provide opportunity for larger, more established breeding flocks to sell their superior genetic breeding stock. The primary source of income for purebred flocks is obtained through sales of rams to commercial sheepmen. Rams are sold at prices ranging from $150 to $500. The enterprise requires ability to select desirable breeding stock, and animal husbandry techniques—including feeding and care to keep animals in desirable condition at all times.

Commercial sheepmen demand high quality, well grown rams. A beginner should not start in the purebred business. It is much better to gain experience with less expensive commercial sheep.

The small ewe flock can be managed either on extensive or limited acreage. With lots of land, sheep do a good job of salvaging roughages. They are well adapted to graze with dairy or feed cattle.

Larry Arehart is Director of the Sheep Production Program at Colby College, Colby, Kans.

Joe Quincy's experience in Iowa is an example of success with sheep grazing behind dairy cattle. His 45 ewes have shown him a substantial gross income.

The ewes are managed with the cows on a 40-acre tract. They glean cornstalks and grass meadows during fall and winter. They have produced a 166 percent lamp crop. During the lambing period the ewes are kept in an old chicken house equipped with water and electricity.

Many producers feel that a small ewe flock more than justifies the expense and time necessary to gain additional income on a few acres. Such flocks can gross $100 per head.

Sheep also have the ability to produce well under confined situations. This type production has become more popular as land costs increase relative to feed costs. High production levels (200 percent lamb crop) are required to achieve a profit.

Profits per ewe could range from $50 to $65 per head depending on genetic input and management. Understanding of marketing conditions is required to be successful.

Feeder Lambs

Lamb feeding is another option. Many producers buy feeder lambs to raise to market weight. This does not require extensive labor. It can be done with limited equipment, and on small acreage.

Lamb feeders buy lambs, treat them for appropriate health problems, and gradually work them to a full feed which is most often dispensed in a self feeder. This enterprise requires good marketing conditions for lambs, a source of feeder lambs, and an understanding of diseases and nutrition.

Good working facilities and a timely disease prevention program are important in sheep production.

A new, innovative way to feed lambs is to keep them on elevated slatted floors. Advantages are reduced space requirements, elimination of bedding, reduction of parasites, and generally increased feed efficiencies. Initial investment, of course, is much greater.

Some small operators produce black wool for home spinning, while others produce lamb for home use and neighborhood locker plants.

Many sheep producers in the Northeast feel the best lamb market is the hothouse or baby lamb market. Special breed and out-of-season breeding is advisable to get the most lucrative price. You can get $60 to $75 for a lamb that is only 8 to 10 weeks old. A few top lambs at the right market may bring slightly higher prices.

Some ethnic groups desire early lambs around their particular religious holidays in early spring. Many lambs can be sold at this time by an alert producer. Buyers will be back year after year and pay good prices. Easter lambs may bring anywhere from $60 to $70 each.

Case Histories

Sheep enterprises begun as hobbies or small supplemental income packages can grow to substantial operations. Mrs. Pierre Knuppel of Grover, Colo., started with 50 head and now manages 800 to 1,000. Mrs. Louise Green of Loveland, Colo., became involved in raising small volumes of black wool for spinning and weaving. She has progressed to major sheep-related enterprises associated with wool for craft and local art festivities.

Others are raising long wool sheep that produce wool which satisfies requirements of the home spinner. Some are even tanning pelts and then making miniature sheep to sell at the local arts and crafts festivals. Producers from the East to the West Coast are involved in these enterprises.

John Paugh began his operation in 1955 in Montana on 80 acres with 200 ewes. His success has encouraged his growth in the sheep industry to 1,000 head which now are the major support for his wife, three sons, and two daughters.

Ten years ago Bob and Esther Hiatt of Hillsboro, Oreg., began with 150 head on 50 acres. They started irrigating their land and have since increased the size of the flock to over 300 ewes. The Hiatts have been able to raise quite a bit of their own feed. In the past year or two they have been feeding the ewes some waste feeds, mainly onions.

About 20 of their 50 acres are irrigated. It has been profitable for the Hiatts. Bob works full-time off the farm.

An intensive operation that began as a small enterprise and has grown to major size is that of Russell and Jonathan May of Timberville, Va. They are presently managing 1,250 ewes.

Francis Chester at Gordonsville, Va., beginning with a small number of sheep, has developed an activity he calls the wool fair. Each spring the fair attracts 2,000 to 3,000 people, mostly from cities or suburbs, to see such things as sheep shearing, wool spinning, and sheep dog trials.

Sheep adapt well regardless of where they are, from the beet and onion fields of Washington and Oregon to the wheat of the Plains states, the cornstalks of the Cornbelt, or the rolling hills of the Northeast.

Pros and Cons

Capital investments are relatively small for sheep. A good commercial ewe can be purchased for $60 to $75 depending on quality, and a purebred for $125 to $150 depending on breed, type, and pedigree. The return for sheep can be high per dollar invested if proper management methods are followed.

Sheep require a minimum amount of labor. Lambing and shearing can be planned to occur when other work is not heavy. Sheep housing can and should be kept inexpensive by using available buildings.

These animals consume home-grown roughages except for a small amount of grain needed at special times. Sheep improve the fertility of the soil.

The high cost of fencing and sometimes high feed costs, along with the nuisance of dogs and predators, are among the disadvantages in sheep production in most sections of the country. A new concept is use of electric fencing to help reduce fencing costs as well as cut dog and predator losses.

Belief that sheep can produce well on scant feed and under poor management is one reason for lack of success with them. Sheep will clean fence rows and eat weeds and brush, but they require care and need specific, high quality feeds at critical times such as before and after lambing.

Internal and external parasites will limit profits unless controlled by good pasture management and easily administered treatments.

Pools, Auctions

Marketing lamb and wool can be a major problem if you are planning to produce in areas where sheep numbers are sparse. The sheep producer should plan production for the market in mind. New concepts of lamb and wool pools can help you obtain a competitive price.

In the past few years, tele-a-auctions also have become important bargaining tools. In Virginia, a tele-a-auction system has raised lamb prices from below the national average to about $1 above. The number of lamb buyers for area farmers has grown from one primary buyer to ten or twelve who negotiate with producers by phone auction hook-up.

As news of the tele-a-auction success is spread, other similar groups are heading for telephone line bargaining. Today, Idaho and Iowa are using such bargaining agents to merchandise their products.

Before starting a sheep enterprise, talk with others who have sheep regarding costs and returns, markets and management. Contact your country agricultural or regional specialist, or your Extension animal scientist at the state university. In some cases, it is helpful to have such people visit your farm and go over the possibilities with you—or you may wish to visit them.

Study your individual case carefully before starting. Then begin in a small way and grow into the enterprise as you become better acquainted with the animals and their management.

For more information on production, these materials are available from the Sheep Industry Development Program of the American Sheep Producers Council, 200 Clayton Street, Denver, Colo. 80206:

The Sheepmen's Production Handbook, covering basic areas of sheep production including genetics, physiology of reproduction, nutrition, health and management. Price $12.50.

The S.I.D. Model Sheep Management System, 10 model systems for different sizes and types of sheep operations, including an operational calendar of management requirements, breeding and feeding recommendations, health and marketing information, facilities and equipment plans, and budget and cash flow projects. Price $3 for set of 10.

Further Reading:

U. S. Department of Agriculture, Lamb Slaughtering, Cutting, Preserving, and Cooking on the Farm, Farmers Bul. No. 2264, on sale by Superintendent of Documents, U.S. Government Printing Office, Washington, D. C. 20402. 1977. $1.40.

Two-a-Day Milkings Kill Joy of Dairying for Most Families

By Robert Appleman and Kenneth Thomas

Part-time dairy farming may involve milking a small herd morning and night in addition to holding a full-time job in business or industry. Or replacement females may be raised to be sold as bred dairy replacement heifers, thereby eliminating both the labor and the equipment required in a milking operation. Some youth projects (4-H or FFA) would fit in this category.

Other operators have a cow, or even a few cows, with the family consuming all or most of the milk produced.

A dairy enterprise may be for fun, food or profit. But in any case it requires intensive management, an inflexible daily labor schedule, and a higher per unit capital investment than most other part-time farming enterprises.

Because of the management skills required in keeping even one cow, and the need to milk her twice daily, a dairy cow will seldom be selected as a family fun project.

Since a dairy cow's milk production varies so much during the year (depending on when she last freshened or gave birth to a calf), balancing a family's relatively constant demand for milk and dairy products becomes difficult. Thus, a dairy cow as an economic source of family food may also become questionable.

Therefore, the major focus of this chapter will be on the part-time dairy operation as a source of profit. An alternative program of raising replacement stock for others is also discussed.

When considering the dairy cow for profit, you need to realize that a reasonable level of production must be achieved just to cover feed and other operating expenses. Such levels of production per cow usually require management skills normally

Robert Appleman is Professor of Animal Science and Extension Dairyman, University of Minnesota, St. Paul. Kenneth Thomas is Professor of Agricultural and Applied Economics, and Extension Economist, at the university.

beyond the abilities of younger family members. In addition, part-time farmers face the further dilemma that the dairy cow is very demanding of their scarcest resource—labor.

Dairy farms can be labor efficient, but this usually requires a size operation beyond the desires and capability of most part-time farmers. The comparative advantages and problems with part-time dairying are outlined in the table.

Breed Selection

The most important consideration in choosing a breed of dairy cattle to establish a dairy farm is the present and future market situation. Most areas favor breeds that produce the largest volume of milk. This undoubtedly explains why 80 to 85 percent of the dairy cattle are Holsteins. There are regions, however, where special milk markets have been developed and other breeds, especially Jersey and Guernsey, are common.

Other factors to consider include:

Personal preference. What breed have other members of the family been associated with?

Resale or salvage value. Larger breed animals are usually worth more when their useful life as milk producers is ended.

Suitability for meat. Half the calves are males. The surplus calves from larger breeds are generally considered superior

---

Pros and cons of part-time dairying

Advantages	Disadvantages
Efficiency Dairy cattle are efficient in converting forage and grassland into milk.	**Confining** Cows must be milked twice daily, every day while lactating, usually about 10 of every 12 months.
Permanency Dairy enterprises will continue to be important. They flourish even in Europe where competition between crops and livestock for limited acreages of available land is much more intense than in the U.S.	**Labor** Each cow requires 60 to 80 hours of labor annually. Part-time dairymen often are short of time.
	Feed Quality To achieve acceptable production per cow, top quality forage is required. This is often difficult to purchase and requires excellent management abilities to produce.
Stability Because it takes 2 years for a heifer to mature, the dairy industry is less subject to large fluctuations in market supplies and prices received. Dairy farming provides a steady, regular income.	
	Capital The per unit capital investment is greater than for most other part-time farming enterprises.
Product Quality There is no satisfactory substitute for fluid milk in the diets of infants and children.	**Regulations** Milk and milk production are highly regulated and often must comply with stringent health regulations.

Common Dairy Breeds

Breed	Estimated % of the population	Typical annual production			Typical weight	
		Milk (lbs)	Fat (lbs)	Fat (%)	Mature cow (lbs)	Newborn calf (lbs)
Ayrshire	less than 2	11,600	450	3.9	1,200	75
Brown Swiss	less than 1	12,700	510	4.0	1,500	95
Guernsey	less than 10	10,300	470	4.6	1,100	75
Holstein	80-85	14,700	530	3.6	1,500	95
Jersey	less than 10	9,500	470	4.9	1,000	60
Milking Shorthorn	less than 1	10,500	380	3.7	1,250	75

meat producers, although tenderness and taste evaluations of meat from the smaller breeds have been outstanding.

Temperament. When young family members must handle cows, consider this factor: Holsteins are generally superior in this trait but all breeds are acceptable if cattle are handled with gentleness.

Calving difficulty. The Jersey breed has fewer problems with difficult calvings.

Dairy farming seldom is a good enterprise for the part-time farmer unless he has an abundance of family labor. Labor requirements can be reduced some with mechanization, but small herds normally operated by part-time farmers seldom justify the expense of much mechanization beyond installing a milking machine and bulk milk cooler. Even then, to keep costs down most part-time dairymen should be adept at purchasing good used equipment.

The approximate number of cows and replacement females that can be maintained by a part-time dairyman are illustrated in a table. A typical dairy herd consists of 55 percent cows (85% in milk), 25 percent heifers over 10 months old, 12 per-

Number of dairy cows and replacements possible, depending on amount of labor available

Man-hour equivalents per day	Number of dairy cows and replacement heifers			
	Purchase all feed [a]		Raise own feed [b]	
	Milk cows	Total (all ages)	Milk cows	Total (all ages)
2	9	16	7	13
3	14	25	10	18
4	18	33	14	25
5	23	42	17	31

[a] Based on 80 hours of labor per cow annually.
[b] Based on 105 hours of labor per cow annually.

cent heifers from 6 weeks to 10 months, and 8 percent calves up to 6 weeks of age.

The primary causes of unprofitable dairy operations include low production per cow, high feed costs, and low production per man-year of labor expended. Since the last factor is almost certain to apply to the part-time dairyman, and it may not be possible to avoid high feed costs, it is imperative that a high level of production per cow be achieved if a profit is to be obtained.

A herd of Holsteins should produce at 14,000 pounds per cow annually, or better, to justify a moderate cost facility and provide the desired returns to labor and capital (see table). If only 11,000 pounds of milk per cow is obtained, the return to labor, facilities, and livestock (line 5) is reduced by $215 when

Estimated costs and returns per Holstein cow, including appropriate share of replacement stock

| | Level of production | |
	11,000 lbs ($)	14,000 lbs ($)
1...Cash income		
Milk hold @ $9.50/100 lb., net	1,045	1,330
Cull cow and calf sales	135	135
TOTAL	1,180	1,465
2...Feed costs		
Corn equivalent (70 or 90 bushel @ $2.25)	158	203
Hay equivalent (8.5 tons @ $60)	510	510
Protein supplement (320 or 520 lbs @ $9)	29	47
Salt, mineral, milk replacer, etc.	19	19
TOTAL	716	779
3...Other costs		
Health (veterinary and drug)	22	25
Breeding	20	24
Power and fuel	25	25
Bedding	27	27
Repairs, equipment	30	30
Supplies, insurance on livestock, record-keeping, insurance on buildings, equipment, etc.	26	26
TOTAL	150	157
4...Variable costs (2 + 3 = 4)		
TOTAL	866	936
5...Return to labor, facilities, and livestock (1 − 4 = 5)		
TOTAL	314	529
6...Fixed costs		
Buildings ($300 @ 10%)	30	30
Equipment ($400 @ 15%)	60	60
Livestock ($800 @ 8%)	64	64
TOTAL	154	154
7...Return to labor and management (5 − 6 = 7)		
TOTAL	160	375
Value per hour of labor [a]	**2.00**	**4.69**

[a] Based on 80 hours of labor per cow annually.

351

compared to that achieved at the 14,000-pound level of production. If the combined debt repayment load exceeds that indicated (line 6), then the value for labor expended is lowered still more (line 7).

Feed Needs

Forages are the basic feed for a dairy cow, supplemented with sufficient grain, protein, minerals and vitamins to support high production.

Forage consists of the whole plant, usually rather high in fiber content. Examples are alfalfa hay, corn silage and pasture.

Dairy cows usually consume forage dry matter at about 2.0 to 2.2 pounds per 100 pounds of body weight. Thus, a 1,400 pound Holstein cow will consume between 31 and 34 pounds of 90 percent dry matter hay, or its equivalent, daily. Youngstock more than 3 months of age will consume equivalent amounts of forage, based on body weight at the time.

Some part-time farmers may choose dairy cattle because they have surplus pasture land available. Lush, growing pasture that is properly fertilized and managed can be an excellent source of nutrients. But its value decreases as it matures, and trampling is a problem, resulting in nutrient waste.

Rotating cattle and maintaining fences around small fields, to reduce nutrient losses, requires more labor and is difficult to accomplish. Unless this is done, a high level of production per cow seldom is achieved from pasture.

Concentrate mixtures (feed grains, protein supplement, minerals and vitamins) contain less fiber and are higher in nutrient content. As production increases, the amount of concentrate mixture fed is increased to meet nutrient needs.

Forage quality varies greatly, depending on when it is harvested, how it is handled and how it is stored. Late cut forage (such as alfalfa in full bloom), weather-damaged forage, and feeds stored where water leaches out much of the nutrients

Guidelines for feeding concentrate mixtures

Breed	Amount of milk daily	Amount of concentrate mixture
Holstein and Brown Swiss	0-40 lbs	1 lb per 4 lbs milk
	40-70 lbs	1 lb per 3 lbs milk
	over 70 lbs	1 lb per 2½ lbs milk
Ayrshire, Jersey, Guernsey and Milking Shorthorn	0-30 lbs	1 lb per 3 lbs milk
	30-60 lbs	1 lb per 2½ lbs milk
	over 60 lbs	1 lb per 2 lbs milk

may result in the grain (concentrate) mix feeding schedule being even higher than indicated in the table.

Harvesting losses frequently amount to 20 percent of the forage grown. Storage and feeding losses may approach another 10 to 20 percent. Thus the total forage requirement (hay equivalent) for a Holstein cow and her share of the replacement youngstock may approach 8.5 tons annually. Similarly, the amount of concentrate mixture needed will approach 2.1 and 2.6 tons per cow producing at the 11,000 and 14,000 pound level, respectively.

The kind and form of forage fed varies tremendously in different regions because of climate, topography and soil type. Each state's land grant university, and the county Extension agent, have the expertise and publications available to provide you with specific cropping and feeding programs. Contact them before investing time and effort in planning this phase of the enterprise.

Land, Housing

Land requirements for the dairy enterprise vary considerably with its production potential and type of crop grown. Since forage yields may range from 1.5 to over 6 tons of hay equivalent per acre, the land required when home-grown forages are utilized may range from 1.3 to 5.7 acres per cow, including replacement stock. An average figure frequently used in the Lake States region is 3 acres per cow for forages and 1 acre per cow for feed grains.

Housing needs for dairy cattle, especially milking cows, differ markedly among the major regions. In the North, cows must be protected against snow, winter winds, and sub-zero temperatures. In the South and Southwest, shade is needed to minimize the effects of high summer temperatures on milk production. Providing an overhead roof is desirable in most regions to divert rainwater, keep cattly dry, and improve working conditions.

The general floor area requirements per cow are 40 to 60 sq. ft. in the housing area, 20 to 30 sq. ft. of paved feeding area, and 12 to 18 sq. ft. of holding area.

In terms of overall barn requirements, a stall barn usually provides 80 to 90 sq. ft. of space per cow. When cows are in loose housing, the covered area provided averages about 50 sq. ft. with another 100 sq. ft. of outside concrete slab for feeding and cattle movement.

In stanchion barns, stalls that are too short make it difficult to keep cows clean. Narrow stalls contribute to teat, udder and leg injuries. Modern standards call for stall platforms 4 ft. wide and from 5-1/2 to 6 ft. long.

When cattle are grouped in loose housing situations, separate areas should be provided for feeding, housing, milking, calving, and maintaining youngstock.

While dry cows and "springer" heifers—those close to calving—are sometimes kept together, separate pens are desirable for the different age groups of youngstock: milkfed calves, calves from weaning to 3 months, calves from 3 to 10 months, heifers from 10 to 15 or 17 months (breeding age), and bred heifers.

Such a division provides the opportunity to vary feeding programs (ration composition), reduce feed costs, avoid competition between large and small heifers, and promote fast growth for maximum profit.

Problem Areas

For the enterprise to be profitable, the dairy manager must be aware of the following potential problem areas, and how to avoid them:

Failure to detect "heat" or time to breed. Observe cattle to be bred for "standing heat" three or more times daily.

Mastitis. Install a good milking system, use good milking procedures, and take the time required to prevent new infections by "dipping teats" with a good disinfectant and treating cows going dry (out of production).

Calf losses. Feed colostrum, provide a dry, well, ventilated calf pen, and observe calves frequently (3 or more times daily).

Poor nutrition—Harvest forage at proper stage of maturity and feed enough concentrate mixture to encourage top production.

Poor genetics. Use genetically superior A.I. (artificial insemination) sires and cull the poor producers.

Disease.—"Pills and shots" will never replace good management. Observe cows frequently and regularly, feed balanced rations, keep housing facilities clean and dry, and practice good sanitation.

Production of quality milk begins on the farm. Each producer is generally regulated by the Grade A Pasteurized Milk Ordinance of the U.S. Public Health Service. Similar but less stringent regulations apply to manufacturing grade milk producers.

Grade A milk must meet specific requirements (different markets vary) for: (1) bacteria, (2) inhibitors—antibiotics, etc., (3) somatic cell count—mastitis, (4) adulteration—added water, and (5) rate of cooling and temperature of holding milk on the farm. Production facilities require an appropriately constructed and maintained milkroom or milkhouse, barn and/or milking area, and a potable water supply. Routine inspection is required.

Some states permit sale of raw milk directly to the consumer provided the consumer purchases the milk at the farm where the milk was produced. Part-time farmers located in a densely populated area occasionally sell milk in this manner to enhance their income potential. Some risk is involved.

Prior to milk pasteurization, many diseases were attributed to consumption of raw milk. Pasteurizing milk destroys any disease-producing bacteria that might be present. Another benefit is that shelf life (storage time) of milk is increased by destroying any spoilage bacteria in the milk.

Home pasteurization of milk can be done several ways. One process requires heating the milk to at least 145° F and holding it continuously at or above this temperature at least 30 minutes. Another process requires a temperature of at least 160° and holding the milk continuously at or above this temperature at least 15 seconds.

Be sure that containers used to store pasteurized milk have been thoroughly sanitized to prevent recontaminating the milk.

Whether milk produced on the farm is consumed entirely by the family or sold directly to another consumer, it is advisable to pasteurize it.

Replacement Stock

A part-time farmer with a limited supply of labor and capital, but an excess of land suitable for forage production, may wish to raise replacement heifers for neighboring dairymen. This approach has special appeal when the part-timer is in a dairy area and is living on a farmstead with surplus buildings.

"Contract" raising of replacement females for another dairyman can incorporate the advantages of the other dairyman raising his own (knowledge of the production potential, minimizing opportunity to introduce disease, and control of growth rate, time of breeding and selection of sire). It minimizes the disadvantages of the other dairyman raising his own (removes competition with the milking herd for labor, space, feed and capital).

Two types of contracts are in general use. In one, the heifer raiser agrees to raise replacement heifers for a dairyman for a set period and at a specified monetary rate, which is either in terms of dollars per month or cents per pound of gain.

The second type of contract includes an option-to-purchase clause, in which calves are sold to the heifer raiser and the dairyman retains first option to buy back any animal prior to freshening (calving).

The amount of forage required to raise replacement heifers from 6 weeks to freshening at 24 to 28 months is between 4.5 and 5.8 tons of hay equivalent (2.5 tons per year). This means that about 3.4 heifers can be grown out on the forage needed to maintain each cow and her share of the replacements.

Thus, 17 "springer" heifers can be raised annually on the same land that will support only 10 cows. At the same time, labor required to feed and care for these animals will be reduced one-third to about 50 or 55 hours annually.

Successful contract raising of heifers depends primarily on locating dairymen interested in supplying calves and purchasing "springer" heifers, ability of the grower to raise the calves satisfactorily, and the nature and completeness of the contract between the cooperating dairyman and heifer raiser.

Heifer raising operations in many regions of the U.S. have a *potentially* higher return per head than beef enterprises. The main disadvantages of raising dairy heifers are:

- The time from purchase of the calf until sale of the grown animal is relatively long.
- Farm family members must "catch" or observe the heifers in heat (ready to be bred by artificial insemination) or use a bull. The first alternative takes time and effort, the latter may be dangerous and expensive.
- Management requirements for raising replacement heifers are higher than for most beef enterprises.

A dairy farm operator needs a wide variety of skills. Experience is the best teacher in both animal husbandry and management decision-making. The part-timer should ask himself the following questions:

- Why do I want to be a dairy farmer?
- Why do I think there is a good future in dairying?
- Do I have the ability to work with animals and get adequate production from them?
- Am I willing and able to work long and inflexible hours?
- Can I obtain animals that will become the foundation of a high producing and profitable herd?
- Can I get sufficient capital to begin a profitable dairy operation?

Dairy Goats Require Lots of Care Just to Break Even

By Donald L. Ace

Dairy goats provide a source of refreshing and nourishing food that can be prepared in many ways: milk to drink, to make into an endless variety of cheeses and cultured products such as buttermilk and yogurt, to churn into butter and, with the addition of a few extra ingredients, to make ice cream and candy.

But a milking goat demands attention if she is to provide food for the table. Someone has to care for the animals twice a day, seven days a week, all year long. Even if the doe is dry she must be fed and watered. When she is giving birth it may mean a trip to the barn every hour through the night to see how she is progressing.

There is an occasional sick animal to care for, perhaps a frozen water pipe to thaw, manure to clean out. Then too there are educational meetings to attend to help you do a better job of management.

Income from a goat project is closely aligned with the producing ability of the doe, the feeding and care provided by the owner, and the local market for milk not needed for family use. For each doe producing over 3,000 pounds of milk per year there are hundreds that produce under 500 pounds.

If each milking doe averages 5 pounds of milk per day and it sells for 35¢ per pound you have $1.75 income. Budget 5 pounds of hay and 1-1/2 pounds of grain per adult per day at a cost of 40¢. Value of product over feed cost is $1.30.

That's not all profit. A buck eats 30¢ worth of feed. Figure 20¢ per day to feed each head of youngstock. Add the cost of upkeep on buildings, fencing, fertilizer for the pasture, plus feeding equipment, milking utensils, taxes, insurance and veterinary care plus an occasional animal lost to death. You will find expenses nearly equalling income.

Donald L. Ace is Chairman, Dairy Science Extension, The Pennsylvania State University, University Park.

Do not view the small dairy goat herd as a profit maker. Regard it instead as a break-even source of nutritious food for the family that can be produced on otherwise idle land, and as a delightful mental diversion that all "hobbies" should provide.

There are byproducts of the hobby too. Chevon is good roasted or barbecued and the hide makes beautiful gloves and jackets.

Breeding Stock

Selecting breeding stock is not a simple task even for an experienced breeder. There are visual signs that forecast an animal's capabilities. Is she big enough for her age? Compare her to other animals the same age in the herd or in other herds. Are her eyes bright, the hair coat smooth and soft?

Do the ribs arch out and downward from the backbone, does she have greater depth in the rear rib area than in the chest area? The spring and depth of ribbing is evidence of body capacity that is so necessary for good forage intake.

View the udder when it is full of milk. It should have full, strong attachments at the body wall and not hang greatly below the hock area. Size of udder is not a reliable indicator of milk production.

View the udder when milked out. It should have a collapsed appearance. Feel it to see there are no hard lumps. It should be soft, thin-skinned and pliable.

Check milk production records. A youngster must be evaluated through its mother's production plus any records available on other daughters of her sire. Place a greater emphasis on records of the dam and sire than on the grandparents. Look for yearly production information rather than daily production. Weigh these records against a 1,500-pound yearly milk record produced over a 305-day period.

Dairy Herd Improvement records are better guides than daily milk weights. Shy away from animals that milked 10 pounds a day when fresh but were dry in less than 200 days. If the doe you are selecting is in milk, her records speak louder than her mothers and sisters.

The goal is to obtain an animal that can produce milk over a 10-month period. A year-round milk supply is difficult to obtain due to the seasonal breeding nature of the goat. It is impossible to obtain with a few does who stand dry an exceedingly long period each year.

Breed a doe after she weighs 80 pounds or at 10 to 12 months of age. Most goats show heat between August and March. Gestation takes 150 days. Twins, triplets are common.

Consider artificial insemination as a viable alternative to keeping a buck. If the herd numbers less than five milking does it will be the more economical alternative, and higher quality sires may be more easily obtained.

Housing Space

Cold loose housing where animals are free to move about on a dry, bedded pack offers the best quarters. Warm loose housing is discouraged because of problems with ventilation.

Provide 25 square feet of space per adult female housed. Buildings constructed of wood are preferred over cinder block. Masonry construction is difficult to insulate, walls are cold and often damp. Construct a 3-sided building with an open side facing away from the prevailing wind.

Keep a bedded pack at least 15 inches deep in the loafing area. Heat is produced inside the pack and animals rest comfortably in these quarters even though the outside temperature drops to zero. Natural air movement will ventilate and remove moisture.

Do not attempt to keep the inside temperature warmer than outside temperatures. Doing so will cause moisture to condense on walls and ceilings, creating a humid environment that goats do not tolerate. Most respiratory problems and stress-related diseases can be traced to wet conditions and inadequate ventilation.

Provide feeding and watering devices in another area of the barn away from the bedded pack. Much hay is wasted if fed on the bedded pack, and the chance for parasite problems increase. Watering devices always offer wet spots to spoil the bedded pack.

Locate 4 by 4 foot pens along one side to house 3 or 4 kids to a month of age. Older youngstock may be housed in 6 by 8 foot pens.

Buck housing must be separate and downwind from the milking animal quarters. A 6 by 8 foot shed with the open side facing south is adequate for each buck. Provide a minimum 10 by 10 foot exercise area.

Plans for goat facilities are available from the Agricultural Engineering Department, Pennsylvania State University, University Park, Pa. 16802. They include:
· *Building a Buck Barn*—Plan No. 728-392.
· *Building a Keyhole Goat Feeder*—Plan No. 728-394.
· *Building a Milking Stand*—Plan No. 728-100.
· *Loose Housing for 20 Goats*—Plan No. 728-102.
These plans are free upon request.

Two methods of fencing have been successful. Use either wooden or steel posts, 6 to 7 feet long, and set them 15 feet apart.

In one method, use 4-foot-high woven wire plus a single electric wire 12 inches above the top of the woven wire. The electrified top wire repels both dogs and goats that attempt to climb fences.

A second method is to use only electric fencing. Attach insulators to each post starting about 12 inches from the ground and every 6 inches thereafter to a height of 36 inches. String wires between insulators from post to post.

Feeding

Most goat owners find machinery costs too high to warrant growing forage. They purchase both hay and grain. An acre of land will provide ample room to house up to 10 milking animals plus the youngstock and offer some pasture area.

If forage is grown, assume a need for 2,000 pounds of hay per year for each mature animal. One acre of good producing land should provide forage to feed 4 milking does. When calculating forage needs allow 3 pounds of hay per 100 pounds of animal. Add an extra 5% for wastage.

Agronomic practices, crops and cropping systems vary greatly.

For assistance contact your county Extension agent who knows the soils, seed varieties, and cultural practices needed for optimum production.

Goats are ruminants and require fiber in their diet. They readily consume twigs and leafy portions of trees and bushes, plus many weeds. This is acceptable food only for animals not producing milk; the strong flavors of many such foods will carry through to the milk. In addition some of the fiber in browse is low in digestible energy and protein, and milk production will fall off on such a diet.

Feed milk animals good quality legume or grass hays. Grain fed to the herd must provide the extra energy and protein the animal needs but does not get from forage. Therefore, with good legume hay a 12% protein grain ration will be adequate. With average quality grass hay an 18% protein level is required.

Vegetable peelings, leaves of cabbage and lettuce, tops of carrots and turnips are readily eaten by the goat but are not good food for the producing doe. If top production is your goal, feed as constant a diet as possible. Sudden changes in types of food may upset digestion, reduce milk flow.

Fastidious Eaters

Goats have fastidious eating habits. They waste hay and grain if it gets soiled. Clean feed bunks and mangers daily.

Locate hay feeders away from the bedded area on a concrete pad that can be scraped clean each day. The keyhole feeder is best to prevent fecal contamination and wasting of hay.

Locate watering devices away from the bedded pack and on concrete. Goats will consume greater amounts of water in winter if the water is warmed to 60° F. Electrically heated float-type stock waterers are available.

Keep in mind several basic points for disease prevention and control:

- Maintain herd isolation. Raise herd replacements. If animals must be purchased, buy them as young as three days of age from locally known, disease-free sources. Insist on health charts for all animals.
- Visitors may bring diseases to your herd. Insist they disinfect their footwear before walking into the barn. Keep them out of feed alleys.
- Practice sanitation at home. Avoid carrying diseases from one animal to another by regularly cleaning and disinfecting maternity quarters and baby kid pens. Isolate sick animals.
- Control internal parasites by sanitary feeding and housing. Separate youngstock from older animals in barns and pastures. Practice pasture rotation and graze youngstock on new pasture whenever possible. No worm treatment program can be effective without sanitary measures that interfere with the parasite's life cycle.
- Draft-free, well-ventilated, well-lighted buildings are needed for goats.
- Pastures and exercise lots should be well drained. They also should be free from trash and sharp objects that could cause injury, especially to the teats and udder.
- Adequate nutrition is important. Drugs, tonics, fancy mineral mixtures and rumen stimulatory substances are not substitutes for good food.
- Use a veterinarian for diagnosis, treatment and advice on health problems. Too often veterinary aid is sought only for dying animals.
- Goats respond to gentle and patient handling. They will not stay healthy or produce well when abused.
- House the buck separately from the does. Provide plenty of opportunity for exercise. Nothing is more damaging to the herd sire than to keep him tied in a dark corner of the barn.

Tools, Equipment

On a bedded pack and very little exercise the toes grow long and the hoof shell uneven. Pruning shears and a farrier's knife are useful tools to keep the foot manicured.

Tattooing is necessary for identification, especially if goats are shown or registered. Cost of the equipment is about $30.

Remove horns as soon as the horn button can be located. Electric dehorners are the most humane. Cost, about $17.

Behind each horn button and toward the midline of the head is a mass of cells which produce an oily musk, a major source of odor in the buck goat. By clipping the hair these yellow cell masses may be seen lying just under the skin. Veterinarians can remove them surgically, or the owner can burn them out at the time of dehorning.

Castrating is necessary for males kept as pets or for slaughter. A veterinarian will do it, or the owner may use an elastrator (cost, $15), Burdizzo Emasculator (cost, $55), or a sharp jackknife. Disinfect all cutting instruments.

For disinfecting the navel, use a teacup partially filled with Iodine to dip both the cord and "belly-button" area.

Pails, Strainers

Stainless steel utensils are unequalled for durability and ease of cleaning, and if milk is sold commercially may be required. Milk pails cost $23, strainers cost $50, and they last a lifetime. Plastic may be used for home-oriented dairies.

Do not use iron, copper, brass, white metal or worn, plated utensils and spoons in handling milk. They cause milk to oxidize and taste like cardboard.

An electric hot water heater and a stainless steel double sink allow utensils to be washed and sanitized as required by many milk markets, and is recommended for all producers. A drying rack for storing clean equipment between milkings is essential to good sanitation.

Pasteurizing is important to protect the quality of milk and extend its shelf-life. The enzyme lipase reacts with butterfat in milk, causing it to turn rancid and develop a goaty flavor. Pasteurization temperature destroys the enzyme.

Two-gallon electric pasteurizing units may be purchased (cost, $95). A more economical approach is to heat milk in a double boiler to 165° F for 20 seconds. Cool immediately. If the temperature goes higher than 165° or is held for longer than 20 seconds a cooked flavor develops in the milk.

An ice water bath or cold running water cools more quickly than air temperature. Therefore cool milk before putting it in a refrigerator. Larger dairies using milk cans may purchase used water immersion type electric coolers from dairy equipment suppliers.

Some producers purchase used 200-gallon bulk milk tanks; this is recommended.

Goats are creatures of habit. Milk the does within minutes of the same time each morning and evening. Wash the udder and teats before milking. Use a disposable paper towel for each animal. Dry the udder thoroughly before milking.

Clipping the hair from the belly and udder in fall and winter will aid in producing a higher quality milk.

Cool the milk immediately to 40° F. Store it at 34° for best preservation of quality and flavor.

Do not expose milk to fluorescent lights or to sunlight.

Pygmy Goats

The Pygmy is a dwarf goat imported in the early 1950's. It is shorter and more compact than other dairy goats. Although friendly with people, the Pygmies are aggressive and should not be corralled with other breeds of goats and sheep.

There is evidence they produce well if fed good forage and grain plus fresh water. Some does give as much as 5 pounds of milk testing 8% butterfat per day. They may produce two litters of kids a year, so the milk-producing period is short compared with other dairy goats. Bucks breed the year around.

The wooden floor and milking stand are okay for home use, but not for commercial milk production. The stainless steel milking pail helps produce clean milk.

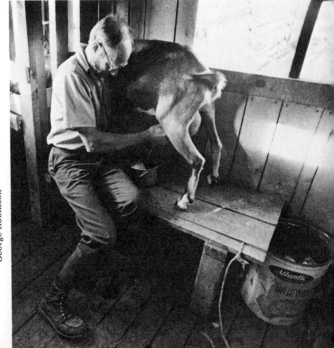

George Robinson

Pygmy prices are on a par with other breeds excepting wethered (castrated) buck kids. They are sold as pets for $50 to $75.

Fencing for Pygmies must be sturdier and higher than for other breeds. In Africa, the Cameroon goat (a type of Pygmy) is called the "tree goat" because it walks up low tree branches to browse. Pygmies found in America are as agile as that too. *(Courtesy Alice Hall, San Bernardino, Calif.)*

Angora Goats

Angoras are kept for the production of a specialty fiber called mohair. Meat production and brush control is second in importance. Never are they raised for milk production.

They are concentrated in areas where the climate is mild and dry. About 90% of this breed in the U.S. are found in the Edwards Plateau of Texas.

Angoras are primarily browsing animals adapted to high, rough, brush land but adapt well to hay, pasture and grain.

They are seasonal breeders with the bulk of kidding extending from late February to mid-April. Multiple births occur in only 10% to 15% of the kiddings.

The fleece of kids is packaged separately and sold at premium prices. Clipping begins at 6 months of age and is repeated every 6 months thereafter. Adult does will shear about 12 pounds of mohair per year; bucks may produce more than twice that amount. Wethers tend to produce a heavier fleece than does.

Fleece production begins to decline at 7 or 8 years of age. Animals are then sold for slaughter. *(Adapted from information provided by American Angora Goat Breeders Association, Rocksprings, Texas.)*

Further Reading:

Ace, Donald L., *Dairy Goats—Correspondence Course 105,* 307 Agricultural Administration Bldg., Penn State University, University Park, Pa. 16802. $5.

Barker, C. A. V., *Dairy Goat Cook Book,* Dairy Goat Journal, P. O. Box 1908, Scottsdale, Ariz. 85252. $3.

Countryside Magazine, Route 1, Dept. D9, Waterloo, Wis. 53494. $9 per year.

Think Twice About Risks of Horse Rental Business

By Robert C. Church

It is unrealistic to consider profit as an incentive for owning horses on a limited, part-time basis. Liability insurance makes it almost prohibitive for even full-time stables with professional help to operate. An element of risk for a horse-oriented business has never been established; therefore, the insurance rate is arbitrarily set.

Volume of business and services such as indoor riding arenas, instruction and training, and professional supervision offered to their clientele keep the full-time stables profitable.

Most rental stables cater to the novice rider who has had no formal instruction and thus is a considerable risk on a horse. A person who keeps horses for public hire must become familiar with the habits, disposition and traits of the horses. An owner who knows a particular horse is apt to be vicious may be held liable for injuries caused by that animal.

The rental horse business is fraught with risk and should be entered into only after considering suitability of the horses involved, insurance costs, availability and safety of trails, and the expertise of those who will supervise riding.

Some people use personal mounts for riding instruction. In most cases a homeowner's policy will cover liability if this enterprise falls within the confines of a casual and not a full-time enterprise. However, many recently written policies do not include this type coverage.

If you have the required skills, training horses on limited acreage as a part-time occupation can provide supplemental income.

But it's not easy to make money from these part-time endeavors. Keep in mind the volume of income that would be required to cover costs involved in order to realize a profit.

Robert C. Church is Associate Professor of Animal Industries and Extension Horse Specialist, University of Connecticut, Storrs.

Without indoor riding facilities, weather very much limits the activities in most areas of the country.

Many small stable operators are in the business as a part-time occupation because they love to work with people and horses.

Selecting a Horse

Horse trading is an old and sometimes not too honorable activity. In no other industry is the point of law *Caveat Emptor* (let the buyer beware) so closely adhered to. Where no warranty is implied or given, the buyer accepts all responsibility unless illegal practices can be proved.

Most dealers in horses are honest businessmen looking for satisfied repeat customers. But too often the sale and purchase of a horse is between private parties with neither a knowledgeable enough horseman to be aware of the horse's faults.

The novice should seek the help of an experienced equine veterinarian or horseman when buying a horse. There are standard examination procedures to determine stable habits, temperament and disposition, and to detect disease and unsoundness. Also there are certain conformation faults that limit the usefulness of the horse and predispose it to specific unsoundness.

Tennessee walking horse cools off in a farm pond. Note that for safety, bridle should have throat latch and girl should be wearing heeled boots.

George Robinson

If you are paying an average or considerable amount for a horse, take steps to ensure a sound investment.

The price of horses varies from slaughter prices to thousands of dollars. Check with several dealers for an idea of the price range. Horses between the ages of 8 to 14 years, well trained, gentle, healthy and sound, make the best beginner horses. The horse should fit your needs and you should fit the horse. A height for ease of mounting and a width to fit your legs makes for a safer, more comfortable ride.

The stallion is not for any but the very experienced horseman. Some mares become difficult to handle when in season. The gelding is a castrated male and is usually more stable than the stallion or mare.

Sources of horses include:

- Public auction barn—usually the dumping ground for less desirable horses, and disease problems. Strictly *Caveat Emptor*
- Horse dealers, as discussed earlier
- Horse farm sales. Some deal in trained horses, but most sell weanlings, yearlings, 2-year-olds, and surplus breeding stock of one of the registered breeds
- Casual sales, as outgrown pony, former 4-H mount, going off to college, or just lost interest. Sometimes dumping an unsuitable or fractious horse.

The former owner of a horse, with no ties to the present owner, will usually tell you the truth about the horse.

Space and Fencing

In most communities, keeping horses in heavily populated areas has precipitated the inclusion of animal ordinances in zoning regulations. Laws governing cruelty to animals, nuisances, health and safety usually are already on the books. Be sure to check all governmental requirements.

Pasturage should be high in fertility of soil and density of desirable forage species in order to provide enough feed for the horse. Your County Extension Farm Agent can give sound advice.

If the horse is to be confined with limited riding, about 800 square feet should be provided for a minimum exercise area. Longeing the horse for a half hour a day can be done on very limited space. (Longeing is having the horse circle you on a long 25- to 30-foot line.)

Fencing for horses usually involves more expense due to the temperament of the horse. Barbed wire, one of the most inexpensive fencing materials, is the least desirable for horses.

Common breeds of horses

Breed	Normal usage	Comments
Arabian	Pleasure (E & W) Park Stock Horse Trail Riding Endurance Riding Parade Pleasure Driving	Exceptional endurance. Foundation breed of most other purebred breeds
Appaloosa	Pleasure (E & W) Stock horse Trail riding Endurance riding Reining Cutting Parade Gymkhana, timed events Pleasure Driving Roping Polo	A color breed; not pure bred
American Saddle Horse	3 and 5 gaited Pleasure Fine harness Pleasure driving	Mostly used in show ring and on bridle path
Morgan	Pleasure (E & W) Park Stock horse Trail riding Pleasure driving Roadster	Descended from one foundation sire (Justin Morgan)
Paint	Pleasure Stock horse Reining, roping Cutting Racing (short) Parade Trail riding Polo	Color breed of quarter horse conformation. Two basic color patterns—Overo & Tobiano
Pinto	Used more extensively under English saddle Polo Parade	Color breed. Conformation may be characteristic of most of the English type breeds. Overo and Tobiano
Palomino	Pleasure (E & W) Stock horse Parade	Color breed. Many double registered, especially Quarter horses

E & W refers to English and Western saddles.

Common breeds of horses

Breed	Normal usage	Comments
Quarter Horse	Pleasure (E & W) Stock horse Reining, roping Cutting Racing (short dist.) Trail riding Gymkhana Timed events Polo Hunters Dressage	Largest in registration of all breeds. Due to its versatility and temperament, a very popular breed.
Standard Bred	Harness Racing—trotters and pacers Roadsters	Used almost exclusively as racing and driving horses
Thoroughbred	Racing Hunters Jumpers Polo Stock horses	Used extensively to add quality, speed and endurance to many of the other breeds
Tennessee Walking	Show ring Pleasure	A characteristic running walk; when developed to the extreme is used mostly in the show ring
Connemara	Medium-sized adult's and children's mounts Hunters Jumpers	Noted for their temperament, hardiness, and jumping ability
Pony of Americas	Children's mount Pleasure (E & W) Timed events	46" to 52" in height and have Appaloosa markings
Hackney Pony	Harness show ponies	Extreme flexion in the knees and hocks. Probably the showiest of all at the trot
Shetland Pony	Children's mount Pleasure driving Harness show Pony (American type)	Old Island type short and blocky, and new American type modeled after the American saddle horse
Welsh Pony	Children's mount Roadster and racing ponies Hunter ponies	A very hardy pony.

The horse is easily frightened and in small pastures is apt to try to run through barbed wire, which frequently causes grave injury.

Suitable types of fences include:

Wood—plank, board, split rail, rail, and buck and rail. Wood is the safest fencing for horses, but expensive to construct and maintain.

Metal—woven wire, chain link, welded pipe, cable, barbless wire and plain wire.

Electric—Use only an approved, safe system. Follow directions for installation. Horses must be trained to an electric fence, which can be used to divide pastures for rotational grazing.

All wire or metal fences should be grounded at least every 800 to 900 running feet to a metal rod driven into the ground to permanent moisture depth.

Fenceposts

Locust, red cedar, and osage-orange are the best wooden posts available. Railroad ties in good condition have long life, are strong, and hold staples well. Pressure-creosoted, discarded telephone poles may also be used.

All other wood posts—as white cedar, oak, hemlock, hickory, pine, spruce, birch, beech and ash should be treated with a wood preservative such as creosote or pentachlorophenol.

Note that with metal posts, a T-shaped driven post is stronger than U or V shaped posts. Concrete posts have long life but are expensive.

Managing Pasture

Many horse pastures are no more than exercise lots. The quantity and quality of feed on these rundown pastures won't meet the horse's nutritional needs.

Proper management through introducing adapted grasses and legumes, and a proper stocking rate, can provide a considerable portion of the horses daily nutritional needs during the grazing season.

Pleasure horses are usually pastured on very limited acreages. Therefore, a balance between the needs of the horse and those of the pasture grasses must be maintained.

The pasture manager has several ways of stimulating the growth of pasture plants. To improve soil fertility he can lime and fertilize. Eliminating brush and weeds will provide more sunlight, nutrients, and water for the desirable pasture plants. Drainage may be required to improve the growth environment. Reseeding to change the plant composition will increase forage productivity.

Large pastures divided into several smaller pastures and rotated during the grazing season provide more forage per acre. Overgrazing and putting horses on pasture too early and leaving them on too late in the season is very damaging to pasture.

Plants grow most rapidly in a 60° to low 70° F temperature range. When the temperature is 20° above or below this range, all growth virtually ceases. Excessive cloudiness and lack of rainfall affect plant growth adversely.

Defer grazing until the new growth is 4 to 6 inches high if the stocking rate is at least 1-1/2 acres per horse. Pastures to be grazed early (2 to 3 inches of growth) should be grazed only 1 to 2 hours a day. Turf injury will occur if horses are turned out on waterlogged pastures.

Pastures going into winter with scant top growth suffer extensive winterkill and do not recover well enough in spring to compete with weeds.

These management practices will help maximize forage production: Drag pastures to break up and scatter droppings. Clip to control weed growth. Remove excess shade trees. Apply lime if a soil test indicates need. Apply phosphate and potash if there is a good stand of clover. Mulch bare areas to prevent erosion.

A horse can provide hours of fun for a loving owner, but lots of work goes along with the fun.

Charles O'Rear

Feed Requirements

Too often through lack of knowledge the horse is overfed and underworked, or overworked and underfed, or given a feed of such low quality that it is not assimilated in the digestive process. These problems can readily be corrected.

Physiological problems that may exist are with teeth (wolf teeth, sharp teeth, long teeth, decayed or broken teeth), parasite infestations, infections of the intestinal tract, allergies, age, spoiled feed, and feed of a wrong consistency (ground too fine).

There are other causes of nutritional deficiencies, but those are the most common.

To satisfy the horse's nutrient requirements, you need to provide: a balanced ration in amounts large enough for growth (if immature); energy (for body temperature regulation, vital body functions and work; the repair of wornout body tissues); development of young (pregnant mare); production of milk (lactating mare).

A balanced ration must consist of correct amounts of:

Protein. Vegetable seed derivatives such as soybean oilmeal furnish more of the essential amino acids.

Energy, fuel for work. Energy is derived primarily from the carbohydrate and fat portion of the ration.

Vitamins—A and D are the most important. Horses kept on pasture or with access to the sun normally assimilate enough vitamin D. Stabled animals should be supplemented. Vitamin A is derived primarily from green leafy roughages (hay) in the form of carotene.

Minerals—Calcium and phosphorus in proper balance (most agree on a ratio of 1.1:1), and salt in proper amounts are vital.

Clean water is essential for digestion and should be provided and accessible at all times (except for hot horses or in other cases where intake must be controlled). A horse consumes an average 10 to 12 gallons per day.

Each bag of commercial feed has a tag attached. The tag lists the percent of crude protein, fiber and fat, and the feed ingredients.

Pelleted feeds are popular with some horsemen because of ease of storage, less dust, less manure and less waste. The big disadvantage is that most horses develop a wood-chewing habit if pellets are fed without some hay.

Good hay must be cut at the proper stage of maturity. It

should have a bright green color, an abundance of firmly attached leaves, and a sweet and pleasant smell.

Tips on Feeding

Savings on feed costs can be realized by prudent buying, proper care of the horse's teeth, elimination of parasites, feeding according to need, reduction of waste, and clean water in sufficient amounts.

Follow a well planned feeding program. Since the horse is a creature of habit, a poor feeding schedule can establish bad habits in the stall. Crowding, kicking, pawing, and biting are some of the most common problems. Giving a horse "treats" by hand can establish a habit of nipping or biting.

Several feed management practices that should be followed are:

Feed each horse as an individual and know its weight and age. Feed at regular times. Use clean feed boxes.

Horses doing hard work need more grain and less hay. Split grain ration into 3 or 4 feedings per day with the biggest portion at night. Split hay into 1/4 morning, 1/4 noon and 1/2 at night.

Use a good quality, nutritionally balanced feed. Never feed moldy, dusty or frozen feed. Feed by weight, not by measure.

Have salt and mineral mix available free choice.

Check teeth regularly.

Check 3 H's—hair, hide and hoof—for condition. They are good indicators of digestive problems.

Check consistency and odor of droppings.

Control parasites, internal and external.

Don't make abrupt changes in feed. This also applies to new cut hay. Don't change abruptly from a grass hay to a hay high in early cut legumes (alfalfa or clovers).

Rule of thumb feed requirements for a mature riding horse

Use	Commercial mix grain 10-12% protein	Hay
Light 1-3 hrs/day	1/3 to 1/2 lbs. per 100 lbs. body weight	1 1/4 to 1 1/2 lbs. per 100 lbs. body weight
Medium 3-5 hrs/day	3/4 to 1 lb. per 100 lbs. body weight	1-1 1/4 lbs. per 100 lbs. body weight
Heavy 5-8 hrs/day	1 to 1 1/2 lbs. per 100 lbs. body weight	1-1 1/4 lbs. per 100 lbs. body weight

Control feed intake of the glutton by using a wide flat feed box and spread grain over the surface, not in a pile. Placing softball size stones in the feedbox also slows down the gorger. Don't feed large amounts of garden produce to horses.

When first turning your horses out to pasture in spring, be sure to give them a full feed of hay prior to turning out and then some hay each day for the first week. The first few times, turn out for only two to three hours until they become accustomed to the new lush growth.

Don't throw clippings from yew bushes in with horses. Remove broken limbs of wild cherry trees that contain wilted leaves. Both are poisonous.

Stabling, Equipment

In the Northeast and areas plagued by rapid changes of weather, shelter should be provided. A three-sided shed open to the south or southeast and protected from the prevailing winds, well-bedded and free from drafts, will provide all the shelter needed.

Provide horses kept outside with a slightly higher plane of nutrition, to meet the greater requirements for body heat.

Horses maintained in show condition are either blanketed and housed in enclosed stables, or heat is provided in the stable. The amount of heating will dictate the amount of ventilation and insulation that must be incorporated in order to control condensation.

Doors, windows, louvers and cracks will usually provide enough ventilation in an unheated stable, but avoid drafts.

An ambient temperature of 32° F will prevent freezing of water pipes and also be very comfortable for the horse. Stale air, condensation, dust and drafts are harmful.

A good stable design with attractive fencing does much to make a horse acceptable in a suburban area. Incorporated into

This stable was built according to USDA Riding Horse Barn Plan No. 5838, available through your State Cooperative Extension Service.

R. C. Church

374

the design should be a rodent-proof feed room, hay and bedding storage, a moisture-proof tack room, and provision for manure storage.

Box stalls or straight stalls may be used for horses or ponies. Provide a 5-gallon water bucket.

Cement is not recommended as a floor material. It is slippery, cold, and hard on the horse's feet and legs.

Planking is used in many areas, but it rots and has to be replaced periodically, creates odors, and provides an excellent environment for mice and rats.

The preference of most horsemen is tamped clay free of rocks. Sandy loam free of rocks can also be used.

Sand is used, but during winter it makes a cold and damp bed for the horse. Then too, horses may develop sand colic through ingesting sand while eating spilled feed.

Many types of bedding are available: straw, sawdust, shavings, peat moss, peanut shells, even leaves and pine needles. Some are more absorbent than others and the price range is large. A diligent sorting of soiled bedding can cut down costs and the amount of wastes to be disposed of.

One window per stall is preferred to provide light and ventilation.

Have one electric lighting fixture over every other stall, and lighting for tack and feed room and alleys. All fixtures should be covered by dust-proof shields; wiring and switches should be out of reach of the horses or protected.

The feed room should be horse and rodent-proof, and provide for tool storage.

Completely enclose the Tack Room. If heated, it should be completely insulated. Keep it clean, and free of dust and moisture which deteriorate leather rapidly. The tack room should include blanket, saddle and bridle racks, a first aid cabinet, a saddle cleaning stand, and equipment storage. Boxes can be provided near stalls for grooming equipment.

Stall feeding facilities

		Mature horse	Pony
Grain box	Length	22″	18″
	Throat height	40″	32″
	Width	14″	10″
	Depth	7″	7″
Hay manger	Length	34″	24″
	Throat height	40″	32″
	Width	22″	18″
	Depth	28″	20″

Use oak lumber.

Training or exercise rings do not require as much space as show rings. Normally a 40 by 80 foot ring will suffice.

Careful planning can create extensive and interesting trails in small wooded areas. Neighbors with horses and land can develop connecting trails for mutual use.

Horse owners with only stabling facilities and no nearby trails or rings available will probably have to invest in a horse trailer. Most owners eventually do.

Tack requirements consist of a halter and lead rope, bridle and saddle (English or Western), saddle pad or blanket. The rider's personal attire will depend on the style of riding.

Health Care

Preventive medicine should be the basis for a sound health care program. Providing the following will do much to insure the health of your horse: Adequate shelter, safe fencing, a balanced ration, good dental and foot care, a sanitary environment, control of parasites, and immunization against diseases.

The second line of defense, the most important, is to pick a good veterinarian. A veterinarian who specializes in equine medicine is probably the best choice; second choice is one who has a large animal practice. Check with local horse owners.

The vet will recommend immunizations for your area. Ask for and follow the vet's parasite control program.

Learn to recognize the seriousness of wounds, know what to do, and when to call the vet. The vet can instruct you in follow-up nursing care. With experience and study many problems can be taken care of by the owner.

A good farrier is necessary for the health of the horse's feet. Most farriers can recognize diseased feet. Horses should have their feet trimmed regularly if kept barefoot. If shod, the shoes should be reset or replaced every 6 to 8 weeks.

Routinely clean out the horse's feet. This plus daily removal of soiled bedding and manure, and replacement with fresh bedding, helps prevent diseased feet.

Normal vital body signs for the horse

Rectal temp. range	Normal pulse rate	Normal breathing rate	Attitude	Hide and hair	Membranes eye, nose, gums
100.5° F 99°-100.8°	32-44	8-16	Alert Content Good appetite	Sleek and pliable	Pink (salmon) colored

Deviations from the above may indicate ill health.

How You Can Attract Birds, Other Wildlife to Your Place

By John York and David Allan

The decision to buy your land was right. You have managed the down payment, rebuilt the house and barn, and planted the garden. The old orchard has been renovated, the county Extension agent has been pumped almost dry, and the Soil Conservation Service has planned, staked, measured, and probed nearly everything that couldn't run away.

The state forester helped with the Christmas trees; the county health services helped with the water and sewage. All in all you are pleased as you settle back on the porch for a cool drink and some peaceful thought.

The scraping of a cricket, the far-off tinkle of a cowbell, someone sawing wood back in the hills all tend to wring out the tension of modern living.

A flicker of color catches your eye—it's a small bird frantically trying to catch a bug which is just as frantically trying to hide. A burst of bird song signals success, or is it failure? No matter. It sounds nice.

Human beings evolved in and with the natural world. As we become more urbanized, more complex, and have more social pressures, we seem to lose touch with nature. We tend to forget just how closely our lives are tied to the other living things around us. But—the evening call of bobwhite, a visit from a friendly squirrel, or the romp of fox pups can ease you out of some very trying days. These interludes also add richness to life on a few acres.

All right! You're convinced. You have the land and you want wildlife. Which ones? Where? How long? When? Why?

Even on small acreages an amazing variety of wildlife can be observed. True, you can't expect to have a herd of deer, elk, or a flock of turkeys, but your land may be part of their range

John York is State Biologist for the Soil Conservation Service (SCS) in Phoenix, Ariz. David Allan is State Biologist for SCS in Durham, N. H.

and they may pay a visit. Too often newcomers to the country have high hopes for large numbers of wildlife, especially if they are willing to improve their land. But with most things, certain rules must be followed in order to succeed.

Rule No. 1—Don't move a rock or a brush pile. Don't cut a tree, plant a bush, dig a hole, level a field, or build a fence until you have surveyed the land and know what resources you have.

Once you know your resources, write out a plan of what, when, and how you will proceed with development of the land. You will make fewer mistakes with a well-thought-out plan to guide you.

An example of this is a landowner who raised Cecropia moths for heart research. This moth thrives on chokecherry trees, normally considered a worthless weed tree in old field land. By inventorying the number of trees, the landowner determined how many moths he could expect to produce and the amount of netting needed for their protection. At the same time, he was also improving nearby habitat for songbirds that normally prey on the moth larvae. By planning the resource he could have both moths and songbirds.

Rule No. 2—Read, ask, and visit. Find out what wildlife *should* be on or around your land and what wildlife is actually there. Zoos, local, state, and national parks, game and fish commissions, universities, and some Federal agricultural agencies will be able to give you habitat requirements for both existing and potential wildlife.

A wealth of informational books about nature is on the market today. Use regional guides to learn about the birds, animals, wildflowers, or trees and shrubs of your area.

Walk over your land. Do this often and during all seasons to see the many changes. It is amazing what new things can be observed with each trip.

Have places where you can sit quietly and watch. This is the best way to see what shares the land with you. Learn to really look at things closely, such as scratch marks on a tree trunk. This will tell you what lives there or uses the area.

Rule No. 3—Manage your wildlife. Certain animals may increase beyond the available habitat. They need to be harvested or they will destroy their habitat or develop into nuisances. Trees and shrubs may grow beyond the reach of wildlife and become troublesome as weeds. They will have to be controlled and managed.

Planning for wildlife is like planning on the rest of the

land. Wildlife, whether a pygmy shrew or a moose, need food, water, and some kind of cover. They set up territories just like humans. These things are part of the animal's "habitat". Each kind of wildlife requires specific items in the habitat. To add, maintain, or attract specific wildlife, you must know what they require. A few examples:

Woodcock need small, open fields where they can go through their mating ritual flights at dusk on a spring evening.

You might put out a log or mound as a loafing and preening area for male ducks who are guarding their territory on a marsh.

Songbirds such as bluebirds respond to certain size holes in birdhouses. This offers them protection from larger birds, like starlings, who are more competitive for homesites.

Food Needs

All wildlife requires some kind of food. Plant eaters—some insects, elk, deer, rabbits, and porcupines, to name a few—convert plant energy to protein and fats. Wildlife such as predatory insects, hawks, owls, bobcats, insect-eating birds, and grizzly bears feed on the plant eaters. This complex web of food links will collapse without plants.

Some wildlife feed on the ground, some in small shrubs. Some feed only on evergreens, while others subsist only on broadleaf trees. Plan your habitat so a wide diversity of seed, berry, and foilage plants will be available. This is especially true if you are planning for a variety of wildlife but no one kind in particular.

Many plants provide both food and cover. On the deserts, some plants provide food, cover, and water. Plant as many multipurpose types as possible.

A well-made birdhouse.

S. P. McClendon, Jr.

Jeff Smith of Hollis, N.H., has successfully done this by converting an old apple orchard into a wildlife area. He plowed and planted strips of sunflowers, sorghum, millets, and small grains as food plots between the orchard rows. Other strips have been seeded to legumes such as clovers, crownvetch, and Lathco flatpea for food. Every few years he plows under the legumes, which have enriched the soil with nitrogen, and re-plants with grain strips.

The old apple trees have been revitalized by severely cutting the tops back and grafting them to small fruited crab-apples that hold their fruit through the winter. The gnarled and broken branches are full of holes that provide homes for birds and animals. Deer bed in the legumes, grouse eat the winter fruit and buds, and in the fall there is a riot of migrating birds feeding in the grain strips.

You can provide food not ordinarily available on a year-round basis to wildlife (feeding your chickens to a fox doesn't count). Birdseed, suet (beef fat), peanut butter, nuts, fruits, bread, sugar water, and the like can be provided in winter. Birds can also be fed in summer; for example, honey water tubes bring hummingbirds to your home, where they can be more easily seen.

Professional biologists are divided on the benefits wildlife receive from supplemental feeding. It is fun and probably the "pros" will balance the "cons". If you do winter feed, you must continue because wildlife become dependent on the source. If it is sporadic or cut off, they suffer and may die in severe weather.

Water, Cover

The importance of water for wildlife depends somewhat on where you live. Water is vital and a major wildlife attractant on the deserts in Arizona. If the land is surrounded by streams, bayous, ponds or springs, water development on the land is relatively unimportant. Some wildlife, such as kangaroo rats, need no free moisture at all. Others need a readily available and constant supply.

Water attracts all kinds of wildlife, from the truly aquatic species to those that just casually use it. A water situation can be as simple as a birdbath or a small backyard pool. It can be a permanent part of the landscape or a temporary pool built of plastic sheeting. Farm ponds or tanks are more perma-nent types of structures. Small marsh developments provide habitat for wetland-loving wildlife.

On the Andrew Wyeth-James land along the coast of

Maine, a shallow 2-acre marsh created in old field land attracted about 40 different species of sea birds and shore birds. They came to wade, feed, drink, or bathe in fresh water.

All natural wetlands such as potholes, marshes, swamps, bogs, ponds, or streams need to be considered in planning your land. Again, this brings us back to the basic rules of meeting the wildlife's needs.

You need to consider two kinds of cover, natural and artificial (or man-made). Both have a place in the plan for wildlife. Natural cover provides resting, roosting, nesting, protection, and foraging areas. This part of wildlife habitat is easily managed by manipulating vegetation. Vegetation can be planted, pruned, thinned, or cleared, depending on what is needed.

Artificial or man-made cover can be brush piles, nesting boxes, piles of rock, birdhouses, log piles, and similar structures. These types of cover can be a valuable addition to natural cover and benefit many kinds of wildlife.

Dispersal of different cover types is important. Cover is needed for protection of wildlife travel routes and for escape.

Developing a Plan

Ideally, the wildlife plan should be developed as an integral part of your overall land use plan. If the land is being farmed, you will find many farming practices can be either good or bad for wildlife.

Windbreaks, field borders, hedgerows, and farmstead landscaping can all contribute cover and food. Some types of drain-

Canada geese nest each
year at this pond.   John W. Anna

ing or the burning of fence rows and roadsides, may destroy habitat. Land not being farmed can benefit from reseeding, mowing, or rotation harvesting of hardwood timber that is beneficial to deer and smaller mammals or birds.

In developing a plan for your land, consider the following:

Soils—Obtain a soils map if possible. This will provide information on what types of vegetation can be grown, where drainage may be needed, or where to build a pond or marsh.

Vegetation—An inventory of the existing vegetation is important. Find out what acreage is in grassland, shrubland, wetland or forest. Locate the different types on an aerial photograph or map. This will give you the pluses and minuses in what you need for good wildlife habitat.

You may be pleasantly surprised to find that you already have all the wildlife plants you need. The vegetation may only need management. Management can be such things as pruning, fertilizing, mowing, or releasing vegetation from competition with other plants.

Water—Determine the quantity of available water and how it might be used. Consider all the possibilities such as wildlife, recreation, irrigation, or personal use. The quality and location is important.

Wildlife—List the wildlife you know are on the place or that have potential to be there. Write out their habitat needs and cross-check these needs against the habitat you have already. Whatever is lacking you need to manage or plant.

Planting plan—Plan the kinds of plants you will need and their location. Be sure they are hardy in your area and adapted to the soils on your land. Determine quantities needed, cultural practices required in establishment, and, most important, their use by the wildlife desired.

Don't try to do everything at one time. Spread the plantings out over a period of time. The seed and shrub catalogues look great during winter, but many good plans fall by the wayside when the time to plant arrives.

Go to some places where these plants are already established, see what they look like, how the plants produce, and if this is what you really want.

Management—Plan the different types of practices to fit your land. These practices could be such things as seeding food plots, cutback borders, shrub border of hedgerow plantings, or forest clearings.

22 Mammals, 60 Birds

David Allan owns eight acres of land in an area of mixed farming and urban homes. An acre and a half is used as home grounds consisting of lawns, a large home garden, dwarf fruit trees, grapes, blueberries, and other fruits. Shrub borders of Cardinal autumn-olive, grey-stemmed dogwood, rugosa rose, Rem-Red honeysuckle, and Pinklady euonymus have been planted along property boundaries and the edge of the woods.

These plants provide food and nesting cover for song-birds, and screen adjoining property. Abandoned fields are managed for wildlife by mowing. This is done with an old power mower in the difficult spots, and a riding mower on level areas.

Wild shrubs such as highbush blueberry, shadbush, roses, and other fruit-producing shrubs are left to form islands in the open field. Native wildflowers are also left for bees and beauty. Milkweeds are encouraged, providing the essential needs of the migratory Monarch butterfly. Mowing is delayed until ground-nesting birds have raised their young.

One part of the field had been stripped of topsoil by a previous owner. While this was initially a loss, rare orchids, sundews, and other native plants began moving in. The sub soil is wet and a marshy area is visited by shore birds and herons.

Ponds draw a variety of wildlife, such as mallard ducks and turtles.

John Haagen

Katherine C. Gugulis

A small spot is planted with wild cranberries for holiday harvest, another spot furnishes fiddlehead ferns for greens, and a small bog has been built for pitcher plants and for star-nosed moles and other bog dwellers. A wet alder swamp has walking trails for bird watching and openings cut for wood-cock.

Cutback borders released old apple trees, native dogwoods, and other shrubs that have responded to the sunlight and pro-duce good cover for birds and small mammals. The woodlands are improved by selective cutting and furnish some firewood. Handtools are used for the exercise.

Old trees are left for dens, and nut trees for wildlife food. Brush is piled up as shelter for cottontails and hare. The maples are tapped for maple syrup as a spring ritual.

A small pond irrigates the garden and fruits, and provides swimming and skating for neighborhood children and a place for kids to catch frogs and turtles. They beat a constant path around it.

The pond is alive with aquatic life such as fish, snakes, turtles, birds, and insects. After construction the spoil banks were planted to native azaleas, tamarack, legumes, and Christ-mas trees for nesting, fruit, and for family use.

Birdhouses are set up to attract birds that help control in-sects, and den boxes placed in the woods for squirrels and owls.

These practices have been successful as the family has observed 22 mammals, from a pigmy shrew not much bigger than a dime to a buck white-tailed deer. About 60 birds are seen, including grouse, woodcock, and the rare yellow rail.

Nine reptile and eight amphibian species complete their wildlife count. This is topped off with wild flower gardens and native plants that have been introduced. The land is harmon-ious with nature and a restful place after the busy workday.

To Sum Up—As you settle back on the porch for that cool drink and some peaceful thought, reflect a moment on what you may have done with your wildlife plan. You may have preserved the last old oak in the county; you may have the only pair of flycatchers; you may have helped wildlife survive in a hostile area.

As Aldo Leopold has written, "wildlife once fed us and shaped our culture. It still yields us pleasure for leisure hours, but we try to reap that pleasure by modern machinery and thus destroy part of its value. Reaping it by modern mentality would yield not only pleasure, but wisdom as well."

You Can Grow Fish for Fun or for Profit

By Robert Waters, H. Dave Kelly, and W. Mason Dollar

You can grow fish for either fun or profit on a few acres—in fact, on fewer acres than you can grow many other crops.

But growing fish is not easy. It requires clean water, a site suitable for a pond, raceway, or other facility, a good bit of planning, and frequently a great deal of knowhow. Some fish-growing operations require lots of managerial ability, a fairly large investment, and a great deal of time, especially operations for commercial purposes.

Yet growing fish on a few acres can be fun. Some operations can furnish high-quality fishing for you and your family and friends at little cost. Other operations can supplement your income without taking a great deal of your time.

Fish farming as a major source of income, however, can be a time-consuming, high-risk business in which you can lose a large investment in a short time. And you can lose it through no real fault of your own. So, before you go into commercial production, give plenty of thought to the venture.

Growing fish can be classified into two broad types—raising fish for your own recreational use, and to sell for profit.

The goal of most noncommercial fish raising is to provide high-quality sport for a few people at minimum effort and cost. Fish for these purposes are raised mainly in farm ponds, of which there are more than 2 million. These ponds generally provide many hours of good fishing.

Resource Appraisal

The first step in establishing a pond for recreational use is to assess your resources. Soil, water, topography, and other resources will influence your success in raising fish and sometimes may limit the possibility of growing fish at all. Get help

Robert Waters, H. Dave Kelly, and W. Mason Dollar
are biologists with the U.S. Soil Conservation
Service at Auburn, Ala.

in determining the adequacy of your resources from the Soil Conservation Service of the U.S. Department of Agriculture.

Your land must contain a site suitable for impounding water. On most farms and ranches, topography is such that water can be impounded at reasonable cost if a dam is properly located.

The soils of your property must be capable of holding water. This is true not only for soils used in building the dam but also for soils on the pond bottom.

In some sandy or limestone areas, seepage through the soil or material under which water is impounded may result in water loss at a rate faster than it is received. Many dry ponds have been constructed because this important factor was not considered.

A good fishpond must have enough spring flow, well water, or runoff to fill the pond in a year or less and to replenish water lost from seepage and evaporation. Also, water quality must be adequate for the species you intend to grow. Most species are adapted to either warm or cold water.

Fish can be grown in ponds of any size; however, desirable fish populations are not easily maintained in extremely small ponds. Warm-water ponds managed for bass and bluegill for recreational purposes should be an acre in size or greater. Trout ponds should be at least one-third acre.

Generally, the time and costs involved prohibit adequate management by individuals on ponds greater than 15 acres.

Where the pond is covered with ice for a month or more and winter fishkills are frequent, water depth may need to be as

Russell Tinsley

A farm pond bass.

much as 20 feet. If your pond depends on seasonal rains for its water supply, water should be at least 10 or 12 feet deep over a fourth or more of the pond area. Throughout most of the South an average depth of 3 to 4 feet is adequate for fish growth; no more fish can be grown by providing additional depth.

The emergency spillway to prevent floodwater from going over the dam should be designed to keep the flow shallow. This prevents big fish from swimming out and unwanted fish from swimming in.

To prevent unwanted fish from entering the pond from downstream, a 24- to 36-inch vertical overfall in the emergency spillway should also be used.

A drainpipe is useful, and required by law in many states. It should be large enough to drain the pond in 5 to 10 days.

An overflow pipe or trickle tube connected to the drainpipe keeps the normal water level a few inches below the spillway. This reduces erosion of the spillway and prevents drowning of its grass cover. A trickle tube also helps prevent excessive loss of fingerlings soon after the pond is first stocked.

A device to take outflow from the bottom rather than the top helps warm the pond early in the spring, may save fertilizer, and may permit fertilizing in spring when the flow is heaviest.

Most trees and brush should be removed from the area to be covered with water. If possible, remove stumps and snags from the pond bottom. This leaves the bottom smooth for seining, which may be required for good management.

Clearing trees and brush from a strip 20 to 30 feet around the pond reduces the amount of leaves that falls into the pond. Leaves discolor water and encourage growth of algae. And decaying leaves may cause oxygen depletion in the water. A cleared strip also provides a grassy bank for fishing.

Shallow water is troublesome because weeds grow in it. In ponds stocked with bass and bluegill, weeds protect small fish from the bass. Shallow water may also cause mosquito problems. A pond is easier to manage if it has no shallow water.

Either deepening or filling eliminates shallow edges, but usually a combination of the two is best. A minimum depth of 2 feet is good, but 3 feet is better.

Stocking Fish

After considering fish species best adapted to your water, select the species to stock. Either a single species (for example, trout) or a species association (for example, bass-bluegill) may be appropriate depending on your resources and objectives.

A number of trout species may be considered for stocking

if your water is suitable. Generally, rainbow trout and brook trout are stocked in ponds because of their availability and because they are relatively easy to manage and to catch. For variety, rainbow and brook trout may be stocked in the same pond.

When stocking trout, exclude all other species of fish. If your pond contains 50 pounds of other kinds of fish, it will likely have 50 pounds less trout.

Largemouth bass and bluegill are the most commonly stocked species in warm-water ponds. Frequently, redear sunfish are mixed with this combination to provide variety and reduce chances of the fish population becoming unbalanced. In this species combination, bluegill and redear sunfish provide food for the bass. If the population is properly managed, it will provide good fishing for many years.

Before stocking, eliminate all wild fish from your pond and drainage area as they will compete with stocked fish for space, oxygen, and food. You can remove these fish by draining the pond or applying a fish toxicant. If you plan to use a chemical to remove undesirable fish, be sure it is approved for such purposes and applied with the supervision of a licensed applicator or other appropriate personnel in your state.

Private and certain state and federal hatcheries provide

Commercial trout ponds, with water running a complete cycle about twice an hour, maintain a fairly constant temperature.

fish for stocking. Information on sources of fish for stocking can be obtained from the Soil Conservation Service, Extension Service, Fish and Wildlife Service, and your state fish and wildlife agency.

Stocking rates depend on the species selected, natural water fertility, and willingness of the pond operator to supplement the pond with either fertilizer or fish feed, or both.

In Alabama, ponds with average natural fertility are generally stocked with 50 bass and 500 bluegill per surface acre.

Ponds maintained at high fertility levels by the periodic ap-

Fish commonly used for recreation and commercial purposes

Species	Recreation	Commercial							Summer Temperature Range
	Sport Fishing	Human Food	Bait Fish	Fee Fishing	Fingerling	Brood Fish	Ornamental	Eggs	
Blue catfish		•		•	•	•			Warm > 80°F
Bluegill	•			•	•	•			Warm > 80°F
Brook trout	•	•		•	•	•		•	Cold < 70°F
Buffalofish		•			•	•			Warm > 80°F
Carp		•			•	•			Warm > 80°F
Channel catfish	•	•		•	•	•			Warm > 80°F
Crappie	•			•	•	•			Warm > 80°F
Fathead minnow			•			•			Warm > 80°F
Golden shiner			•			•			Warm > 80°F
Goldfish			•			•	•		Warm > 80°F
Hybrid sunfish	•				•				Warm > 80°F
Largemouth bass	•				•	•			Warm > 80°F
Mudkellunge					•				Cold < 70°F
Northern pike					•				Cold < 70°F
Rainbow trout	•	•			•	•		•	Cold < 70°F
Redear sunfish	•				•	•			Warm > 80°F
White catfish		•			•	•			Warm > 80°F

> means greater than. < means less than.

plication of commercial fertilizer are stocked with 100 to 150 bass and 1,000 to 1,500 bluegill per surface acre.

In cold-water areas, ponds with average fertility usually produce enough natural food to support only 500 half-pound trout per surface acre annually. But with supplemental feeding the same waters can produce 2,000 half-pound trout in a year.

Management Levels

Fishing success in your recreational pond will be determined largely by the management of your water and by your fish population.

Intensive management can produce exceptional fish production. However, the cost of fertilization, feeding, chemical controls, supplemental stocking, and other management is often more than most people are willing to pay for a few hours of good fishing. The amount of effort involved in carrying out intensive management is also prohibitive in many cases.

Before you select a species for stocking and decide on stocking rates, determine a level of management best suited to your time and pocketbook.

One way to increase fish production is fertilization. Fertilize before the growing season for warm-water fish. In the extreme South, fertilizer may be applied year-round; however, the fish growing season decreases progressively toward the North.

Effectiveness of fertilization is greatly reduced in the northern States and may not be recommended as a fish management practice. The increased organic production resulting from fertilization may contribute to oxygen depletion and fishkills, especially in areas with long periods of ice cover.

Where fertilization is applicable, ponds need refertilization any time during the growing season that a white object can be seen 18 inches or more below the water's surface. If the pond is fertile enough, the resulting algal bloom will obscure the object at that depth. As many as 12 fertilizer applications per year are needed for proper fertility.

A mineral fertilizer is best for fishponds—organic fertilizers often stimulate growth of undesirable algae and vascular aquatic plants. The amount depends on the fertilizer analysis. In warm-water ponds managed for bass and bluegill, 8 pounds of nitrogen, 8 pounds of phosphate, and 2 pounds of potash per surface acre (100 pounds of 8-8-2 fertilizer or equivalent) is commonly recommended for each application.

Fish production can also be increased by supplemental feeding. In trout ponds, production can be raised from 100 pounds per acre to 1,000 to 2,000 pounds per acre by supplemental feeding.

Other management which may influence the ability of your pond to produce harvestable-sized fish include waterweed control, maintaining a balanced fish population through either harvest or partial removal of fish by drawdown or chemicals, controlling fish diseases, and ensuring good water quality.

Fish for Profit

The most popular commercial enterprises involve raising fish for human food, bait, stocking, and fee fishing. Other enterprises produce brood fish, eggs, and ornamental fish.

As a rule, commercial enterprises require a larger investment and more know-how than raising fish for your own recreation. Generally it's best to start on a small scale, expanding the operation as you gain knowledge and experience. In most operations, fish are raised at densities far exceeding what nature provided for.

A dependable supply of good quality water is essential for raising commercial catfish and trout. Water from wells, springs, streams, or runoff ponds is suitable if you take necessary precautions.

Most commercial trout are raised in raceways that have a reliable year-round flow from springs, wells, or streams. The number of trout that can be raised each year is determined by the volume of water available. A minimum flow of 450 gallons per minute is needed.

The best source of water is a well. Using well water avoids problems of unwanted fish, flood hazard, pesticides, and muddiness. Your well should provide enough water to fill the ponds in a week or so, replace water lost through evaporation, and

Fishing for channel catfish in a fee-fishing pond.

Robert E. Waters

supply water needed to replenish oxygen. Where underground water sources are unknown or questionable, drill a test well.

Springs are a good source of water but may contain undesirable fish. The flow in dry seasons must be known to determine spring adequacy. Oxygen level is usually adequate for your fish, but it's a good idea to check. If oxygen is not adequate, aerate the water. Kill undesirable fish with an approved fish toxicant or remove them by filtering the water.

Water taken from a pond, stream, bayou, canal, or other surface source usually contains undesirable fish. They will get into ponds and compete with desirable fish unless you take steps to keep them out.

If your operation requires a pond, your land must contain a site suitable for impounding water. The topography must be such that water can be impounded at reasonable cost and the soil must be capable of holding water.

If raising fish in raceways is your goal, the site should be such that a raceway can be built and operated at reasonable cost.

Put your raceway near the water supply. Whenever possible locate it so water can flow by gravity from the source throughout the raceway.

Soil Conservation Service technicians can help you decide if your water supply is of the quality and quantity required by your operation and by the fish you plan to grow.

Labor, capital, electrical power, and many other resources may be needed. Consider these needs carefully.

Design and Layout

Proper design of the facility can often reduce costs and make your operation more efficient—for example, designing raceways in which water flows by gravity, thus reducing or eliminating pumping costs.

Facilities should be no more elaborate than required to do the job efficiently and economically. Storage space for fertilizer and feed is called for in many cases. Earthen fishing piers are desirable in large ponds for fee fishing. All-season roads may be needed.

Other facilities and equipment that may be needed include feed bins, automatic feeders, pumps, concession stands, scales, mechanical aerators, fish-cleaning facilities, large freezers, thermometers, water-analysis kits, cages, bait, fishing tackle, and holding vats. Some of these are expensive.

Species of fish for your operation will be determined

mainly by the quality and quantity of your water, the region where you live, and your type of operation.

You may stock more than one species of fish, especially when growing fish in a pond. In many cases, results are satisfactory only when several species are stocked.

Sometimes an additional species is stocked to provide variety or increase fishing success. White catfish are stocked frequently in fee-fishing ponds with channel catfish, because the whites bite better during daytime than channel catfish.

In many enterprises, one or more species are stocked for purposes other than direct fish production. Largemouth bass and fathead minnows frequently are stocked in commercial catfish ponds, especially in ponds likely to become infested with unwanted fish or fish that prevent maximum catfish production. The bass feed for the most part on the minnows and grow large enough to control unwanted fish. Thus, they help ensure maximum production of catfish.

Stock only the species and numbers recommended in your area for your kind of enterprise and for the quantity and quality of your water. Exclude all others.

Managing Water and Fish

Management of water and fish will depend upon the size of your operation, the species you plan to grow, the density at which you stock the fish, and the source, quantity, and quality of your water.

A dependable year-round well is an excellent source of water. But well water frequently contains little dissolved oxygen and large amounts of carbon dioxide and nitrogen— a combination deadly to fish. You can dispel the harmful gases and add oxygen by splashing well water over a hard surface or by spraying it into the air before it enters the water where the fish are raised.

In most cases it is easier and less expensive to prevent unwanted fish from entering your operation than to control them after they enter. Here are some ways to prevent their entrance:

When possible, use a dependable, year-round source of water that is free of fish. If surface water must be used, filter it before it enters your operation.

Locate, layout, and construct your operation and facilities so overflows from streams and other bodies of water cannot enter. Entrance of such water is a sure way of introducing unwanted fish and, frequently, diseases and parasites.

When you drain a pond or raceway, let the bottom dry out

completely before refilling. This, of course, eliminates fish that are present. It also destroys many parasites and disease-causing organisms.

Diseases and parasites can be real problems, especially when fish are stocked at high densities in impoundments. You can avoid many disease and parasite problems by taking these precautions:
- Use water that is free of parasites and disease-causing organisms
- Bring only fish that are free of diseases and parasites into your operation. Make sure they are treated before stocking
- Use no feed material which may introduce diseases and parasites

Supplemental feeding is required in most commercial operations. Use only the kind and amount of feed recommended in your area and for the density at which your fish are stocked. Check frequently to see that all feed is being eaten.

Oxygen Deficiency

Oxygen deficiency is a major cause of fishkills, especially in operations stocked at high densities.

Some ways to help prevent oxygen deficiencies are:

Avoid excessively deep water. Use a dependable source of high-quality water. Stock only the recommended number and species of fish. Feed only the recommended amounts of feed.

If fertilizer is needed, use only inorganic fertilizer. Allow no runoff water from feedlots, poultry houses, pig parlors, and dairy barns to enter your enterprise. If fish are stocked at high densities and a supplemental feed is applied, check the dissolved oxygen content frequently, especially at daybreak during warm weather.

The time and method of harvesting will be determined to a great extent by your operation, the species of fish you grow and the volume of your business.

Fish may be harvested during any season if facilities are available for handling them. Frequently, facilities for dressing and storing large volumes are required. Usually it is best to harvest fish from a commercial enterprise during cool weather, especially if large quantities are involved.

Restocking is another key part of management. Some operations need complete restocking every year, others periodically. Still others require almost continuous restocking with adult fish during the fishing season.

Other important aspects of management include care of

the watershed, livestock exclusion, treatment for diseases and parasites, weed control, and controlling nuisance wildlife such as beaver and muskrat.

Market Demand

Economics and market demand can make or break you. Give them a good bit of thought before investing your money and labor in fish production. Unfortunately, less information is available on these factors than on other aspects of growing fish. Here are a few things to keep in mind:

Your initial costs in commercial enterprises may include ponds, raceways, wells, drainpipes, fences, roads, seines, boats, motors, special trucks. Some of these costs are high.

Yearly maintenance is another cost. Generally, the bigger the operation, the greater the cost of maintenance. There are yearly production costs, too. These may include pumping, stocking, feed, fertilizer and chemicals, labor, taxes on land, telephone, interest on operating capital, and harvest costs.

Generally, the cost per unit of production (usually per fish or per pound of fish) is lower in enterprises that produce large quantities of fish. If there is little or no demand for large quantities, you can lose money in a hurry even though your cost per unit of production is low.

Your best bet, of course, is to produce a high-quality product at the lowest possible cost per unit. Then, have a guaranteed market for the fish you produce, especially a market that ensures a reasonable return on your investment and labor.

Market demand is influenced a great deal by location. Fee-fishing enterprises should be near population centers, preferably within an hour's drive.

High-volume commercial operations, especially those producing fish for human food, should be near plants equipped to process and store large volumes of fish.

You can obtain technical assistance and information from many sources, among them the Cooperative Extension Service, Soil Conservation Service, your state game and fish agency, and the U.S. Fish and Wildlife Service.

Personnel from these agencies can give information you need in deciding whether your water, land, and location are suited for a particular operation. They also can give you the latest information on managing your enterprise. Turn to them first for advice.

Other information sources are trade magazines and private organizations. Some universities and colleges have experts in fish production on their staffs—contact them.

Vacation Farms Attract Half a Million Each Year

By Malcolm I. Bevins

If you looked around the countryside at vacation farms, you would be struck by their great diversity. Some vacation farms have animals. Some have guest rooms in the farmhouse, others have cottages. Some offer fully prepared meals, others provide cooking facilities. Some have swimming pools and tennis courts, others simply offer the "ole swimming hole." The rural environment with open space and clean air is common to all.

Most recent estimates indicate that about half a million Americans currently take a farm vacation—less than 1 percent of our total population. Over half of these vacationers come from the nine largest U.S. cities. Two out of three adult farm vacationers were raised or spent some time on farms or ranches in their childhood.

The typical farm or ranch vacationer can be described as highly educated, and holding a professional or managerial position with relatively high earnings. Most typically, this vacationer also has a larger than average family.

A recent survey indicates the desire for peace and quiet is the single, most important reason for taking a farm vacation. Secondly is the desire to give a special experience to the children. Less than one-fifth take such a vacation for reasons of economy.

Generally, farm vacationers do not seek a highly organized program of recreational activities, but they desire to participate in a variety of activities. Horseback riding, hiking, and swimming all rank high on the list of desired activities. A single activity, like observing farm animals, is not enough. Urbanites seek to place themselves temporarily in the clean lifestyle of rural America.

Malcolm I. Bevins is Extension Economist and Associate Resource Economist, Vermont Agricultural Experiment Station, University of Vermont, Burlington.

A vacation farm operator and his family must like working with people, often as much as 18 hours a day. This calls for time, patience, and understanding. A certain amount of privacy is lost. But a better understanding of people, places, and things is almost sure to rub off on the farm family.

Farm vacationers need private quarters—their own dining room, lounging area, and toilet facilities. Cleanliness is an absolute necessity, not only to attract and keep guests, but also to meet health department requirements. Accommodations should be cozy; there is nothing like a crackling fire on a cool or damp night.

Water is a key word in more ways than one. An abundant supply of pure water under pressure is needed, between 100 and 150 gallons a person per day. Additionally, clean water for recreation is highly desirable. Swimming is a top recreational activity during summer months. A pool often is preferred over a pond, and quality is easier to control.

Good, old-fashioned, home-cooked meals loaded with variety and imagination keep folks happy. Guests are on a holiday schedule, so meal hours must be flexible.

Clientele Goals

Early in the planning process a decision must be made as to whether or not to seek special clientele. Some vacation farm operators choose to cater to young families, others prefer older couples without children. Although some operators cater to both groups, this can lead to problems. Older couples like a quiet setting, children like lots of activity.

A vacation farm catering to young families has a limited season, tied to summer vacations. On most farms, this coincides with the busiest harvesting schedule. Potential conflict between both activities must be carefully evaluated.

Operators will also have to decide whether to accept children not accompanied by parents. The degree of responsibility is quite different.

The final decision on clientele should reflect the entire farm family's personal attitudes. Although the joys associated with working with youngsters can be most rewarding, sharing experiences with older persons can be equally satisfying.

Legal liability for accidents or injuries represents another basic consideration. Some activities pose greater risk than others. Coverage will cost more for horseback riding than for swimming. Discuss these matters carefully with your insurance agent. Incorporation becomes an important consideration as a technique for limiting liability.

Fringe benefits will likely accrue if the vacation farm enterprise is established. A tennis court or swimming pool developed for the vacationers can be used during off-peak periods by the farm family.

How to Charge

Nationally, vacation farms are relatively few in number and diverse in character. For this reason there is no established food and lodging price range that can be considered appropriate for all. However, a few guidelines do apply.

Business should be attracted on the basis of recreational opportunity, not low cost.

Some budgeting of potential income and expense at several different prices will be helpful. Income must cover all costs and return a reasonable profit for time spent.

Income on the vacation farm is not restricted to food and lodging charges. Farm products like maple syrup can be sold directly to your guests at retail prices, as can fishing tackle and bait to the fishermen. Boat rentals will also be profitable.

The camper needs a variety of services such as ice, firewood, bottled gas, and food supplies. It is not uncommon for campers to spend more money on services than on camping fees, but don't "nickel-and-dime" your guests excessively. Vacationers are willing to pay a reasonable total price but object to a constant bombardment of small charges.

Several features of a vacation farm give it greater appeal than standard resorts. Prices are usually lower because the farm operator has less invested than the traditional resort operator. Farms are less crowded than resorts. And country living has an attractive new image, especially among young people.

In the future, energy conservation may trigger an increase in the popularity of vacation farms. Why travel 2,000 miles to Yellowstone National Park when a fun-filled vacation can be had much closer to home?

The RV Market

The recreation vehicle (camper, trailer, or motor home), so popular today, must be considered in the total picture. Currently, recreation vehicle (RV) manufacturers are shipping over half a million units annually to dealers.

An estimated 6 million portable camping units across the country are in use today.

RV owners seek a place to park their unit where electricity and water are available, and clean toilet facilities conveniently located. Catering to this group is easy for the farm operator—no sleeping accommodations are needed.

The potential for vacation farms looks good, but will never be fully realized without a significant marketing effort. Consumers are bombarded by advertising and promotional programs from other recreation suppliers looking for a share of the market.

Vacation farm operators might well look to their old friend, the cooperative, for help. As often demonstrated, a group of farmers advertising together can achieve results that would be impossible for one alone.

Don't make any decisions about this business without first visiting at least one on-going vacation farm. Go as a regular guest—a great deal more can be learned.

While you rub elbows with other guests, find out their likes and dislikes. Talk to the management about problems and opportunities, and stay objective during this evaluation process.

Take part in recreational activities offered, both familiar and unfamiliar. This can be fun and you may uncover useful ideas.

Keep notes on your thoughts and impressions and take plenty of pictures. These can help you evaluate your experience when you get back home.

State Standards

Before you start your business, visit the state health department or licensing authority. Find out about state standards. If your planned facilities are substandard, how much will it cost to bring them up to the mark?

Contact the local office of the Soil Conservation Service and discuss the physical attractiveness of your area to vacationers. Give some thought to conservation practices that might improve your farm's appeal.

Drop in at your county Extension office. Specialists there and at the State university can help you evaluate the feasibility of a vacation farm and lead you to other sources of assistance.

If financial help is needed, talk with your banker as soon as possible. Perhaps you will qualify for financial aid from either the Small Business Administration or the Farmers Home Administration.

Finally, gather the entire family and discuss, in frank terms, potential impact of the venture on family life. Is everyone ready for this change? Carefully weigh all benefits against all costs, both monetary and nonmonetary. Take a vote. If the "ayes" have it, welcome aboard. Good luck!

Dude Ranches Capitalize on 3 Current Trends

By J. Hugh Winn and Tom Davis

In the West, vacation farms are called dude ranches. A working dude ranch usually is an enterprise that has one or more agricultural operations and provides recreational facilities and services to guests for a fee. The agricultural operations involved are usually livestock-oriented.

Three current trends are expanding demand for dude ranch vacations: A desire to participate in the lifestyle of the old West, energy-conscious Americans seeking single-destination vacations, and heightened environmental awareness of the public creating the urge for recreation in natural areas.

A dude ranch in the West requires 40 to 80 acres as a headquarters. It is essential that this area be near a large tract of publicly-owned land available for recreation use such as backpacking, trail riding, and cross-country skiing.

Scenic values are important to dude ranch guests. Mountains, forests, sparkling rivers, streams with beaver ponds, all are part of the desired ranching experience as well as wildlife and livestock viewing.

Access to dude ranches need not be easy. Remote locations add to the desirability of the ranching experience, but may be a hazard for the ranch operator or guest. Arrangements must be made for providing emergency services in case of accident or illness.

A dude ranch operator must combine characteristics of a rancher, businessman, hotel operator and recreation director. The operator must know livestock, horsemanship, ranch maintenance, business operations, and public relations, as well as how to get along with people and to cope with a variety of situations. The operator's family must be willing to become part of

J. Hugh Winn is Extension Professor and Agricultural Economist, Colorado State University, Fort Collins.
Tom Davis is Extension Assistant Professor, Recreation Resources, at the university.

the enterprise and to sacrifice some privacy. Above all, they must enjoy working with and living among other people.

Income potential from dude ranching depends primarily on the desires and needs of the ranch operator, who must determine how much time, capital and energy to commit. A balance must be struck between the demands of the agricultural enterprise and the demands of providing visitor services.

Other factors which may affect income potential are management skills of the operator, availability of capital, and competition from other dude ranches or recreational enterprises.

Initial Steps

Before going into dude ranching, visit several operating dude ranches to observe and ask questions. This can give you ideas on needed facilities, equipment, and recreation programs, and problems you may face.

A beginner should start small and add activities as business picks up. This way you will not have invested too heavily if you decide not to continue the operation.

Clientele may support a campground, souvenir stand, or horse breeding program. Local handicrafts or ranch products might be a profitable sideline.

Contact with the clientele makes it easy to respond immediately to needs and desires of the "guest". The owner has daily meals with guests, and can share experiences from the various program activities.

The day of the "mud on the boots" type operation has passed. Proper planning, organizing, directing, coordinating and controlling need to be accomplished by even the smallest vacation-ranch.

Scheduling of guests, activities and events must be precise. Intermediate and long-range plans should reflect anticipated needs of guests.

Meeting Regulations

All States have health, sanitation and safety requirements for any business providing lodging, food service, and various activities. Different regulations may apply to a dude ranch, depending on the size and intensity of the operation.

If the enterprise is small, with lodging in the home, bunkhouse, or cabins, with guests and the operator's family all eating together, it may be classified as a boarding house. Health and sanitation requirements for boarding houses apply. If the operation is larger, with a motel and walk-in restaurant, requirements would have to be met for these establishments.

Potential dude ranch operators should contact the local representative of the State health department for information and advice.

Usually county and state licensing apply to a dude ranch. Municipal licensing requirements must be met if the operation is within municipal boundaries.

At the municipal or county level, a license to do business may be needed. Some counties, however, do not require this. If new facilities are planned, county building codes must be met and permits obtained. Check with your county officials for information.

State licenses required depend on the size of operation, facilities, and services offered. They might include a sales tax permit, lodging and restaurant licenses, beer or liquor licenses, dance hall license, and guide's license.

Permits may be needed to use State and Federal lands or waters in the dude ranch operation. Requirements vary with type of use and agency involved. Get information from the agency managing these lands.

Many States have some zoning regulations for unincorporated areas. Dude ranches may be permitted to operate in open zones—where you can do about as you please, agricultural zones, business zones, tourist or amusement zones, and perhaps in some recreational zones.

Contact the county planning commission or the county zoning administrator for zoning regulations.

Tax Requirements

Taxes required from dude ranches vary among States and for different types of operations. For example, taxes on business property may be assessed at a different rate than on agricultural land.

Special taxes may apply depending on the operation or programs. Examples would be county or State sales taxes, lodgings tax, amusement tax, and vending machine tax.

The dude ranch operator will be subject to withholding and unemployment taxes for employees. Some form of workmen's compensation payment may be required.

The variety and cost of these taxes must be included when considering the feasibility of a dude ranch enterprise.

To control the operation you need certain key indicators to show how well you are doing at any time. Good business records are vital.

The profit and loss statement and balance sheet are the most widely understood and used by all businesses. Monthly cash flow projections are essential to decision-making and must be used.

Develop a record program that separates out and allocates expenses to the correct enterprises. Good records help you have a proper balance of assets. Land, livestock, machinery and equipment inventories must show the best return related to the economic activity. Gross earnings need to be identified with enterprise cost.

Dude ranch operations may have many hazards for people unfamiliar with livestock, ranch equipment, horses and outdoor activities. It is essential to have adequate liability insurance, besides other forms of coverage.

Statutes and court decisions concerned with limits of liability vary. Legal consultation is advised for prospective operators.

Insurance rates may vary among companies. Contact different companies for coverages offered and rate schedules.

Repeat Business

It is your job to make the guests' experience as enjoyable as possible. This can pay off in repeat business. Otherwise you will have to find new customers each season.

The dude ranch operator should be involved in efforts to promote the enterprise at local, regional, state and national levels. This may mean active membership in a local Chamber of Commerce, a regional tourism association, a guest ranch association, and a national organization to promote recreation or tourism.

Like other businessmen, you must advertise. Identify the type of visitor you are most comfortable with and select media which will reach these people. Advertising budgets vary from 0 to 15 percent of gross revenues. Of course, the best advertising is also the least expensive: word-of-mouth by a satisfied guest.

The Small Business Administration and the Farmers Home Administration are two Federal agencies that may be able to help you financially and technically in developing a working dude ranch. Other assistance can be obtained from State fish and game agencies, the Forest Service, the Bureau of Land Management, your State land-grant university, and the county Extension office.

You Can Start Producing Earthworms in a Washtub

By Ronald Abe, William Braman and Ocleris Simpson

Earthworm farming has become popular in recent years due primarily to demand for the worms as fish bait and to supply brood stock to the ever increasing numbers of new growers. Worms also are sought by organic gardeners for natural tillage of the soil.

Since earthworms are great consumers of decaying organic matter and micro-organisms, they may be used in future systems to decompose and recover nutrients from large quantities of organic wastes—such as manure, garbage, and pulpwood sludge.

There also is the possibility of processing large quantities of earthworms into products to be used as a protein supplement for aquatic and farm animal production. The live worm contains 10% protein and if made into a dried product, 60% protein.

Cultural practices recommended in this chapter are for the small English worm which is also called "red worm", "red wiggler", or "red hybrid". It is the most popular farmed worm because of its hardiness and tolerance of high concentrations of decaying organic matter.

With some experience, many modifications in the cultural practices may be made for economic reasons without sacrificing yield. For cultivation of varieties native to your own area, observe the environmental conditions under which large numbers of worms are found and duplicate those conditions.

If ideal growth conditions can be sustained at all times, worms may be raised in mounds or windrows of organic soils without the use of a container or enclosing structure. Since we

Ronald Abe is Professor of Animal Science, Division of Agriculture, Fort Valley State College, Fort Valley, Ga. William Braman is Director of Research, Murphy Products Company, Inc., Burlington, Wis. Ocleris Simpson is Research Coordinator, Fort Valley State College.

are humans and subject to making errors in growing earthworms, it is recommended that a container be used to prevent a massive escape of the worms during unfavorable conditions.

Use any container that is at least 10 inches deep and free of toxic chemicals. Make drain holes of 1/2 to 1 inch in diameter at the bottom. Smaller holes may become plugged with use over an extended period of time.

A three-inch layer of coarse gravel in the bottom will assure good drainage and prevent loss of soil from the container. A layer of sand may be placed on the gravel to fill the large spaces between the gravel and further improve drainage.

A galvanized washtub 2 feet in diameter and 10 inches deep, as an example of a container for a beginner, could be used to produce 4,000 to 6,000 mature size worms a year.

Building a Bed

Construct rectangular beds for large volume production of worms. Limit width of the bed to about 4 feet for easy access to the center during harvesting. Sides of the bed may be made of lumber, concrete block, brick or concrete slab.

Provide adequate drainage. To protect against flooding in areas of high rainfall or in culturing practices using liquefied feeds, use a perforated septic tank drainage line in conjunction with layers of gravel and sand.

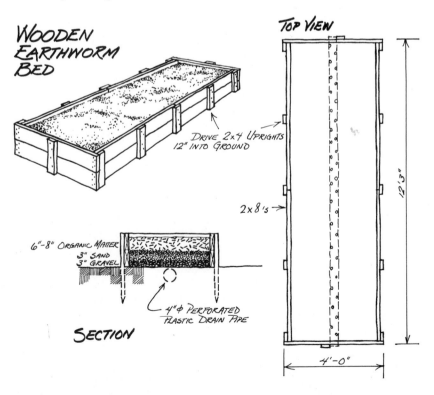

WOODEN EARTHWORM BED

TOP VIEW

DRIVE 2x4 UPRIGHTS 12" INTO GROUND

2x8's

12'3"

4'-0"

6"-8" ORGANIC MATTER
3" SAND
3" GRAVEL

4"⌀ PERFORATED PLASTIC DRAIN PIPE

SECTION

Have walkways between beds for ready access to all areas of the beds.

Shade will help control moisture content of the bed during the hot summer months. A shed or building is the ideal location for the beds; however, the added cost may not justify the assurance against losses from heavy rains in certain areas.

A time-controlled automatic sprinkling system can be used to provide moisture and cooling in summer in the absence of shade. Water frequently and in small amounts.

In filling the bed use soil with a high content of decayed organic matter. Mix about two-thirds top soil and one-third decayed organic matter such as leaf compost, manure, sawdust, ground peanut hulls, peat moss, and other plant byproducts available in your area.

Test the mixture in a small container such as a flower pot or a clean gallon can with at least 25 worms. Make drain holes on the bottom, add a layer of gravel, and fill the container three-fourths full with the soil mixture.

Saturate the soil with water and let excess water drain, then place the worms on the surface. A 24-hour test will give you an idea about the porosity, drainage and other environmental conditions needed by the worms.

If the worms crawl out of the container, move to the gravel layer, or die from shrinking or swelling, try another mixture or seek help from your county Extension agent.

Fill the worm bed with a 6-inch layer of tested soil. Thoroughly water the bed with a spray nozzle or sprinkler, but not to the point of being soggy.

Stocking Rate

Stock the bed with 100 to 500 adult size "red wigglers" per square foot of bed surface area. Breeders for the first bed may be purchased from any bait and tackle store or from worm farms in the area. Some worm farms sell old bed materials already containing thousands of small worms and egg cocoons to start your bed.

To a large degree the stocking rate will determine the mature size of the worms produced and the amount and frequency of feeding during the growing seasons. Overcrowded beds produce small-sized adult worms. Sparsely populated beds with sufficient feed produce large adults.

The "red wiggler" may be fed a variety of feeds such as fresh manure, kitchen waste, or poultry mash feeds. Feeds containing 10 to 12 percent protein on the dry matter basis seem ideal for good growth and reproduction.

Fresh cattle manure, particularly from animals fed high hay or grass rations, makes excellent feed for the worms. Dilute it with about two volumes of water. Pour onto the surface of the bed to form a thin layer.

The manure will form a mat of particles that also provides shade and helps lower moisture evaporation from the soil.

When feeding dry ground feedstuffs, sprinkle them lightly and evenly on the surface. Always water the bed immediately after applying feed. Keep the bed moist throughout at all times. During hot weather, the bed may need light watering several times a day to replace moisture that evaporates from the surface.

When holes appear on the moist crust of feed it is time to apply feed again, usually every 3 to 5 days in vigorously growing cultures. Larger feed particles will disappear more slowly because they must first be fermented by micro-organisms into products that can be ingested and digested by the worms.

Over-Feeding

Over-feeding will produce an extreme amount of fermentation activity which results in a mass escaping and killing of worms. Also, the worm bed will contain too much acid and the bed medium will undergo an unfavorable osmotic shift and cause the worms to shrink.

Should this occur, remove as much feed from the surface as possible and water the bed heavily to leach out the dissolved solids, which subsequently lowers the heat of fermentation.

Over-feeding may also cause heavy infestations of mites on the surface of the bed. If this happens, hold off feeding until the mite population diminishes to small numbers. Then regulate feeding to prevent another "bloom" of mites. Mites compete with the worms for the valuable feed. It is better to sacrifice a few days of worm growth than to waste feed on the mites.

The earthworm is hermaphroditic, which means each worm has both male and female organs.

Copulation occurs through the clitellum, the glandular swelling located about one-third the length of the worm from the head.

In mating, the worms overlap each other with their heads going in opposite directions. As the clitella of the two worms meet, large quantities of mucus are secreted that binds the worms together. Each worm acts as a male and gives off seminal fluid which is stored.

After the worms have separated, a slime tube which is formed by the clitellum of each worm is worked forward over the body and collects a few eggs from the oviducts and receives sperm from the seminal receptacles where they have been stored. The slime tube is gradually slipped off over the head, closing to form a greenish-yellow cocoon.

The cocoon of "red wigglers" is about an eighth of an inch in diameter and lemon-shaped. Under favorable conditions, copulation and production of cocoons take place once every three to five days.

Each cocoon usually produces two to ten worms after an incubation period of about three to five weeks. Under unfavorable conditions, however, the eggs may remain dormant and fertile for months.

The young worms grow rapidly and within a month from the time of hatching they may be able to reproduce. However, it can take six months for the worms to attain full size. Earthworms have a long lifespan and may live more than ten years.

Harvesting, Cleaning

In the growing season, earthworms feed near the surface during the night. Early morning is the best time to harvest them, before they move deeper into the bed.

In harvesting, remove the first two to three inches of bedding material from the surface by hand or with a pitchfork and place it in a tub or bucket. Spread the material loosely in the container to form a one- to two-inch layer and place it in a lighted area. As the worms move to the bottom, remove the soil by hand. Most of the soil can be removed from the mass of worms within a few minutes.

Since many cocoons and small worms will be in the soil, return it to the worm bed.

For larger operations, a commercially produced mechanical earthworm separator may be purchased. These separators seem to work fine with most types of worm bedding materials.

Do not store worms in large numbers in containers without bedding material. Worms on the bottom will quickly die from suffocation, particularly during hot weather.

Worms may be kept alive for at least two weeks in containers filled with moist peat moss.

Worms intended for use as feed or food for pets, fish, livestock or even humans may be thoroughly cleaned by following the illustration shown in steps 3 to 6. Most of the gastrointestinal contents will be removed from the worms in this process.

Disposing
of Property

PART 5

Selling Property: Brokers, Title, Closing, and Taxes

By J. W. Looney

Property transfers are complicated by local restrictions, state rules, federal regulations and tax laws. The seller often must turn to professionals for guidance in accomplishing the orderly transfer of possession and title to a new owner.

After deciding to sell, the property owner must determine whether to employ a real estate broker or attempt to sell the property without one. Important decisions must also be made about what to include in the sale, and the price the property will be sold for. It is on this latter point that the real estate broker can be of great assistance. In addition, services of an appraiser may be desirable to help determine the property's current value.

If the seller enlists the aid of a real estate broker, the seller and broker must agree on terms of the broker's employment. These terms normally are set out in a "listing" contract.

Typically, brokers in a given area will use standardized listing contracts. The seller and broker may agree to include items that are not part of the standard form contract. If so, these should be specifically detailed. Basic types of listing contracts include open listing, net listing, exclusive listing, and exclusive right to sell.

Open listing may be either an oral or written contract. It is a simple agreement in which the seller agrees to pay a stated commission if the broker obtains a buyer to sign a purchase contract agreeable to the seller. This does not preclude the owner from making the sale nor does it preclude contracts with other brokers.

Although there are certain advantages to the seller, this type contract is not favored by most brokers and may have

---

J. W. Looney is Assistant Professor, Agricultural Law, Department of Agricultural Economics, Virginia Polytechnic Institute and State University, Blacksburg.

disadvantages for the seller as well. The broker may not be interested, knowing other brokers will also be authorized to obtain buyers for the property. And a broker who does sign up is less likely to promote and advertise the property.

Net listing is the second type of listing contract. Here the owner sets a base price below which the property is not to be sold. Generally the real estate broker is authorized to add the commission or fee over and above this base amount.

Many owners prefer net listing because they can be assured of receiving the base price. Most brokers do not favor this type contract and many refuse to sign one. Although the arrangement may be oral or written, it is advisable to put all terms into writing—particularly the net price expected by the owner.

Exclusive Listing

A more common type of listing contract is the exclusive listing, preferred by most real estate people. One broker is appointed to act as agent for the seller for a set time—often three or six months.

The broker who obtains a buyer during this period is assured of a commission. Thus, brokers are more likely to promote and advertise the property.

A similar arrangement is exclusive right to sell. But here the broker is entitled to a commission if the property is sold at any time during the contract term even if the owner arranges the sale.

As a practical matter, most brokers prefer this arrangement. However, many sellers often have contacted potential buyers on their own before the listing. So it is fairly common practice to modify the exclusive right to sell contract by providing that the broker gets no commission if a sale is concluded with a buyer previously contacted by the seller.

No matter which type listing contract is used, seller and broker should agree on a number of essential points. Particularly important is that the sale price and terms of sale be specified, and any personal property to be sold with the real estate be designated.

Both parties should understand the basis on which the real estate commission will be determined. Generally a broker's commission is based on the selling price, and may vary from five to ten percent depending on the type of property involved.

Other provisos of the listing contract may specify the amount of "down payment" or "'earnest money" expected from a potential buyer, and what is to be done with this money until final transfer of the property.

Multi-Listing

In some areas, brokers have a "multi-listing" service under which a number of brokers can be authorized to sell the same property at the same time.

Many multi-listings provide for sharing the commission between the real estate office obtaining the listing and the office which arranges the sale. Such arrangements often are advantageous to the seller who is assured of the widest possible efforts to obtain a buyer.

Once an interested buyer has been found, the real estate broker negotiates a sales contract between seller and buyer. In most areas, brokers are authorized to prepare the sales contract. It may also be prepared by a lawyer, and in some states local law requires a lawyer.

Negotiations between the parties often are handled exclusively by the broker with no direct contact between buyer and seller until after the sales contract is completed.

The typical contract includes price, financing arrangements, closing date, possession date, type of deed to be transferred, insurance requirements, title examination, agreements on taxes, and other matters the parties wish included. The contract should contain a legal description of the property and specify any personal property involved in the sale.

Many contracts will also include arrangements for a survey of the property and inspections such as termite, plumbing and heating, or electrical.

The contract may include determining who bears the loss if the property is destroyed before the closing date.

Once the parties reach an agreement, the buyer usually deposits earnest money toward purchase of the property. Typically this goes to the real estate agent who puts it into an escrow account in a bank. Earnest money of ten percent of the purchase price ordinarily is required. This will be applied toward the total purchase price once the sale is closed.

Title Examination

Most real estate sales contracts provide for the seller to deliver "marketable title" to the property. That means title can be transferred free of reasonable doubt as to its validity. The purchaser is thus assured that involvment in a lawsuit related to title is unlikely after purchase.

Once the sales contract is agreed upon and earnest money deposited, the buyer gets a specified time to obtain a title examination of the property. The buyer has an opportunity to obtain professional assistance in this examination.

If no title defects appear, the sale can be concluded. But should examination of title reveal defects that raise questions about its validity, the seller is given time to correct the defects. If the defects cannot be corrected within a reasonable time, the buyer may be relieved from the purchase contract.

Much misunderstanding arises in the title examination process because of its complexity. The process requires a thorough search of courthouse records in the county where the real estate is located.

In some states the search is made by professional abstractors who furnish a certified summary of their findings. This summary, called abstract of title, can be examined by the attorney for the buyer. In other states the search is conducted by the attorney directly.

The person conducting title search must examine each transfer of property in the records which relate to the property being sold. It is customary to go back 60 or more years to determine the exact "chain of title." In some areas the chain of title may be traced to the original government land patent 100 or more years back.

Besides examining records of the property itself, title search includes examining other public records which may affect title.

After all records have been checked, a title report is submitted to the potential buyer. This report gives full information regarding title to the property. An attorney may be asked to give an opinion on "marketability" of the title.

Title Insurance

In many areas, particularly with urban property, it is customary for the buyer to obtain title insurance. This insurance policy does not insure against every loss that might occur, but insures against defects that would normally appear in the records.

Title insurance is available from a number of companies and may be obtained through most attorneys. Normally the buyer pays the cost. However, the sales contract may provide that the cost be paid by the seller or shared between seller and buyer.

During the period from the time of the sales contract until closing date of the transaction, the buyer will arrange necessary financing for the purchase.

Certain types of financing arrangements require the seller to be directly involved. For example, the seller may personally finance the sale by accepting from the buyer a promissory note securing the unpaid balance on the purchase price with payments to be made to the seller in installments.

In other arrangements, the buyer may take over or assume an existing loan which the seller has against the property. In some situations the seller may be relieved of any future obligation. However, in many States the original maker of the obligation—the seller—will continue to be liable for the payment unless the lender releases the seller of this obligation.

Under some financing arrangements the seller may be required to bear the expense of "points," a charge assessed by the lender in addition to any interest rate paid by the buyer.

Points often will be charged to the purchaser. However, in some financing the seller must pay points. On FHA and VA loans, for example, if points are involved they will usually be paid by the seller.

Points fluctuate according to availability of money for lending, and may range from one to nine. Generally one point is the equivalent of one percent of the mortgage amount, two points equivalent to two percent, and so forth.

The parties themselves may agree in the sales contract on who will pay the points.

Settlement

A closing date will be specified in the sales contract. Settlement varies depending on type of property involved and complexity of financing and transfer arrangements. At the time of closing the title transfer occurs, and the purchaser pays the remaining portion of the purchase price.

Often closing will be handled in the presence of buyers, sellers, attorneys for both parties, real estate brokers,and perhaps representatives of the lender. Normally the attorneys prepare all legal documents before the closing date. At closing, the broker commission is paid as well as all fees connected with the transaction.

Again, the sales contract can detail responsibility of each party regarding payment of expenses at closing. But generally the seller is expected to pay the cost of drafting the deed, transfer taxes, the real estate commission, the attorney's fee for representating the seller, and an apportioned share of taxes, insurance, and utility bills up to the point of closing.

The buyer, in turn, generally pays the title examination fee, the cost of drafting financing instruments, recording fees, appraisal fee, cost of survey, title insurance, and the attorney's fee for representing the buyer. Usually possession by the buyer is allowed shortly after closing.

Tax Implications

Sale of property may involve important income tax implications for the seller, particularly if the sale results in gain to the seller. Gain is the excess of net selling price of the property over the tax basis.

The tax basis for real estate is the owner's original investment plus cost of any improvements on the property minus any depreciation previously claimed. The amount of the tax basis may be recovered by the seller without payment of tax. However, any amount the seller receives above the tax basis is subject to tax as a capital gain.

Capital gain, under present tax law, is taxable at a lower rate than ordinary income. As an example, assume the property owner paid $50,000 originally for a parcel of land. Say the owner made improvements costing $10,000 and claimed no depreciation in previous tax years. The owner's tax basis would be $60,000.

If sales price for this parcel is $100,000, the gain would be $40,000 ($100,000 — $60,000 = $40,000). This amount is taxed as a capital gain. The taxpayer adds half of the capital gain to his other income to determine tax due.

The problem for many sellers is that adding the gain to other income results in a high tax in the year of sale.

The Internal Revenue Code allows deferral of tax on a portion of the gain where the purchase price is payable over two or more years. This is the "installment method" of reporting gain.

To qualify for the installment method, the initial payment in the year of sale must be no more than 30% of the total selling price. If at least two payments are received in two or more years and the initial payment is less than 30% of the selling price, the taxpayer is allowed to pay tax on the amount of gain in the year payment is received rather than in the year of sale.

For example, assume in the earlier case that the seller agreed to receive only $20,000 of the total $100,000 selling price in the year of sale, and to accept the remaining $80,000 in installments over a ten-year period.

The ratio of gain, $40,000, to contract price, $100,000, is 40%. Thus, 40% of the amount received in the year of sale is taxable as gain (40% × $20,000 = $8,000). In subsequent years, 40% of each year's payment on the balance will also be subject to tax as gain.

The obvious advantage of such an arrangement is that it will be less costly for most taxpayers to spread the tax over a number of years rather than paying it all in the year of sale.

Under other tax law provisions a person who sells a residence at a gain is not taxed if the proceeds are reinvested in another residence within 18 months of the date of sale, or if another residence was purchased within 18 months before the sale.

Where a new home is built it must be completed within 18 months before or two years after sale of the old residence. Construction must begin within at least 18 months of the date of sale of the old residence.

If a new residence costs less than the old, then some gain must be recognized.

Where taxpayers over 65 sell a principal residence, the first $35,000 is excluded from income. If the sales price is more than $35,000, a part of the gain is excluded.

Besides possible tax consequences, the seller must consider local laws and regulations which may affect the property's sale potential.

Local zoning ordinances may regulate size and nature of parcels sold. Some counties set minimum lot sizes. Others require surveying, platting, and reporting of the potential sale of residential lots. Some specify establishment of essential services such as water and sewer, streets, and utilities.

Note: Legislation under consideration in Congress when this chapter was written may change taxes on capital gains.

Providing for Your Heirs—
Non-Sale Property Transfers

By Donald R. Levi

Most non-sale real estate transfers made by owners of small acreages relate to a desire to pass on their estate to their heirs. Occasionally an owner may exchange his acreage for other real estate of like kind in order to avoid income taxes associated with an outright sale, but the typical non-sale transfer is to family members.

Basically, these transfers fall into two general classes— lifetime gifts and testamentary bequests. Motivations for each may differ, partly depending on legal and tax considerations. These motives are better understood by considering the estate planning objectives of most persons.

Most persons have three basic estate planning objectives. First, they want to make sure they are taken care of during retirement. Second, they often desire to leave certain assets to specific relatives or friends.

Third, once the first two objectives are achieved, most want to minimize the expenses and taxes associated with transferring property to the next generation. Minimizing these costs maximizes the value of property transferred to heirs.

Most non-sale transfers of property are related to these estate planning objectives, so it is useful to briefly review them.

Retirement Security. Before Social Security, retirement security often was achieved through owning real estate. It was sold at retirement, with proceeds serving as the retirement nest egg.

Some still use this method. Usually they sell to non-relatives, but sometimes to heirs under a "family annuity" arrangement.

A real estate tract is transferred to children in exchange

Donald R. Levi is Associate Director, Texas Real Estate Research Center, Texas A&M University, College Station.

for the children's promise to pay the parents a fixed monthly sum for the remainder of their lives.

Many real estate owners prefer this to the regular life insurance annuity because a premature death would financially benefit their heirs rather than an insurance company. But it can result in both federal gift and income tax liability, so competent tax advice should be obtained.

Retirement security may also be obtained, singly or in combination, by maximizing Social Security benefits, or by participating in employer, self-employed, or "individual retirement account" retirement programs authorized by the Internal Revenue Code.

Distribution Preference. Once retirement is reasonably assured, most turn their attention to the desired distribution of property among children, grandchildren and others. Obviously, the preferred distribution is a highly personal matter, so it is difficult to make generalizations which apply to everyone.

However, a couple of observations are in order.

First, an equal distribution among children is not necessarily a fair distribution. The personal, health and financial situation existing for children and their families may affect distributional choices. So may the extent to which relatives and friends have been available and of assistance during times of need. Only the person concerned can make judgments on these matters.

Second, lifetime gifts are less popular than testamentary bequests with most elderly persons, because the former involves a loss of control before death. Thus, lifetime gifts take away assets which could have been used for retirement security. This, together with the fear that inflation may erode purchasing power of funds set aside for retirement, has the effect of discouraging lifetime gifts for the less wealthy.

One major advantage of lifetime gifts is that one knows the desired property distribution has been achieved, and that it will not be upset by a family dispute over validity of a will.

Legal fees and taxes associated with transferring property to heirs are referred to as "intergeneration transfer costs". They consist of legal and other fees involved with probating an estate, state inheritance taxes, and the unified federal gift and estate taxes.

Generally speaking, our goal is to minimize these costs while also achieving retirement security and desired asset distributional patterns. Clearly though, our major concern should be to minimize the total of all intergeneration transfer costs.

It is illogical to set up wills so as to save one dollar in probate expenses if this costs an additional ten dollars in federal gift and estate taxes. One reason for working with a competent estate planner is to avoid this trap.

Probate has several functions. It assures that the deceased person's debts are paid, and also makes sure that debts owed the deceased will be collected. It keeps title clear to real estate and other assets by specifically identifying all heirs and designating which ones inherit each asset. It assures that the deceased's wishes are carried out in distributing assets.

Probate Expenses

While there are some filing fees and advertising expenses in administering an estate, usually the major expenses are fees paid to the estate's personal representative (the executor or administrator) and the attorney hired to assist.

The personal representative and attorney normally receive fees of about equal size. In many states their compensation is set by state law and is based on a sliding scale—that is, the percentage they are entitled to decreases as the size of estate increases. In other states the statutory compensation called for is a "just and reasonable" fee, which presumably depends on the amount of work actually required to administer the estate.

Other indirect costs may be associated with probate, such as lost profits suffered by a family business because probate constraints and red tape prevented sales and/or purchases to be made optimally.

An attorney can advise and assist you in preparing wills, trusts, and other legal instruments so as to avoid or minimize both direct and indirect probate expenses.

Most states have either an inheritance or an estate tax.

An inheritance tax is a tax on the right to receive property, and generally is based on the amount (value) of property inherited by each person. Many states provide tax-free exemptions for some specific value of inheritance.

Both the exemption size and the tax rate may vary, depending on the degree of blood relationship. That is, in several states lower inheritance taxes are due when property is inherited by close rather than distant relatives. To this extent these taxing schemes serve as an incentive to keep property in the family.

An estate tax is a tax on the right to transfer property. Therefore, it is based on the total value of property owned (or controlled) by the decedent. In general, the amount of taxes collected under state inheritance and estate taxes are not greatly different.

Federal Taxes

Generally, the tax bite is much more substantial at the federal level.

Since 1976, lifetime gifts and transfers occurring at death have been "unified" and are now cumulative (added together) for tax computating purposes.

That is, the value of taxable lifetime gifts made this year is added to taxable gifts made in previous years. The total tax due is computed on this sum.

Gift taxes paid in previous years are subtracted from the tax so computed, and the difference is the amount due and payable for this year. Thus, as additional taxable gifts occur over time, they are subject to progressively higher tax rates.

It is important to remember that the person who makes the gift, and not the person who receives it, must pay any federal tax due.

Similarly, federal estate taxes due after death on one's estate must be paid out of the estate. Further, property transferred at death is called the decedent's "last gift", and its value is added to any previous taxable lifetime gifts in order to determine the additional federal tax due at death.

Thus, all non-sale transfers of property—both lifetime and testamentary—are combined to determine the total federal tax liability assessed on the privilege of transferring property to the next generation.

The dollar figure used to compute the gift and estate tax due is fair market value. There is just one exception to using fair market value, and it is available only on transfers made at death—never on lifetime gifts.

'Special Use' Value

This exception involves real property used in closely-held businesses, including family farms and ranches. If all qualifying requirements are met, then land may be valued at its "special use" (agricultural) value rather than fair market value.

Particularly in these areas where there is a large difference between agricultural and fair market value, qualifying for this special tax treatment may save many tax dollars. Such large differences most often are found near large urban areas because it is here that the fair market value of land may be driven up by a large number of buyers seeking to own a few acres near their jobs.

However, not all of these owners will qualify for this lower federal estate tax valuation. The basic intent of Congress was to make the use of agricultural value available only for

420

family businesses which are going to be continued as a family business by the next generation.

Several technical legal tests have been developed to determine who is really a farmer. And total or partial recapture of any tax savings resulting from using agricultural value is required when the farm or ranch is transferred within 15 years of death to someone other than a qualified heir.

Unified federal gift and estate taxes are computed separately for husbands and wives. Thus, a simple way to begin minimizing taxes is to have ownership roughly divided equally between husband and wife, taking into account differences in life expectancies and recognizing that there are financial advantages to deferring tax payments until later.

To see that this division will help minimize taxes, just think of the assets held by a married couple as being represented by one large box. If all assets are owned by one of the spouses, the taxes due on transferring the property to the next generation will be assessed on contents of the entire box. However, if the box can be divided into two parts with each spouse computing and paying taxes on their respective part, it is possible to stay down in lower tax brackets.

Splitting asset ownership into "two boxes" is encouraged by a preferential federal gift tax treatment (called a martial deduction) that is available. It lessens or avoids gift taxes on non-sale transfers between spouses, so that the "two box" scheme may be achieved at lower tax cost.

Gifts to charities are encouraged by federal gift and estate tax law since they are entirely exempt from tax. In addition, such gifts may also qualify for an income tax deduction.

Planning Tools

With this background, let's discuss some legal instruments available for non-sale transfers. These instruments are often called the "tools" of estate planning.

Wills, trusts, joint ownerships, gifts, sales, family annuities, and other arrangements may be used to achieve estate planning objectives, depending on family circumstances.

Wills. Basically, anyone owning property or having children needs a will. Furthermore, each spouse should have a will, particularly if there are minor children, because generally only the longer surviving spouse may legally name a guardian for them.

Obviously, wills may be used to achieve the desired property distribution. But they also may:
· Minimize death taxes and probate expenses

- Keep title to property out of the hands of children until they are mature enough to manage it
- Name guardians for children and property
- Select the person or corporate body who will administer your estate in probate; and
- Help the family business operate relatively free of probate red tape during the transitional period following death.

Non-sale transfers often are used to create co-ownership arrangements. Most states have two basic kinds of co-ownership—a tenancy in common and a joint tenancy. A third available in some states, called a tenancy by entireties, is similar to a joint tenancy but can be used only by married couples. A quite different concept, community property, exists in eight states.

Tenancy in common co-ownership interests can be transferred by will or deed.

Joint tenancy co-ownership interests can be conveyed by lifetime deed, but not by will. A joint tenancy has a "right of survivorship", which automatically transfers the deceased co-owners' interests to the surviving co-owners (often without going through probate).

Holding property with family members in joint tenancy is sometimes called a "poor man's will", because it achieves the desired property distribution. (It does not provide the other functions of a will.) However, joint tenancies can cause adverse tax consequences if an estate is subject to federal gift and estate taxes.

Types of Trusts

A trust involves a non-sale transfer of property to a trustee, who invests and manages it for the benefit of the beneficiaries. These beneficiaries can be anyone selected by the trust creator (called a grantor).

Selection of a trustee may range from a family member to a bank or trust company.

The selection may vary depending on the kind of property in the trust and the expertise of alternative trustees in managing different types of property. Some trustees may be adept at managing a diversified stock portfolio; others are skilled in farm management.

There are two basic kinds of trusts, testamentary and lifetime.

A testamentary trust is created in one's will. An example might be one created for the benefit of minor children, distributing income to them for support and maintenance but retaining title until they are mature enough to wisely handle ownership and management decisions.

Lifetime trusts, on the other hand, may be either of two distinct types—revocable or irrevocable.

Under a revocable trust the grantor retains the power to terminate it and take the property back. If the trust still exists when the grantor dies, the trust property need not go through probate and may be distributed outside the public view.

The fact that trust distributions are not a matter of public record (while probate matters are) is thought by some to be an advantage. However, the existence of the power to revoke is sufficient to include the value of trust property in the grantor's estate when computing federal estate taxes due.

An irrevocable trust is one the grantor cannot revoke. Since it constitutes a completed gift, its creation may be a taxable gift for federal gift tax purposes.

Trusts cost money to create and operate (trustees are paid for their services).

On the other hand, depending on the specific provisions they contain, they may help minimize or avoid death taxes and probate expenses, relieve others of ownership and management obligations, assure high level management, minimize income taxes, and achieve desired property distribution at your preferred point in time.

Photos

William E. Carnahan served as visual coordinator in rounding up and screening photos for the color section in the front of the book. He also was a major contributor of color photos, traveling in several regions to take them. He contributed black and white photos as well. Carnahan is with the *Science and Education Administration.*

Another important photo contributor is George Robinson, a former U.S. Department of Agriculture (USDA) photographer who now lives and works in New England.

In the color credits below, USDA photographers are listed by name only. Some details are given on others. In one case only the company name is known. Numbers in the color credits match numbers printed with the color photos.

Photographer's names, if known, appear with the black and white photos. In the black and white credits, only non-USDA photographers and their affiliation (if known) are listed. Where the photographer's name is unknown, the source is listed, along with the page the photo appears on.

Color Photo Credits

Robert C. Bjork, 33

William E. Carnahan, Cover photo, 13-15, 20, 22, 23, 35-37, 40, 41, 43-45, 47-49, 51, 54, 55, 58, 60

Scott Duff, Oregon Cooperative Extension Service, 25

Malcolm W. Emmons, Ohio Cooperative Extension Service, 16

Norman E. Gary, University of California, Davis, 39

Tom Gentle, Oregon Cooperative Extension Service, 21

Greg Gerber, Lincoln University Cooperative Extension, 26, 50

Lyn Jarvis, Vermont Cooperative Extension Service, 46, 59

Kenneth Lorenzen, University of California, Davis, 38

Ron Rivers, Oklahoma State University, 57

George Robinson, Jericho, Vt., first photo in color section, 12, 17-19, 27-32, 34, 42, 53, 56

Troy-Bilt Roto Tillers, Garden Way Manufacturing Co., Inc., 24

Black and White Photo Credits

Vincent 'Abbatiello, Rutgers University, Brunswick, N.J.

R. C. Church, University of Connecticut

The Dakota Farmer, Aberdeen, S.D., 55

Thomas DeFeo, Cincinnati, Ohio

Scott Duff, Oregon State Extension Service

Norman E. Gary, University of California, Davis

Susan Griffin, Montgomery County (Md.) Cooperative Extension Service

Kevin Hayes, Pennsylvania State University

International Harvester, 162

Barry W. Jones, Texas Agricultural Extension Service

Kenneth Lorenzen, University of California, Davis

John Messina

Michigan Consolidated Gas Co., 157

Mississippi Cooperative Extension Service, 74

Charles O'Rear, Manhattan Beach, Calif.

C. R. Roberts, University of Kentucky

George Robinson, Jericho, Vt.

Rosemary's Herb and Pottery, 211

Jared M. Smalley, Area Extension Agent, Wadena, Minn.

Russell Tinsley, *The National Future Farmer*

Wallaces Farmer, Des Moines, Iowa, 274

Janet Yeary, Fresno, Calif.

How to Get Help From USDA

If you want your garden soil tested or if you have questions about gardening, livestock, insects, plants or animal diseases or other questions about living on a few acres, contact your county **Cooperative Extension Service.**

Some pesticides are for restricted use only, and you must be certified before you can purchase or use them. Your county Extension agent can tell you which pesticides are restricted and can also suggest alternative chemicals or methods for controlling pests. In addition, your agent can tell you how to become certified if that is necessary.

County Extension offices are usually located at the county seat in the post office, the courthouse, or in a building shared by other USDA agencies.

County Extension offices are listed in the telephone book under the name of the land-grant university in your state, or as (Cooperative or Agricultural) Extension Service under your county government.

Help is also available from the Cooperative Extension Service at the land-grant university in your state.

The **Soil Conservation Service** (SCS), working through the local soil and water conservation district, can help you find out what your land and water resources are like and how to improve them in order to grow better crops, support more wildlife, boost the supply and quality of water, and produce more income.

SCS also can help you and your neighbors solve flood, drought, and other resource problems and provide information to help you make decisions about land uses.

For more information about assistance available, contact the Soil Conservation Service field office in your county. In most cases it is in the county seat and is listed in the telephone directory under "U.S. Government, Department of Agriculture." You also can write to U.S. Department of Agriculture, Soil Conservation Service, P.O. Box 2890, Washington, D.C. 20013.

The **Farmers Home Administration** (FmHA) makes loans to farm and non-farm families alike for home ownership and improvement in rural areas.

In the countryside and towns of up to 20,000 population that are classified as rural, FmHA makes improvement loans that will bring existing houses to a standard of adequacy, including the installation of wells, waste disposal facilities, and inside plumbing. Loans also are made to improve home insulation.

Loans directly from FmHA are available only to families of low and moderate income who cannot obtain conventional credit (the agency may be authorized at times to guarantee home loans by commercial lenders to above-moderate-income families). Loans are not made for landscaping an existing home, for gardening on a non-commercial scale, or for improving acreage that is extraneous to the essential homesite.

Young people of student age—minors—may get loans on their own signature for income-producing projects carried out as part of a youth organization (4-H, Future Farmers of America) or school programs, such as a garden cultivated for purposes of selling the produce.

Business loans are available to rural families for straightaway business enterprises. Farm loans are made to those who can qualify as bona fide farmers producing products for sale and relying on farm earnings for a substantial part of their living.

The Farmers Home Administration has 1,825 county offices. Some in more lightly populated areas serve more than one county. Locations are found in local telephone directories under U.S. Government or the "F" alphabetical listings. FmHA offices also can be located by asking at other agricultural agency offices or other lending institutions in rural localities, or by writing to the Farmers Home Administration, U.S. Department of Agriculture, Washington, D.C. 20250.

Contact the **Agricultural Stabilization and Conservation Service** (ASCS) if you have questions about participation in Federal farm programs, cost-sharing assistance for soil and water conservation practices, or loans for farm storage facilities. When disaster strikes, contact ASCS regarding emergency feed for livestock, restoration of farmland, payments for field crop losses (for farm program cooperators), and indemnity payments for pesticide losses to beekeepers.

You can find this agency in your phone book listed alphabetically by itself, or listed under U.S. Government, or under the county name.

Index

LC Card No. 78-600097

For sale by the Superintendent of Documents, U.S. Government Printing Office
Washington, D.C. 20402
Stock Number 001-000-03809-5
Catalog Number A1.10:978

U.S. GOVERNMENT PRINTING OFFICE: 1978 O—251-000